RF PHOTONIC TECHNOLOGY IN
FIBER LINKS

In many applications, radio-frequency (RF) signals need to be transmitted and processed without being digitalized. These analog applications include CATV, antenna remoting, phased array antenna and radar amongst others. Optical fiber provides a transmission medium in which RF modulated optical carriers can be transmitted and distributed with very low loss. With modulation and demodulation of the optical carrier at the sending and receiving ends, the optical fiber system functions like a low-loss analog RF transmission, distribution, and signal processing system. RF photonic fiber technology has particular advantages in that it is more efficient, less complex, and less costly than conventional electronic systems, especially at high microwave and millimeter wave frequencies. Analog signal processing of RF signals can be achieved optically while the signal is being transmitted along the optical carrier. Examples of such processing techniques include up- and down-conversion of RF frequencies, true time delay of RF signals, and optical distribution of RF clocks.

This volume presents a review of RF photonic components, transmission systems, and signal processing examples in optical fibers from the leading academic, government, and industry scientists working in this field. It discusses important concepts such as RF efficiency, nonlinear distortion, spurious free dynamic range, and noise figures. This is followed by an introduction to various related technologies such as direct modulation of laser sources, external modulation techniques (including lithium niobate modulators, polymer modulators and semiconductor electroabsorption modulators), and detectors. In addition, several examples of RF photonic signal processing technology, such as the phased array, the optoelectronic oscillator, and up and down RF frequency conversion and mixing, are presented. These will stimulate new ideas for applications in RF photonic signal processing.

RF Photonic Technology in Optical Fiber Links will be a valuable reference source for professionals and academics engaged in the research and development of optical fibers and analog RF applications. The text is aimed at engineers and scientists with a graduate-school education in physics or engineering. With an emphasis on design, performance, and practical application, this book will be of particular interest to those developing novel systems based on this technology.

WILLIAM CHANG pioneered microwave laser and optical laser research at Stanford University between 1957 and 1959, whilst working as a lecturer and research associate. He subsequently joined the Ohio State University and established quantum electronic research there between 1959 and 1962. He became Professor and Chairman of the Electrical Engineering Department at Washington University, St. Louis in 1965, where he initiated research into guided wave and opto-electronic devices. In 1979 Professor Chang joined the University of California at San Diego and founded the well-known electronic device and materials group, as well as chairing the Department of Electrical and Computer Engineering from 1993 to 1996. He was made an Emeritus Professor of the department in 1997. Professor Chang has published more than 140 research papers on optical guided-wave research which in recent years have focused particularly on multiple quantum well electroabsorption modulators in III–V semiconductors.

RF PHOTONIC TECHNOLOGY
IN OPTICAL FIBER LINKS

Edited by

WILLIAM S. C. CHANG
University of California, San Diego

CAMBRIDGE
UNIVERSITY PRESS

CAMBRIDGE UNIVERSITY PRESS
Cambridge, New York, Melbourne, Madrid, Cape Town, Singapore, São Paulo

Cambridge University Press
The Edinburgh Building, Cambridge CB2 8RU, UK

Published in the United States of America by Cambridge University Press, New York

www.cambridge.org
Information on this title: www.cambridge.org/9780521803755

First published 2002
This digitally printed version 2007

A catalogue record for this publication is available from the British Library

Library of Congress Cataloguing in Publication data

RF photonic technology in optical fiber links / edited by William S. C. Chang.
        p.   cm.
Includes bibliographical references.
ISBN 0 521 80375 6
1. Optical communications.   2. Fiber optics.   3. Radio frequency modulation.
4. Photonics.   I. Title: Radio frequency photonic technology in optical fiber links.
II. Chang, William S. C. (William Shen-chie), 1931–
TK5103.59 R47 2002
621.382′75–dc21   2001052858

ISBN 978-0-521-80375-5 hardback
ISBN 978-0-521-03708-2 paperback

Contents

Contributors

Garry E. Betts
MIT Lincoln Laboratory
Lexington, MA 02173-9108, USA

John E. Bowers
Department of Electrical and Computer
* Engineering*
University of California Santa Barbara
CA 93106, USA

William B. Bridges
California Institute of Technology
M/S 136-93, 1200 E. California Blvd.
Pasadena, CA 91125, USA

William K. Burns
Naval Research Laboratory
Washington DC, 20375-5000, USA

William S.C. Chang
Department of Electrical and Computer
* Engineering*
University of California San Diego
La Jolla, CA 92093-0407, USA

Charles Cox
Research Laboratory of Electronics
Massachusetts Institute of Technology
Cambridge, MA, 02139, USA

Roger Helkey
Chromisys Corp.
25 Castilian Drive
Goleta, CA 93117, USA

Marta M. Howerton
Naval Research Laboratory
Washington DC, 20375-5000, USA

Ronald T. Logan
Phasebridge Inc.
859 S. Raymond Avenue
Pasadena, CA 91105, USA

Xiaolin Lu
Morning Forest, LLC
Highlands Ranch,
CO, USA

Willie Ng
HRL Laboratories, LLC
M/S RL64, 3011 Malibu Canyon Road
Malibu, CA 90265, US

Stephan A. Pappert
Lightwave Solutions, Inc.
8380 Miralani Drive, Suite B
San Diego, CA 92126-4347, USA

Joachim Piprek
Department of Electrical and Computer
* Engineering*
University of California Santa Barbara
CA 93106, USA

Ron Stephens
HRL Laboratories, LLC
M/S RL64, 3011 Malibu Canyon Road
Malibu, CA 90265, US

Greg Tangonan
HRL Laboratories, LLC
M/S RL64, 3011 Malibu Canyon Road
Malibu, CA 90265, US

Timothy Van Eck
Lockheed Martin Space Systems
  Company
3251 Hanover Street
Palo Alto, CA 94304, USA

Ming C. Wu
Department of Electrical Engineering
University of California Los Angeles
Los Angeles, CA 90095-1594

X. Steve Yao
General Photonics Corporation
5228 Edison Avenue,
Chino, CA 91710, USA

Daniel Yap
HRL Laboratories
M/S RL64, 3011 Malibu Canyon Road,
Malibu, CA 90265, US

Paul K. L. Yu
Department of Electrical and Computer
  Engineering
University of California San Diego
La Jolla, CA 92093-0407, USA

Introduction and preface

RF technology is at the heart of our information and electronic technology. Traditionally, RF signals are transmitted and distributed electronically, via electrical cables and waveguides. Optical fiber systems have now replaced electrical systems in telecommunications. In telecommunication, RF signals are digitalized, the on/off digitally modulated optical carriers are then transmitted and distributed via optical fibers. However, RF signals often need to be transmitted, distributed and processed, directly, without going through the digital encoding process. RF photonic technology provides such an alternative. It will transmit and distribute RF signals (including microwave and millimeter wave signals) at low cost, over long distance and at low attenuation.

RF photonic links contain, typically, optical carriers modulated, in an analog manner, by RF subcarriers. After transmission and distribution, these modulated optical carriers are detected and demodulated at a receiver in order to recover the RF signals. The transmission characteristics of RF photonic links must compete directly with traditional electrical transmission and distribution systems. Therefore the performance of an RF photonic transmission or distribution system should be evaluated in terms of its efficiency, dynamic range and its signal-to-noise ratio.

RF photonic links are attractive in three types of applications. (1) In commercial communication applications, hybrid fiber coax (HFC) systems, including both the broadcast and switched networks, provide the low cost network for distribution of RF signals to and from users. RF photonic technology has already replaced cables in commercial applications such as CATV. (2) At high frequencies, traditional microwave and millimeter wave transmission systems, using coaxial cables and metallic waveguides, have extremely large attenuation. Electrical systems are also complex and expensive. Other advantages of RF photonic methods include small weight and size, and immunity to electromagnetic disturbances. RF photonic systems offer an attractive alternative to traditional electrical systems at high frequencies. However, much of the RF technology for high frequency application is still in the research and development stage. It is important to understand the operation of each new development and to assess the implication of each new component before any application of the photonic link. (3) Once the RF signal is carried on an optical carrier, photonic techniques may be used to process the RF signals. An obvious example is the frequency up- or down- conversion of the RF signal. Therefore, photonic RF signal processing represents a potential attractive application area for new applications. However, in photonic RF signal processing, the system performance will have additional requirements than just the requirements for bandwidth, efficiency and dynamic range. For example, the phase noise of the RF signal becomes an important consideration in sensor applications.

System design consideration and the choice of the technologies and components to be used in RF photonic links, as well as the evaluation of their performance characteristics, are very different for analog links than for digital optical fiber links. For example, the "on–off" threshold switching voltage of a modulator is important for digital communication systems while the slope efficiency is the important figure of merit for analog modulation. A thorough understanding of the analog system issues and component requirements is necessary for a successful system design.

This monograph describes, in detail, the various key components and technologies that are important in analog RF links. The components are evaluated in terms of their potential contributions to the RF links, such as RF efficiency, bandwidth, dynamic range and signal-to-noise ratio. Since the modulation of an optical carrier is much smaller than its bias for analog links, a special feature of the analyses presented in this book is the use of small signal approximations with emphasis on the reduction of nonlinear distortions.

The objectives of this book are: (1) to present to the reader various key technologies that may be used in RF photonic links: (2) to assess the significant aspects of various technologies; (3) to explore extant and potential applications of such technologies; (4) to illustrate specific applications of RF photonic links.

The analyses of basic RF photonic links are presented in Chapter 1. The analyses show clearly the important figures of merit of various components and the system objectives of analog RF photonic links. Chapter 2 describes the role of RF subcarrier links in commercial local access networks. Modulation and detection techniques are of particular importance in RF photonic links, because they determine the nonlinear distortion, the bandwidth, the efficiency, and, in certain cases, the noise of such links. Chapters 3 to 7 describe various modulation techniques, including the direct modulation of semiconductor lasers, the $LiNbO_3$ external modulators, the traveling wave modulator, the polymer modulator and the electroabsorption semiconductor modulator. The basic materials and principles of operation, the performance expectation and the advantages and limitations of each modulation technique are presented. In Chapter 8, a description of the key features of various detectors is presented. In the next three chapters, Chapters 9 to 11, three novel techniques, photonic frequency up- and down-conversion, integration of antenna and modulators at high millimeter wave frequency and optical generation of high RF frequency oscillation are discussed. They may offer hitherto unavailable opportunities for applications of RF photonic technology. Since RF modulation of optical carriers can be transmitted via fibers over long distance and with true delay, RF photonic technology can be used for antenna remoting and RF signal processing. Chapter 12 illustrates an important application of RF photonic technique to antenna remoting and to the phased array antennas.

University of California, San Diego   *William S. C. Chang*

1

Figures of merit and performance analysis of photonic microwave links

CHARLES COX

MIT, Cambridge, MA

WILLIAM S. C. CHANG

University of California, San Diego

1.1 Introduction

Microwave links serve important communication, signal processing and radar functions in many commerical and military applications. However, the attenuation of microwave RF signals in cables and waveguides increases rapidly as the frequency of the signal increases, and it is specially high in the millimeter wave range. Optical fibers offer the potential for avoiding these limitations for the transmission of RF signals.

Photonic microwave links employ optical carriers that are intensity modulated by the microwave signals[1] and transmitted or distributed to optical receivers via optical fibers. Since the optical loss for fibers is very low, the distance for photonic transmission and distribution of microwaves can be very long. When the modulation of an optical carrier is detected at a receiver, the RF signal is regenerated. Figure 1.1 illustrates the basic components of a simple photonic microwave link.

Since the objective of a photonic microwave link is to reproduce the RF signal at the receiver, the link can convey a wide variety of signal formats. In some applications the RF signal is an unmodulated carrier – as for example in the distribution of local oscillator signals in a radar or communication system. In other applications the RF signal consists of a carrier modulated with an analog or digital signal.

From the point of view of input and output ports, photonic microwave links function just like conventional microwave links. A common example is to transmit or to distribute microwave signals to or from remotely located transmitting or receiving antennas via optical fibers. Another common example is to distribute cable TV signals via optical fibers. Not only RF signals can be transmitted in this manner; microwave functions such as mixing (up-conversion or down-conversion) of signal frequencies can also be carried out photonically. We will discuss the

---

[1] Intensity modulation is in universal use today. Modulation of other parameters of the optical wave, such as its frequency, is in the research stage.

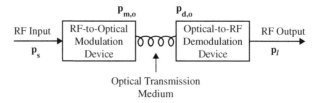

Figure 1.1. Basic components of a fiber optic link: modulation device, optical fiber and photodetection device.

figures of merit and performance measures of photonic RF transmission links in this chapter. The details of devices for the conversion between the RF and optical domains are discussed in other chapters of this book.

The objective of RF photonic links is clearly to achieve the same functions as conventional microwave links with longer distance of transmission, reduced cost, better performance, higher operating frequency, or reduced complexity and size. For this reason, their performance will be evaluated in terms of the criteria for microwave links. For microwave links – which are passive – typical performance criteria include just RF loss and frequency response. There is no nonlinear distortion or additional noise unless the signal is amplified. In photonic microwave links however, additional noise, such as the laser noise, can degrade the noise figure, and nonlinearity of the modulation process can reduce the spurious free dynamic range. Therefore, for microwave transmission and distribution using photonic links, which are more like active RF components, the important performance criteria are: (1) the RF gain and frequency response, (2) the noise figure (*NF*), and (3) the spurious free dynamic range (*SFDR*). These topics will be discussed in Sections 1.2, 1.3, and 1.4, respectively. For systems involving frequency mixing the figures of merit and the performance will be discussed in chapters presenting these techniques.

From the optical point of view, the magnitude of the RF intensity modulation is much less than the intensity of the unmodulated optical carrier. For this reason, one can analyze the link in the small signal approximation. Consequently, at a given DC bias operation point, the time variation of the optical intensity can be expressed as a Taylor series expansion about this DC bias point as a function of the RF signal magnitude.

To demonstrate clearly the system impact of such photonic link parameters as the CW optical carrier intensity, the noise, and the nonlinear distortion, we have chosen to discuss in this chapter the properties of just an *intrinsic* RF photonic link. In an intrinsic link no electronic or optical amplification is employed. An RF signal, at frequency ω, is applied to a modulation device to create intensity modulation of an optical carrier. The modulated carrier is transmitted to the receiver by a single fiber. The RF signals generated by the photodetector receiver constitute the RF output of the link. Only direct detection of the optical intensity (without amplification or any noise compensation scheme such as coherent detection) is considered for our discussion. Distribution of the optical carrier is limited to a single optical receiver. RF gain, *NF* and *SFDR* of such a basic link are analyzed in this chapter. Analysis of complex links involving amplification, distribution, and schemes for noise or distortion reduction, can be extended from the analysis of the intrinsic link.

There are two commonly used methods to create an RF intensity modulation of the CW optical carrier; one is by direct RF modulation of the laser and the second one is by RF modulation of the CW optical carrier via an external modulator. When the distinction between these two modulation methods is not germane to the discussion, we refer to them collectively as the modulation device. The RF gain, the *NF*, and the *SFDR* for both methods are discussed here.

1.2 Gain and frequency response

Conversion efficiencies, especially of the modulation device, are typically less than 10%, which leads immediately to photonic link losses of greater than 20 dB! This situation has motivated considerable work on efficiency improvement techniques. Conceptually there are two routes one can pursue: (1) improve the modulation device, and (2) improve the circuit that interfaces the modulation signal source to the modulation device. In this chapter we set up the analytical framework that shows how each of these approaches affects the solution. We also present an introductory investigation of how the interface circuit can improve performance.

The linear RF gain (or loss) of the link, g_t, is defined as the ratio of "the RF power, p_1, at frequency ω delivered to a matched load at the photodetector output" to "the available RF power at the input, p_s, at a single frequency ω and delivered to the modulation device." That is,

$$g_t = \frac{p_1}{p_s}. \tag{1.1}$$

Frequently the RF gain is expressed in the dB scale as G_t, where $G_t = 10 \log_{10} g_t$. If G_t is negative, it represents a loss.

The RF gain of a photonic link is frequency dependent. For links using simple input and output electrical circuits consisting of resistances and capacitances, their RF gain generally decreases as the frequency is increased. To quantify the *low pass* frequency response, the RF bandwidth f_B is defined as the frequency range from its peak value at DC to the frequency at which G_t drops by 3 dB. For a *bandpass* frequency response, which requires a more complex driving circuit, the G_t peaks at a center frequency. In this case, the bandwidth is the frequency range around the center frequency within which G_t drops less than 3 dB from its peak value.

There are three major causes of frequency dependence. (1) The directly modulated laser or the external modulator for the CW laser may have frequency dependent characteristics. (2) The voltage or current delivered to the modulation device may vary as a function of frequency due to the electrical characteristics of the input circuit. (3) The receiver and the detector may have frequency dependent responses.

Let $p_{m,o}$ be the rms magnitude of the time varying optical power at frequency ω (immediately after the modulation device) in the fiber. Let $p_{d,o}$ be the rms magnitude of the optical power at ω incident on the detector. Here, the subscript o denotes optical quantities, subscript m designates modulated optical carrier at ω at the beginning of the fiber and subscript d designates modulated optical carrier at ω incident on the detector. Quantities that do not have the o subscript apply to the RF. Then

$$p_{d,o} = T_{M-D} p_{m,o},\tag{1.2}$$

where T_{M-D} is the total optical loss incurred when transmitting the optical modulation from the modulation device to the detector. This quantity includes the propagation loss of the fiber and the coupling loss to and from the fiber. Since $p_{m,o}$ is proportional to the current or the voltage of the RF input, $p_{m,o}^2$ is proportional to p_s. Since the RF photocurrent at ω generated by the detector is proportional to $p_{d,o}$ and since the RF power at the link output is proportional to the square of the photocurrent, $p_{d,o}^2$ is proportional to p_{load}. Substituting the above relations into Eqs. (1.1) and (1.2) we obtain:

$$g_t = \left(\frac{p_{m,o}^2}{p_s}\right) T_{M-D}^2 \left(\frac{p_{load}}{p_{d,o}^2}\right)\tag{1.3a}$$

or

$$G_t = 10\log_{10}\left(\frac{p_{m,o}^2}{p_s}\right) + 20\log_{10} T_{M-D} + 10\log_{10}\left(\frac{p_{load}}{p_{d,o}^2}\right).\tag{1.3b}$$

Strictly speaking, the definition of gain requires an impedance match between the modulation source and the modulation device. A typical RF source can be represented by a voltage source with rms voltage v_s at a frequency ω in series with an internal resistance R_S. As we will see below few, if any, modulation devices have an impedance equal to R_S. There are clearly myriad circuits that can accomplish the impedance matching function. The selection for any particular application depends upon many criteria – primary among them are cost, bandwidth and efficiency. We have chosen a few illustrative examples below; a more comprehensive discussion is included in the book by Cox [1].

The simplest – and least expensive – form of matching is to use a resistor. However a resistor is a lossy device, so this form of matching is not very efficient. Thus lossless impedance matching (an ideal situation that can only be approximated in practice) offers greater efficiency but at increased circuit complexity, and hence cost. We will present one example from both categories in the discussion below.

As will become clearer in the sections to follow, $p_{m,o}^2/p_s$ is an important figure of merit because it contains the combined effects of the modulation device impedance and slope efficiency. To get a feel for $p_{m,o}^2/p_s$, we will present explicit formulations for several representative cases. In each case we will need to derive two expressions: one for the relationship between the modulation source power, p_s, and the modulation voltage or current; and one relating the modulation voltage or current to the modulated optical power, $p_{m,o}$. Thus the defining equation for $p_{m,o}^2/p_s$ will be different for different methods used to obtain the RF modulation of the optical carrier.

1.2.1 The $p_{m,o}^2/p_s$ of directly modulated laser links

The optical power output of a semiconductor laser is produced from the forward biased current injected into its active region. Figure 1.2 illustrates the instantaneous optical power $p_L(t)$ $(p_L(t) = p_L + p_{l,o})$ as a function of the total injected current $i_L(t)$, $(i_L(t) = I_L + i_l)$. Here, the bias current is I_L, and i_l is the small-signal RF modulation, i.e., $i_l \ll I_L$. If the modulation current is $\sqrt{2}i_s \cos \omega t$ where i_s is the rms magnitude of the RF current, then $p_{l,o}$ is the rms magnitude of the laser power

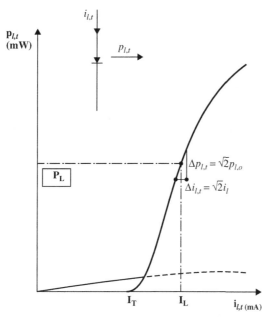

Figure 1.2. Representative plot of a diode laser's optical power, $p_{l,t}$, vs. the current through the laser, $i_{l,t}$, with the threshold current, I_T, and a typical bias current, i_L, for analog modulation.

at ω. Clearly, there is an approximately linear dependence of the change of $p_L(t)$ on the change of $i_L(t)$ in the range from the lasing threshold to the saturation of the laser output.

Direct modulation of semiconductor lasers (i.e., direct modulation of $i_L(t)$) is simpler to implement than external modulation. Hence it is the most commonly used method to achieve intensity modulation of the optical carrier, primarily because it is less expensive. However, the useful bandwidth of modulation is limited to the range from DC to the laser relaxation reasonance, which is typically a few tens of GHz. There is a detailed discussion of the modulation bandwidth, the noise, the nonlinearity, and the modulation efficiency of semiconductor lasers in Chapter 3 of this book.

The slope efficiency of the laser s_1 is defined as the slope of the $p_L(t)$ vs. $i_L(t)$ curve at a given laser bias current I_L in Fig. 1.2,

$$s_1 = dp_L(t)/di_L(t)|_{I_L} = p_{1,0}/i_1. \tag{1.4}$$

Here, s_1 includes the power coupling efficiency of the laser output to the guided wave mode in the fiber.

The i_1 is determined from the analysis of the circuit driving the laser. The laser can be represented electrically by an input impedance consisting of a resistance R_L in parallel with a capacitance C_L.

It is well known in circuit theory that the maximum RF power is transferred from the source at voltage v_s to the load impedance when the load impedance is matched to the source impedance, R_S. This maximum available power, p_s, is $v_s^2/(4R_S)$. Therefore the design goal of any RF circuit driving the laser is to provide an impedance match from R_L and C_L to R_S.

Let us consider first a simple resistively matched driving circuit illustrated in Fig. 1.3. This is applicable for frequencies where the reactance of the capacitance, $X_L = 1/j\omega C_L \gg R_L$. For in-plane lasers, typically $R_S \gg R_L$; thus a simple way to satisfy the matching condition at low frequencies is to add a resistance R_{MATCH}

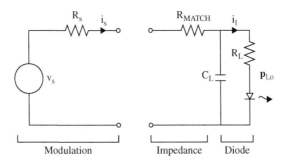

Figure 1.3. Circuit for a resistive magnitude match between a source and a diode laser whose impedance is represented by the parallel connection of a capacitor and a resistor.

in series with the laser so that

$$R_L + R_{MATCH} = R_S. \tag{1.5}$$

Ohm's law permits us to write the expression for the laser current, which in conjunction with Eq. (1.4) yields an expression for the modulated laser power, viz.:

$$i_1 = \frac{v_s/2}{(R_L + R_{MATCH})}, \qquad p_{l,o}^2 = \frac{s_1^2}{R_L + R_{MATCH}} \cdot \frac{(v_s/2)^2}{R_s}. \tag{1.6}$$

Dividing the second equation in (1.6) by the maximum power available to the laser at this terminal, p_s, $(v_s/2)^2/R_s$, yields an expression for $p_{l,o}^2/p_s$:

$$\frac{p_{l,o}^2}{p_s} = \frac{s_1^2}{R_s}. \tag{1.7}$$

The frequency variation of $p_{l,o}^2/p_s$ depends on the frequency dependencies of both i_1 and s_1. The frequency dependency of s_1 will be discussed in Chapter 3; for the purposes of the present discussion it will be assumed to be a constant, independent of frequency.

We can get a feel for the frequency dependency of i_1 from the simple circuit of Fig. 1.3. At frequencies higher than DC, i_1 drops because of C_L. From the analysis of the circuit shown in Fig. 1.3, we obtain

$$i_s = \frac{v_s}{R_S + R_{MATCH} + \dfrac{R_L}{1 + R_L(j\omega C_L)}},$$

$$i_1 = \frac{i_s}{1 + R_L(j\omega C_L)},$$

$$|p_{l,o}| = s_1|i_1| = s_1 v_s \left| \frac{1}{2R_S + (R_S + R_{MATCH})(j\omega C_L)R_L} \right|,$$

$$\frac{|p_{l,o}|^2}{p_s} = \frac{s_1^2}{R_S} \left[\frac{1}{1 + \left(2 - \dfrac{R_L}{R_S}\right)^2 \left(\dfrac{R_L \omega C_L}{2}\right)^2} \right]. \tag{1.8}$$

In addition to the frequency variation of s_1, this matching circuit will cause a 3 dB drop of $|p_{l,o}|^2/p_s$ whenever

$$\frac{R_L \omega C_L}{2} = \frac{1}{2 - \dfrac{R_L}{R_S}}. \tag{1.9}$$

The simple circuit illustrated in Fig. 1.3 may be replaced by any one of a number of more sophisticated circuits that matches R_L and C_L more efficiently to R_S.

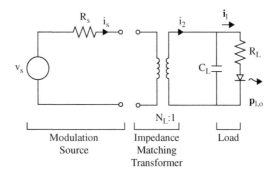

Figure 1.4. Circuit for an ideal transformer magnitude match between a source and a diode laser whose impedance is represented by the parallel connection of a capacitor and a resistor.

Consider here a matching circuit that can be represented symbolically as an ideal transformer as shown in Fig. 1.4. To achieve the desired match, the turns ratio of the transformer is chosen such that

$$N_L^2 R_L = R_s, \quad \text{and} \quad i_2 = N_L i_s. \tag{1.10}$$

We obtain from conventional circuit analysis:

$$i_1 = \frac{v_s N_L}{2R_s + R_s R_L (j\omega C_L)},$$

$$\frac{|p_{1,o}|^2}{p_s} = \frac{s_1^2 N_1^2}{R_s \left[1 + \left(\frac{R_L(\omega C_L)}{2} \right)^2 \right]}. \tag{1.11}$$

At low frequencies, $j\omega C_L$ is negligible, and we obtain

$$\frac{p_{1,o}^2}{p_s} = \frac{s_1^2}{R_s} N_L^2. \tag{1.12}$$

Thus transformer (lossless) matching, as show by Eq. (1.12), provides a $p_{1,o}^2/p_s$ that is N_L^2 times larger than the value obtained from resistive (lossy) matching, as shown by Eq. (1.7). The frequency response of this driving circuit will cause a 3 dB drop in $p_{1,o}^2/p_s$ when

$$\frac{R_L \omega C_L}{2} = 1. \tag{1.13}$$

Comparing Eq. (1.13) with Eq. (1.9), we see that the 3 dB bandwidth of the trans-former match is a factor of $2 - \frac{R_L}{R_s}$ larger than for the resistively matched circuit. This is an unusual result in that normally one trades increased response for decreased bandwidth.

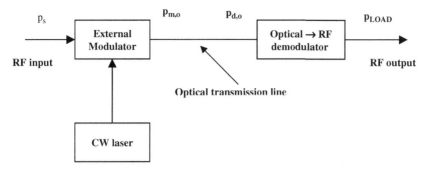

Figure 1.5. Illustration of an externally modulated fiber optic link.

Given the improvement in gain and bandwidth that is obtained in going from a resistive to a transformer match, it is natural to ask: how much further improvement is possible with some other form of matching circuit? Alternatively, one may be interested in using a matching circuit to provide a large i_1, i.e., a large gain g_t, over only a narrow band. The answers to these questions are provided by the Bode–Fano limit. A companion book, *Analog Optical Links*, by Cox [1], also published by Cambridge University Press, discusses such topics extensively.

The discussion in this subsection is applicable to different types of lasers including edge emitting and vertical cavity surface emitting lasers.

1.2.2 The $p_{m,o}^2/p_s$ of external modulation links

For external modulation, the laser operates CW and the desired intensity modulation of the optical carrier is obtained via a modulator connected in series optically with the laser. Figure 1.5 illustrates such an external modulation link. Similar to the case in directly modulated lasers, in external modulation links the frequency variation of $p_{m,o}$ depends on the specific modulator used and on the RF circuit driving the modulator. Different from directly modulated lasers, the bandwidth of state-of-the-art external modulators is approximately five times that of diode lasers. Consequently, the bandwidth of $p_{m,o}^2/p_s$, even when a broadband matching circuit is used, is determined in most applications primarily by the properties of the circuit driving the modulator.

In comparison with directly modulated laser links, the major advantages of external modulation links include (1) the wider bandwidth, (2) higher $p_{m,o}^2/p_s$, (3) lower noise figure and (4) larger *SFDR*. The disadvantages of external modulation links include (1) the additional complexity and cost of optical connections, (2) the necessity of maintaining and matching the optical polarization between the laser and the modulator,[2] and (3) the nonlinear distortions induced by external modulators.

---

[2] Although polarization independent modulators have been investigated, to date all these modulators have a significant reduction in sensitivity. Thus they are rarely (never?) used in practice.

For more than 10 years the most common type of external modulator that has been considered for RF photonic links is the Mach–Zehnder modulator fabricated in LiNbO$_3$. More recently the polymer Mach–Zehnder modulator as well as the semiconductor Mach–Zehnder and electroabsorption modulators have all also received some consideration for these applications. Each of these modulator types can have one of two types of electrodes: lumped element for lower modulation frequencies and traveling wave for higher frequencies. These different types of modulators are discussed in detail in Chapters 4, 5, 6, and 7 of this book.

The early dominance of the lithium niobate Mach–Zehnder modulator led to the use of the switching voltage for this type of modulator, V_π (to be defined below), as the standard measure of modulator sensitivity. However, with the increasing popularity of alternative types of external modulators – primary among them at present is the electroabsorption (EA) modulator – there is the need to compare the effectiveness of completely different types of modulators. Further, there is also the issue of how to compare the effectiveness of external modulation with direct.

One way to meet these needs, that has proved quite useful from a link perspective, is to extend the concept of slope efficiency – discussed above for direct modulation – to external modulation. This is readily done by starting with the external modulator transfer function, which is often represented in the literature by an optical transmission T of the CW laser power as a function of the voltage v_M across the modulator, $T(v_M)$. To incorporate $T(v_M)$ into the external modulation slope efficiency we need two changes in units: modulation voltage to current and optical transmission to power. When these conversions are substituted into Eq. (1.4) we obtain

$$s_m \triangleq \left. \frac{dp(i_L)}{di_L} \right|_{I_M} = R \left. \frac{dp(i_L)}{dv_M} \right|_{V_M} = R P_L \left. \frac{dT(v_M)}{dv_M} \right|_{V_M}. \tag{1.14}$$

Here R is the modulator impedance, P_L is the CW laser power at the input to the modulator and V_M is the modulator bias point.

Let us assume first that, within a given bandwidth of ω, T responds instantaneously to v_M. In other words, T is a function of the instantaneous V, independent of the time variation of v_M (this compares to the lumped element electrode case). For example, in Mach–Zehnder modulators whether fabricated in LiNbO$_3$ or polymers, it is well known that

$$T = \frac{T_{FF}}{2} \left[1 + \cos \left(\frac{\pi v_M}{V_\pi} \right) \right], \tag{1.15}$$

where T_{FF} is the fraction of the total laser power in the modulator input fiber that is coupled into the modulator output fiber when the modulator is biased for maximum transmission. In a balanced modulator, one of the transmission maxima occurs

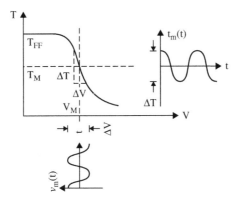

Figure 1.6. Illustration of the transmission T of an external modulator as a function of V. The $V(t)$ is illustrated in the inset below the $T(V)$ curve. It consists of the bias voltage V_M and the RF voltage $v_m(t)$. The transmittance T as a function of t is shown in the inset on the right. It consists of a bias transmittance T_M and a time varying transmittance $t_m(t)$.

at $v_M = 0$. V_π is the value of v_M which is required to shift the modulator from maximum to minimum transmission; ideally, at minimum transmission $T = 0$. In electroabsorption modulators,

$$T = T_{FF} e^{-\alpha(v_M)L}. \tag{1.16}$$

Here, L is the length of the modulator, and α is the absorption coefficient.

The relationship between T and v_M is illustrated in Fig. 1.6. At $v_M = V_M$, the bias transmission is T_M. The total applied voltage v_M consists of the sum of a DC bias V_M and a RF voltage $v_m(t)$. For a single frequency RF input, $v_M(t) = V_M + v_m(t)$ and $v_m(t) = \sqrt{2}v_{rf} \cos \omega t$, where v_{rf} is the rms RF voltage applied to the modulator. The resultant T is $T(t) = T_M + t_m(t)$. The resultant $t_m(t)$ and the input $v_m(t)$ as a function of time are illustrated in the two inset figures on the right hand side and below the $T(V)$ figure. The optical carrier power in the optical fiber transmission line, immediately after the modulator, is $P_L T$, where $P_L T = P_L T_M + P_L t_m(t)$. In general, T is not a linear function of v_M, and t_m is a distorted $\cos \omega t$ function.

As seen by Eqs. (1.15) and (1.16), we often have an analytic expression of $T(v_M)$ for an external modulator. In such cases, in the small-signal approximation ($v_m \ll V_M$), T can be expressed as a Taylor series expansion with $(v_m \cos \omega t)^n$ terms. The higher order terms give rise to distortions to the modulating signal and will be discussed again in Section 1.4.

In the first-order approximation (neglecting $n > 1$ terms), we can apply Eq. (1.15) to Eq. (1.14) to yield an expression for the slope efficiency of Mach–Zehnder modulators,

$$s_m = -\frac{\pi T_{FF} P_L R}{2 V_\pi}. \tag{1.17}$$

Applying Eq. (1.16) to Eq. (1.14) yields the analogous expression for electroabsorption modulators,

$$s_m = -\frac{\pi T_{FF} P_L R}{2 V_{\pi,eq}},\qquad(1.18)$$

where $V_{\pi,eq}$ is defined in Chapter 6.

From Eqs. (1.17) and (1.18), it is clear that to increase the slope efficiency, one wants a low value of V_π; in other words one wants as sensitive a modulator as possible. It is also clear from these equations that to maximize the slope efficiency it is equally important to have the product of the average optical power and the optical transmission as high as possible. Thus one wants a modulator with as little optical loss as possible and a laser with as high optical power as feasible.

It is important to note that all three of these parameters – V_π, T_{FF} and P_L – can be chosen independently. Therefore the upper bound on s_m for external modulation is not set by fundamental limits but by practical considerations. This is in contradistinction to the direct modulation case where there is a fixed upper bound to these quantities that is set by conservation of energy.

At high frequencies, T is no longer just a function of V. T depends on the time variation of the RF signal. In other words s_m is dependent on ω. This case corresponds to traveling wave modulators, two examples of which will be discussed in Chapters 5 and 6.

Similar to the case for directly modulated lasers, in external modulation links the frequency dependence of $p_{m,o}$ is a product of the frequency dependence of s_m and v_m. Both the magnitude (e.g., T_{FF} and V_π, or $V_{\pi,eq}$) and the frequency dependence of s_m will be different for different types of modulators such as the LiNbO$_3$ and polymer Mach–Zehnder modulators and electroabsorption modulators, and for different electrode designs such as lumped element and traveling wave. The s_m properties of various modulators will be discussed in Chapters 4 to 7. In the remaining discussion of this subsection we will assume s_m to be independent of ω.

It is important to note that s_m and the bandwidth of the modulator cannot be optimized independently. The V_π or $V_{\pi,eq}$, i.e., s_m, is directly related to the C_M of the modulator. For example, in a lumped element Mach–Zehnder modulator V_π is inversely proportional to the length of the electrode L, while C_M is proportional to L. Therefore there is a direct trade-off between s_m and the bandwidth of $p_{m,o}^2/p_s$. The trade-off between s_m and the bandwidth for electroabsorption modulators is discussed in Chapter 6. The common practice for broadband applications is to determine the largest C_M that will provide the desired bandwidth and then design the modulator electrodes, i.e., L, to maximize s_m under that constraint.

As was the case with matching to the diode laser, the magnitude and the frequency dependence of v_m will depend on the modulator input impedance, the driving circuit design, and the modulator electrode configuration – lumped element or traveling wave. Further, the matching circuit design approach will be the same as for the direct modulation case, however the details will be different because external modulators are voltage controlled devices with high input impedance while the lasers are current controlled devices with low input impedance.

1.2.2.1 Impedance matching to lumped element modulators

In all lumped element external modulators, v_m, the time dependent voltage across the waveguide, is considered to be independent of the position along the electrode. Electrically, the modulator electrode behaves like a lumped capacitor in parallel with a leakage resistance. Although more sophisticated representation of some modulators may include additional resistance or reactance, we will base our discussion here on an equivalent circuit of a capacitance C_M in parallel with a resistance R_M. In comparison, the equivalent circuit for forward biased diode lasers has a small R_L, whereas here R_M is a high resistance.

As we did for the diode laser, we will present expressions for $p_{m,o}^2/p_s$ with resistive and transformer type impedance matches to the modulator. Figure 1.7 shows a circuit using a resistance R_{MATCH} for matching. Since $R_M \gg R_S$, R_{MATCH} is placed in parallel with R_M. From circuit analysis, we obtain

$$v_m = v_s \frac{\left(\dfrac{R_E}{R_E + R_S}\right)}{1 + \dfrac{R_S R_E}{R_S + R_E}(j\omega C_M)}, \qquad \text{where } R_E = \frac{R_M R_{MATCH}}{R_M + R_{MATCH}}. \quad (1.19)$$

For maximum transfer of RF power at low ω, R_E is designed to be equal to R_S.

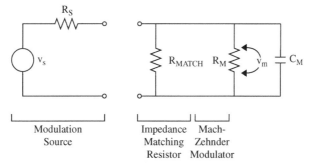

Figure 1.7. Circuit of a resistive magnitude match between a resistive source and the parallel connection of two lumped elements – R and C – which represent a first order impedance model for a Mach–Zehnder or an electroabsorption modulator.

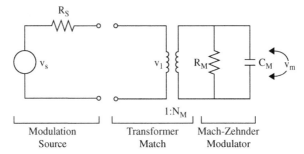

Figure 1.8. Circuit for an ideal transformer magnitude match between a source and the *RC* impedance model for a Mach–Zehnder or an electroabsorption modulator.

Therefore,

$$\frac{p_{m,o}^2}{p_s} = \frac{s_m^2}{R_S} \frac{1}{1 + (\omega C_M R_S/2)^2}. \tag{1.20}$$

Higher efficiency matching can again be obtained using a lossless circuit that can be represented as a transformer matching the input impedance to R_S. Figure 1.8 illustrates the use of an ideal transformer to match the input impedance of a modulator to R_S. Since $R_M > R_S$, we need a step-up transformer for matching, with $N_M^2 R_S = R_M$ and $v_m = N_M v_1$. From the circuit analysis we obtain

$$v_m = N_M v_s \frac{R_M}{(R_M + N_m^2 R_S) + N_m^2 R_S R_M(j\omega C_M)} = N_M v_s \frac{1/2}{1 + [N_M R_M(j\omega C_M)/2]}$$

and

$$\frac{p_{m,o}^2}{p_s} = \frac{s_m^2 N_M^2}{R_S} \frac{1}{1 + \left(\dfrac{N_M^2 \omega R_S C_M}{2}\right)^2}. \tag{1.21}$$

Comparing Eq. (1.20) with Eq. (1.21), we conclude that the transformer matching yields a low frequency gain that is N_M^2 times larger than the resistive matching. However we also note that the transformer match bandwidth is N_M^2 lower than the resistor match bandwidth. This represents the more typical situation where the gain increases by the same factor that the bandwidth decreases such that the gain–bandwidth product remains constant.

1.2.2.2 Impedance matching to traveling wave modulators

In Chapters 5 and 6, it will be shown that the electrodes of traveling wave modulators form an electrical transmission line. When the electrical propagation loss is negligible, the transmission line has a real characteristic impedance Z_0 which is independent of ω. The transmission line is usually terminated with a load resistance

R_L, where $R_L = R_0$ so that the input equivalent impedance of the terminated modulator is just a pure resistance R_0. In this case, the results obtained in Eqs. (1.19) to (1.21) are applicable, except that $C_M = 0$ under this condition. The equations predict and infinitely wide bandwidth. If $R_0 = R_S$ (R_S is usually 50 ohms), we have an ideal match to the modulation source without the use of any matching circuit. Although it is desirable to design Z_0 to be close to R_S from the impedance matching point of view, Z_0 is often considerably lower than R_S to obtain a large s_m. The trade-offs between s_m and Z_0 for Mach–Zehnder and electroabsorption modulators are discussed in Chapters 5 and 6. In reality, if Z_0 is significantly different than R_S, it is difficult to obtain a broadband match.

1.2.3 The $p_1/p_{d,o}^2$ of photodetectors

The photodiode generates an RF current i_d which is proportional to the optical power at the frequency ω incident on the photodiode, $p_{d,o}$. In other words, $i_d = s_d p_{d,o}$, where s_d is the slope efficiency of the photodiode in units of A/W. Let the photodiode be represented electrically by R, C_D, and R_D, as shown in Fig. 1.9. R is the leakage resistance of the reverse biased photodiode. It is typically very large compared to R_D and R_{LOAD}. The RF output power from the photodiode is delivered to the load shown as R_{LOAD}, where $R_{LOAD} > R_D$. To maximize the power transferred to the load, we need to match R, C_D, and R_D, to R_{LOAD}.

As an example, we have chosen the transformer matching circuit shown in Fig. 1.9. In this case,

$$i_{LOAD} = N_D i_2,$$

$$i_{LOAD} = \frac{N_D}{(R_D + N_D^2 R_{LOAD})(j\omega C_D) + 1 + \dfrac{R_D + N_D^2 R_{LOAD}}{R}} i_d. \qquad (1.22)$$

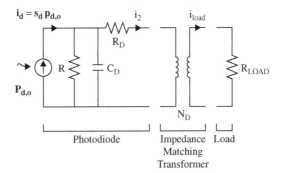

Figure 1.9. Schematic of simple, low frequency photodiode circuit model connected to an ideal transformer for magnitude matching the photodiode to a resistive load.

For a practical transformer turns ratio, $N_D^2 R_{LOAD}$ cannot match R. Thus $R \gg R_D + N_D^2 R_{LOAD}$, and we obtain:

$$\frac{p_1}{p_{d,o}^2} = \frac{i_{LOAD}^2 R_{LOAD}}{(i_d/s_d)^2} = \frac{N_D^2 s_d^2 R_{LOAD}}{\left(R_D + N_D^2 R_{LOAD}\right)^2 \omega^2 C_D^2 + 1}. \tag{1.23}$$

Equation (1.23) allows us to explore the changes in gain and bandwidth as a function of N_D. Let us assume that s_d is independent of ω within the frequency range of interest, then

$$\text{Gain increase}|_{low\,\omega} = \frac{p_1}{p_{d,o}^2}\bigg|_{N_D} \bigg/ \frac{p_1}{p_{d,o}^2}\bigg|_{N_D=1} \cong N_D^2. \tag{1.24}$$

$$\text{Bandwidth increase} = \frac{R_D + R_{LOAD}}{R_D + N_D^2 R_{LOAD}} \cong \frac{1}{N_D^2}. \tag{1.25}$$

Equations (1.25) and (1.26) express the same result we obtained for the modulator, that the product of gain and bandwidth is a constant. Consequently detector circuits (i.e., the N_D) can be designed to provide the highest gain within a desired bandwidth.

In reality the slope efficiency s_d also has a bandwidth dependence. Detectors with large s_d have generally small bandwidth. Figure 1.10 shows the slope efficiency of some commonly used photodetectors versus the 3 dB frequency. Chapter 8 discusses in detail the s_d for various photodetectors.

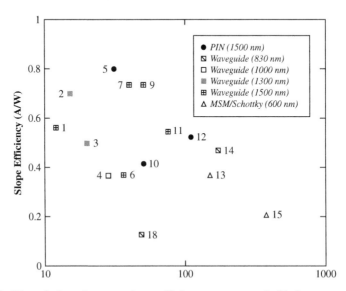

Figure 1.10. Plot of photodetector slope efficiency vs. upper 3 dB frequency for various photodetectors.

1.2.4 General comments on link gain

We can now combine the specific expressions of the previous sections into the general expression for link gain, Eq. (1.3a). For the low frequency gain of a direct modulation link we insert Eqs. (1.12) and (1.23) – the latter with $\omega \to 0$ – into Eq. (1.3a) to obtain:

$$g_{t-DM} = \frac{N_L^2 s_l^2}{R_S} s_d^2 N_D^2 R_{LOAD} = s_l^2 s_d^2 \big|_{R_S = R_{LOAD}; N_L = N_D = 1} \tag{1.26a}$$

The analogous expression for an external modulation link is obtained by substituting Eqs. (1.20) and (1.23) into Eq. (1.3a). We continue to assume that $\omega = 0$ in Eqs. (1.21) and (1.23). The result is

$$g_{t-DM} = \frac{N_M^2 s_m^2}{R_S} s_d^2 N_D^2 R_{LOAD} = s_m^2 s_d^2 \big|_{R_S = R_{LOAD}; N_M = N_D = 1} \tag{1.26b}$$

In comparing Eqs. (1.26a) and (1.26b) we see one of the advantages of expressing the link gain in terms of slope efficiency: a simple expression which can be applied to both types of modulations.

In all early links – and most links today – $g_t < 1$, which means the links have loss. This is consistent with the fact that the slope efficiencies of conventional diode lasers and PIN photodetectors are constrained by energy conservation to be < 1. By inference, it was assumed that the slope efficiency of external modulators was also constrained to be < 1.

There are two general approaches to reducing the link loss, which may be used separately or in conjunction. For narrow band links, it is possible to trade excess bandwidth for reduced loss. To achieve gain in a directly modulated link, for example, requires $R_D/R_L > s_l^2 s_d^2$. This has been demonstrated experimentally by Ackerman *et al.* [2]. The Bode–Fano limit defines the extent of this trade-off [3].

The other general approach to reduce the link loss is to increase the slope efficiency. This is a broadband approach that does not involve a gain–bandwidth trade-off. Although in principal it would be equally effective to improve the slope efficiency of either the modulation device or the photodetector, in practice all attention has focused on improving the modulation device.

The first theoretical study – and experimental verification – of improved slope efficiency was by Cox et al. [4] for external modulation. This work demonstrated a slope efficiency of 2.2 W/A and a net link power gain of 1 dB out to 150 MHz. More recently Williams *et al.* [5] have applied this technique to produce an external modulation link with 3 dB gain over 3 GHz.

It is also possible to improve the broadband slope efficiency of direct modulation. One technique for doing so is the cascade laser – as proposed and demonstrated by Cox *et al.* [6]. Using discrete devices, a link gain of 3.8 dB has been achieved over a bandwidth of < 100 MHz. Further improvements are anticipated as integrated versions are developed.

1.3 Noise figure

Noise plays important roles in setting the minimum magnitude signal that can be conveyed by the link and in contributing to the maximum *SFDR* for RF photonic links. The *SFDR* of RF photonic links will be discussed in detail in Section 1.4. We will focus our discussion in this section on noise sources and the noise figure.

For RF photonic links the noise power at the output of the link n_{out} is related to the noise at the input of the link n_{in} by the noise figure, *NF*, which is defined as,

$$NF = 10 \log_{10} \left(\frac{s_{in}/n_{in}}{s_{out}/n_{out}} \right) = 10 \log_{10} \left(\frac{n_{out}}{g_t n_{in}} \right),$$

$$\text{where} \quad n_{in} = kT_o \Delta f \quad \text{and} \quad T_o = 290 \text{ K.} \tag{1.27}$$

Here, s_{in} and s_{out} are the RF signal power at frequency ω at the input and the output of the link, respectively, k is Boltzmann's constant and Δf is the noise bandwidth. When the input and output terminals are matched, s_{out}/s_{in} is the gain of the link g_t, which we learned how to analyze in the previous section.

The two forms of Eq. (1.27) make clear a couple of facts about the noise figure: (1) when there is no noise added by the link, the minimum *NF* is 0 dB, and (2) *NF* does not depend on the signal power s_{in}.

Clearly, n_{out} will be larger than $g_t n_{in}$ because of the additional noise contributions from the laser, the detector and the circuit elements. Therefore, we can rewrite Eq. (1.27) as

$$n_{out} = g_t n_{in} + n_{add} \text{ and } NF = 10 \log_{10} \left(1 + \frac{n_{add}}{g_t n_{in}} \right). \tag{1.28}$$

From an alternative point of view, the effect of additional noise n_{add} in Eq. (1.28) can be considered as an equivalent additional noise at the input which is n_{add}/g_t. Equation (1.28) can also be generalized to include noise added at any point in the link $n_{add,i}$,

$$NF = 10 \log_{10} \left(1 + \frac{\sum_i n_{add,i}/g_i}{n_{in}} \right). \tag{1.29}$$

Here g_i is the gain between the link input and the additional noise source.

NF is commonly used to specify the noise property of any RF component. Equations (1.27) and (1.28) allow us to express n_{out} in terms of the NF of the link when $n_{in} = kT_o \Delta f$. The primary objective of this section is to discuss how to find the *NF* of a link.

1.3.1 Noise sources and their models

There are three dominant noise sources in photonic links: thermal, shot and relative intensity noise. All noise sources are statistically independent, so the total noise power from all these sources is simply the sum of the independent noise powers. In the following subsections, we will show the mean square current representation of these noise sources.

1.3.1.1 Thermal noise

Whenever any resistor is used in a circuit, it generates thermal noise, which is also known as the Johnson noise. It is a white noise, meaning that the noise power per unit bandwidth is a constant, independent of the frequency. To limit the noise power contributed by such a broadband noise source, there is usually an electrical filter of bandwidth Δf put in the circuit. Only the signals and the noise within this band need to be considered in the link. The noise contributed by a physical resistance R is commonly treated as a mean square current noise generator in parallel with a noise-free resistor R. The mean square noise current of the thermal noise is

$$\overline{i_t^2} = \frac{4kT\Delta f}{R}, \tag{1.30}$$

where, T is the temperature in kelvin. For each mean squared current representation, it is well known in circuit theory that there is also an equivalent mean squared voltage representation of the same noise source. Figure 1.11 shows that, for terminals A and B, a $\overline{v_{noise}^2}$ in series with R is completely equivalent to the $\overline{i_{noise}^2}$ in parallel with R, provided $\overline{v_{noise}^2} = R^2\overline{i_{noise}^2}$.

For any complex impedance, thermal noise is generated in the resistive part of the impedance. When there is a load matched to R, the noise power delivered to the load is $(\overline{i_t^2}/4)R = kT\Delta f$.

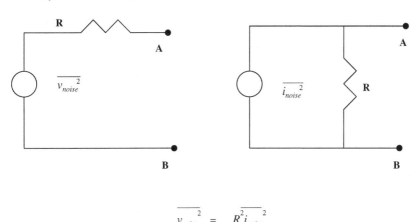

$$\overline{v_{noise}^2} = R^2\overline{i_{noise}^2}$$

Figure 1.11. The equivalent mean squared voltage and current source of noise.

1.3.1.2 Shot noise

Shot noise is generated whenever an electrical current with average value $\overline{I_D}$ is generated via a series of independent random events, e.g. the photocurrent in a detector. Shot noise is also a white noise source. Within a filter bandwidth Δf, the shot noise is represented by a mean square shot noise current generator (placed in parallel with a noiseless current generator I_D),

$$\overline{i_{sn}^2} = 2q\overline{I_D}\Delta f, \tag{1.31}$$

where q is the charge of an electron. Note that the shot noise is linearly proportional to I_D. Thus the shot noise can be larger than the thermal noise for sufficiently large I_D. It represents one of the disadvantages of using large P_L in photonic links. For example, the shot noise at 1 mA of I_D is equal to the thermal noise of a 50 ohm resistor at room temperature (290 K).

1.3.1.3 Relative intensity noise

There are fluctuations of laser intensity caused by random spontaneous emissions. These fluctuations are known as the relative intensity noise (*rin*). The *rin* is defined as

$$rin = \frac{\overline{\delta p_l^2}\Delta f}{\overline{P_L^2}}. \tag{1.32}$$

Here $\overline{\delta p_l^2}$ denotes the mean of the squared spectral density intensity fluctuations, and $\overline{P_L^2}$ is the average laser power squared. In general *rin* is a function of $\overline{P_L^2}$. It reaches maximum just above the lasing threshold and it decreases as the laser is above threshold. Relative intensity noise of a laser is usually specified in terms of *RIN*. *RIN* is related to *rin* by

$$RIN = 10\log_{10}(rin). \tag{1.33}$$

The *RIN* spectrum is not flat and hence this is not a white noise source. However, for simplicity, most link analyses assume that *RIN* is a constant within the bandwidth of interest. *RIN* also differs for diode and solid state lasers, and for single mode and multimode lasers. For example, single mode solid state lasers may have a *RIN* of -170 dB for $\Delta f = 1$ Hz, whereas diode lasers typically have a *RIN* of -145 dB for $\Delta f = 1$ Hz.

Both the $\overline{P_L^2}$ and the $\overline{\delta p_l^2}$ will produce corresponding currents squared in the load resistor after detection. Since the same detector and circuit will be used for $\overline{P_L^2}$ and $\overline{\delta p_l^2}$, the ratio of $\overline{\delta p_l^2}/\overline{P_L^2}$ is the same as $\overline{i_{rin}^2}/\overline{I_D}^2$. Hence the relative intensity noise can be represented as a current generator with a mean square current as

$$\overline{i_{rin}^2} = rin \cdot \overline{I_D}^2 \cdot \Delta f. \tag{1.34}$$

Notice that $\overline{i_{rin}^2}$ is proportional to $\overline{I_D}^2$, whereas the shot noise is only linearly proportional to I_D. Therefore the *RIN* noise dominates at high laser average power.

1.3.2 Noise figure analysis of representative links

To find the noise figure of a link, we need to calculate either the n_{add} at the output or the equivalent noise at the input, $\Sigma n_{i,add}/g_i$. In Figs. 1.3, 1.4, 1.7 and 1.8, we have shown examples of the circuit for the RF source driving the laser or the modulator. In Fig. 1.9 we have shown an example of the circuit in which the photodetector is matched to the load. In all links for the transmission of RF, the $p_{d,o}$ input to the photodetector is proportional to $p_{m,o}$, i.e., the modulated output of the modulation device. If we now add mean square noise current generators at various locations of the elements (such as the laser, the detector, and the resistance, that generate the noise) to the driving and detector circuits, we can calculate the resultant n_{add} or the $\Sigma n_{i,add}/g_i$ from circuit analysis. In such a calculation, all circuit elements, except the noise sources, will be considered noiseless elements.

Clearly, the *NF* of a given link will depend on the noise sources, their locations and the circuit configuration of the link. Fortunately, when the n_{add} is caused primarily by a dominant noise source, other noise sources can be neglected. Whether the dominant noise mechanism is thermal, shot or *RIN* noise will depend both on the operating conditions – such as the bias laser power and the average detector current – and on the circuit. We will now discuss how to calculate the *NF* for two cases that occur frequently.

1.3.2.1 The NF *of a* RIN *noise dominated link using a directly modulated laser*

Figure 1.12 shows the equivalent circuits of a transformer matched directly modulated laser link including the noise sources, which is based on the laser and detector circuits without the noise sources that have been shown in Fig. 1.4 and Fig. 1.9. For the convenience of calculation, the thermal noises due to the resistances R_S and R_L are represented by the equivalent noise voltage source in series with the resistance. The *RIN* noise is shown as a current noise source in parallel with the photodetector:

$$\overline{v_{ts}^2} = 4kTR_s\Delta f, \qquad \overline{v_{ti}^2} = 4kTR_L\Delta f,$$

$$\overline{i_{tLOAD}^2} = 4kT\Delta f/R_L, \qquad \overline{i_{rin}^2} = rin\overline{I_D}^2\Delta f. \qquad (1.35)$$

The *RIN* of a typical diode laser is < -150 dB/Hz and the average detector current is 1 mA. Under these assumptions $\overline{i_{rin}^2} = 1.0 \times 10^{-21} > \overline{i_{sn}^2} = i_t^2 = 3.2 \times 10^{-22}$ A^2/Hz. Consequently we can obtain a reasonably accurate value for the *NF* based on only two contributions to n_{add}. One is the thermal noise of R_L. Since the

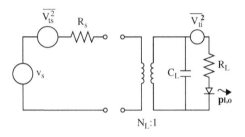

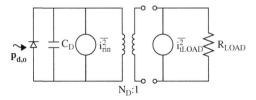

Figure 1.12. The equivalent circuit of a RF link for *NF* calculation with *RIN* dominated noise and a directly modulated laser.

noise source is located in the same loop as the modulation source, it is easy to verify that this source has the same gain from input to output as the modulation source. Therefore, its contribution to n_{add} is $g_t kT \Delta f$. The other contribution to n_{add} is the *RIN* noise. Since it is located at the link output before the transformer, its contribution to n_{add} can be written by inspection to be $N_D^2 \overline{i_{rin}^2} R_{\text{LOAD}} \Delta f$.

Consequently, we obtain

$$NF = 10 \log_{10}\left(1 + \frac{g_t kT + \overline{i_{rin}^2} N_D^2 R_{\text{LOAD}}}{g_t kT}\right) = 10 \log_{10}\left(2 + \frac{\overline{i_{rin}^2} R_{\text{LOAD}}}{N_l^2 s_l^2 s_d^2 kT}\right).$$

(1.36)

Figure 1.13 is a plot of Eq. (1.36) with $s_l = 0.2$ W/A, $s_d = 0.8$ A/W, $T = 290$ K, and $R_{\text{LOAD}} = 50$ Ω. It confirms our intuition that as the *RIN* increases so does the *NF*. For an $\overline{I_D}$ of 1 mA, the shot noise would be equal to the *RIN* noise at -168 dBm. Therefore the link would be shot noise dominated for *RIN* less than -168 dBm. and the above analysis would need to be modified. Notice also that, even for every small *RIN* and shot noise, the *NF* never decreases below 3 dB [$= 10 \log_{10}(2)$]. The theoretical limit of the noise figure for any RF photonic link with passive matching at the input is discussed in more detail in Section 1.3.3.

1.3.2.2 The NF *of a shot noise dominated link using an external modulator*

Figure 1.14 shows the equivalent circuit for calculating the *NF* of a RF link dominated by shot noise and using an external modulator. Solid state lasers, which typically have RIN < -175 dB/Hz, have often been used as the CW source for

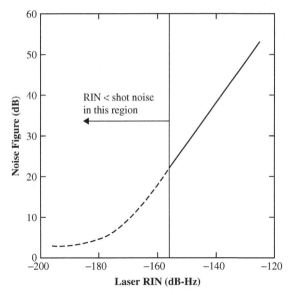

Figure 1.13. Plot of the *NF* of a *RIN* dominated directly modulated laser RF link as a function of *RIN*.

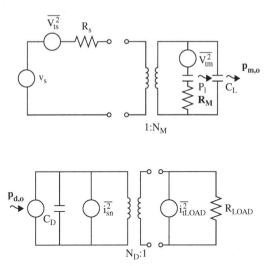

Figure 1.14. The equivalent circuit of an RF link for *NF* calculation with external modulator, dominated by shot noise.

external modulation. The typical $\overline{P_{\mathrm{L}}} = 10$ mW and $T_{\mathrm{FF}} = 3$ dB, resulting in an $\overline{I_{\mathrm{D}}} = 4$ mA (for $s_{\mathrm{d}} = 0.8$ A/W). Consequently,

$$i_{\mathrm{sn}}^2 = 1.28 \times 10^{-21} \ \mathrm{A}^2/\mathrm{Hz} > i_{\mathrm{t}}^2 = 3.20 \times 10^{-22} \ \mathrm{A}^2/\mathrm{Hz} \ (\text{for } R = 50 \ \Omega)$$
$$> i_{rin}^2 = 5.06 \times 10^{-23} \ \mathrm{A}^2/\mathrm{Hz}. \tag{1.37}$$

Transformer matching is used for both the driving circuit and the detector circuit. Similar to Fig. 1.12, the thermal noise sources for R_M, R_{LOAD}, and R_S are shown in the figure. Similar to Eq. (1.37), we obtain

$$NF = 10 \log_{10} \left(2 + \frac{2q I_D}{s_m^2 s_d^2 R_M k T} \right). \tag{1.38}$$

Notice once more that, like the previous example, the theoretical limit of the *NF* is 3 dB. However, unlike the *RIN* dominated case, the *NF decreases* as the shot noise (or I_D) *increases*.

The reason for this apparently anomalous result rests on the combined effects of the *NF* definition and the dependence of gain on optical power. Recall from Eq. (1.29) that the *NF* moves all noise sources throughout the link to equivalent noise sources at the input. In this case to move the shot noise at the output to the input requires dividing by the link gain. In an externally modulated link this gain is proportional to the square of the average optical power. However the shot noise is only linearly proportional to the average optical power. Together these facts imply that the gain is increasing faster than the shot noise. In turn this means that although the shot noise is increasing, the effect of this noise – when translated to the input – is decreasing.

1.3.3 Limits on noise figure

In Sections 1.3.2.1 and 1.3.2.2 we saw that the minimum noise figure for large gains is 3 dB. It can be shown (see for example Cox *et al.* [7]) that the general expression for the noise figure limit for photonic microwave links is simply

$$NF_{min} = 10 \log_{10} \left(2 + \frac{1}{g_t} \right),$$

which means that, as $g_t \to \infty$, $NF \to 3$ dB.

In the low gain case, i.e., $g_t \ll 1$, then the $NF \to -G_t$. This limit has been referred to as the passive attenuation limit, since it is analogous to a passive attenuation in which the *NF* equals the attenuation. Link noise figures approaching this limit have been demonstrated: for example, Cox *et al.* [7] measured $NF = 4.2$ dB. However, in general link noise figures are typically 15–30 dB.

It is important to emphasize that there are three conditions that need to be satisfied for the 3 dB limit to be applicable. The source must be losslessly and passively matched to the modulation device. Removing any one of these conditions invalidates the limit. Active matching, such as with an electronic preamplifier, is one well known and widely used technique for achieving link noise figures less than 3 dB. If the match is not lossless, then the minimum noise figure is simply increased by the amount of this loss.

A passive way to circumvent the 3 dB limit is to introduce an intentional impedance mismatch between the source and the modulation device. Using this approach Ackerman *et al.* [8] demonstrated a $NF = 2.5$ dB. In a traveling wave modulation device, a variation of the mismatch approach is available to circumvent the 3 dB limit. For modulation frequency/electrode length combinations where one-half the modulating wavelength covers the electrode, then any modulation signal traveling on the electrodes in the opposite direction of the optical propagation will introduce zero modulation. This would include noise generated by the traveling wave termination resistance. For this frequency, the minimum noise figure would be 0 dB. To realize this in practice would require that the electrodes themselves generate no external noise – i.e., they would need to be lossless. To date $NF < 3$ dB in traveling wave modulation devices has not been demonstrated experimentally because of insufficient slope efficiency.

1.4 Distortions in RF links

In Section 1.2, we defined the linear gain of the link. The gain relates the RF power at the fundamental frequency ω delivered to the load R_{LOAD} to the input RF power at ω. However, careful examination of the output will include not only the signal at the fundamental signal frequency ω but also at frequencies that are harmonically related to the fundamental. In other words, there will be nonlinear distortion of the fundamental signal. If a link is to be useful in linear applications, then for sufficiently low input power, the harmonics need to be smaller than the noise. As the input RF power at the fundamental increases, the output power at the harmonic frequencies $n\omega$ will increase faster than the signal power at ω for reasons that will become clear below. Thus harmonic distortion power will eventually surpass the noise power. In this case, the *SFDR* will be limited by the nonlinear distortion at high input RF power. Discussion of the effect of noise and nonlinear distortion on *SFDR* will be presented in the next section. Here, we focus on how to evaluate nonlinear distortions.

Let us consider the total photocurrent $i_{T,LOAD}$ delivered to the load resistance R_{LOAD}. Here, $i_{T,LOAD}$ consists of a DC bias photocurrent, $I_{B,LOAD}$, plus the RF signal and distortions. For distortion analysis, the effect of the noise is neglected. Mathematically, the voltage input to the RF link consists of a DC bias voltage V_B plus the RF signal, $v_{rf}(t)$, i.e., $V = V_B + v_{rf}(t)$. We can express the link output photocurrent's relationship to the link input voltage though the link transfer function, h, viz.:

$$i_{T,LOAD}(t) = h[V_B + v_{rf}(t)] \qquad (1.39)$$

Recall from Section 1.2 that the RF part of $i_{T,LOAD}$ can be analyzed in the small signal approximation. In terms of Eq. (1.39), this means that we can approximate h by its Taylor series expansion about its value at V_B,

$$i_{T,LOAD} = h(V_B) + v_{rf} \frac{\partial h}{\partial V}\bigg|_{V=V_B} + \frac{v_{rf}^2}{2!} \frac{\partial^2 h}{\partial V^2}\bigg|_{V=V_B} + \frac{v_{rf}^3}{3!} \frac{\partial^3 h}{\partial V^3}\bigg|_{V=V_B}$$

$$+ \frac{v_{rf}^k}{k!} \frac{\partial^k h}{\partial V^k}\bigg|_{V=V_B} + \cdots$$

$$= I_{B,LOAD} + h_1 v_{rf} + h_2 v_{rf}^2 + h_3 v_{rf}^3 + \cdots + h_k v_{rf}^k + \cdots \qquad (1.40)$$

where,

$$h_k = \frac{1}{k!} \frac{\partial^k h}{\partial V^k}\bigg|_{V=V_B}.$$

For $V = V_B + \sqrt{2} v_s \cos \omega t$ or $v_{rf}(t) = \sqrt{2} v_s \cos \omega t$,

$$v_{rf}^2 = \frac{1}{2}(\sqrt{2} v_s)^2 (1 + \cos 2\omega t),$$

$$v_{rf}^3 = \frac{1}{4}(\sqrt{2} v_s)^3 (3 \cos \omega t + \cos 3\omega t),$$

$$v_{rf}^4 = \frac{1}{8}(\sqrt{2} v_s)^2 \left(\frac{6}{2} + 4 \cos 2\omega t + \cos 4\omega t \right),$$

$$\cdots \qquad (1.41)$$

Therefore, the h_k term contributes to the kth harmonic distortion at the frequency $k\omega$. It also contributes to the $(k-2)\omega t$, $(k-4)\omega t$, ..., $(k-2m)\omega t$, ... lower frequency terms, provided $k > 2m$. In view of Eq. (1.41), we can rewrite Eq. (1.40) as,

$$i_{T,LOAD} = I_{B,LOAD} + \frac{1}{2}h_2(\sqrt{2} v_s)^2 + \frac{6}{16}h_4(\sqrt{2} v_s)^4 + \cdots$$

$$+ \left[h_1 \sqrt{2} v_s + \frac{3}{4}h_3(\sqrt{2} v_s)^3 + \frac{10}{16}h_5(\sqrt{2} v_s)^5 + \cdots \right] \cos \omega t$$

$$+ \left[\frac{1}{2}h_2(\sqrt{2} v_s)^2 + \frac{4}{8}h_4(\sqrt{2} v_s)^4 + \cdots \right] \cos 2\omega t$$

$$+ \left[\frac{1}{4}h_3(\sqrt{2} v_s)^3 + \frac{5}{16}h_5(\sqrt{2} v_s)^5 + \cdots \right] \cos 3\omega t + \cdots$$

$$= I_{B,LOAD} + \sqrt{2} i_{LOAD} \cos \omega t + \sqrt{2} i_{2,LOAD} \cos 2\omega t$$

$$+ \sqrt{2} i_{3,LOAD} \cos 3\omega t + \cdots . \qquad (1.42)$$

There are several important observations about Eqs. (1.40) to (1.41) that need to be discussed. (1) For the signal at the fundamental frequency, the linear relationship between $i_{\text{T,LOAD}}$ and v_s (or i_1) expressed in Section 1.2 is correct only when contributions from h_2 and h_3, etc. are neglected. (2) Usually, $h_1 > h_2 > h_3$, and so on. Therefore within a reasonable range of v_s, the fundamental is much larger than the 2nd harmonic, which is larger than the 3rd harmonic, etc. It follows from this that the 2nd is larger than the 4th, the 3rd larger than the 5th, etc. Thus for the majority of systems for linear use, we only need to consider h_1, h_2, and h_3. (3) The largest coefficient for the cos $n\omega t$ term contains a v_s^n. Therefore the power of the nth harmonic increases at least n times faster than the power of the fundamental as v_s. (4) There is also a slight shift of DC bias from $I_{\text{B,LOAD}}$ as a function of h_2. However, $I_{\text{B,LOAD}}$ is typically much larger than the terms involving higher orders of v_s. (5) Strictly, Eq. (1.40) is valid only if h_n is time independent and frequency independent. In practice, we consider h_n to be time and frequency independent over a given bandwidth. Thus, similar to s_m and s_1, $h(V)$ may be different in different ranges of ω. (6) The time averaged RF power is the time average of $i_{\text{T,LOAD}}^2 (t) R_{\text{LOAD}}$. The time average of the cos $m\omega t$ cos $n\omega t$ cross product terms is zero. Therefore, the time averaged RF power contained in the fundamental and in each harmonic can be calculated separately. The output power of the fundamental, p_1, is $i_{\text{LOAD}}^2 R_{\text{LOAD}}$. The output power of the nth harmonic, $p_{n\text{HM}}$, is $i_{n,\text{LOAD}}^2 R_{\text{LOAD}}$.

Although in principle a single sinusoid for v_{rf} is sufficient to characterize completely the nonlinear properties of a link, practical and system considerations often dictate the use of two or more closely spaced sinusoids to characterize distortion.

For simplicity we only consider here the two-equal-tone case, $v_{\text{rf}} = \sqrt{2}v_s(\cos\omega_1 t + \cos\omega_2 t)$, $\omega_1 \approx \omega_2$. Calculation of v_{rf}^n out to 3rd order, similar to the technique shown in Eq. (1.41), yields various mixed frequency terms, in addition to the harmonics of each tone:

Second order: $\omega_1 \pm \omega_2$, $2\omega_1$, $2\omega_2$; Third order: $2\omega_1 \pm \omega_2$, $2\omega_2 \pm \omega_1$, $3\omega_1$, $3\omega_2$.

For applications where the bandwidth is less than an octave, the 2nd order distortion terms fall out of band and can be filtered out. Of the remaining 3rd order terms, the sum-terms can also be filtered out. This leaves only the $2\omega_1 - \omega_2$ and $2\omega_2 - \omega_1$ terms to limit linearity, which are commonly referred to as the 3rd order intermodulation (3IM) distortion. To demonstrate the relation between the 3rd order harmonic (3HM) and the 3IM distortion, consider the 3IM case when 5th and higher order terms are negligible. The coefficient of $\cos(2\omega_1 - \omega_2)t$ and $\cos(2\omega_2 - \omega_1)t$ is $\frac{3}{4}h_3(\sqrt{2}v_s)^3$. Therefore the amplitude of 3IM photocurrent for the two tone input is simply three times the 3rd harmonic photocurrent for a single tone input. Consequently, the output power of 3IM, $p_{3\text{IM}}$, is nine times the output power of 3HM, $p_{3\text{HM}}$.

Calculation of $h(V)$ for every component in the link is long and tedious. We usually identify a particular component that causes most of the nonlinear distortion

and treat the remaining link components as perfectly linear. For example, in a
link using an external modulator, the photodetector is much more linear than the
modulator. Therefore, we obtain

$$v_m = Av_s, \qquad i_{\text{T,LOAD}} = Bs_d p_{\text{T,d}},$$
$$p_{\text{T,d}} = P_{\text{L}} T_{\text{M-D}} T(V_{\text{B}} + v_{\text{rf}}),$$
$$h \cong Bs_d P_{\text{L}} T_{\text{M-D}} T(V_{\text{B}} + v_{\text{rf}}),$$
$$h_k = Bs_d P_{\text{L}} T_{\text{M-D}} \frac{1}{k!} \frac{\partial^k T}{\partial V^k}\bigg|_{V=V_{\text{B}}}. \tag{1.43}$$

Here, A and B are obtained from the circuit analysis as shown in Section 1.2, and
$p_{\text{T,d}}$ is the total instantaneous optical power incident on the photodetector.

1.4.1 A graphical illustration of SFDR

Figure 1.15 shows the various output powers in dB – such as the fundamental output
at ω, the output power of various harmonics at $n\omega$ and the output noise power – as
a function of the input RF power p_s in dB at ω. The noise power at the output is
determined from the *NF* as discussed in Section 1.3. Since the noise power N is
independent of p_s, it is shown as a horizontal line. In Eq. (1.42), we have shown that
i_{LOAD} is proportional to v_s when higher order contributions can be neglected. Since
the fundamental output p_{LOAD} is approximately proportional to p_s, it is shown as
a line with a slope of one. The position of the line depends on parameters such as
$s_1, s_m, s_d, P_{\text{L}}, v_m/v_s, i_1/v_s, i_{\text{LOAD}}/i_d$, etc. Similarly, since $i_{2,\text{LOAD}} \propto v_s^2$ when higher
order contributions are neglected, the 2nd harmonic output is shown as a line having
a slope of two. Since $h_2 \ll h_1$ the second harmonic output, p_{2HM} starts at a value
much lower than the fundamental. Similarly, both the 3HM and the 3IM output lines
have a slope of three. However, the 3IM output is $10\log_{10} 9$ dB larger than the 3HM
output. As the h_n coefficients change, the vertical positions of these lines change.

When the fundamental output is less than the noise output, the signal cannot be
detected. When the signal is greater than the noise, the signal-to-noise ratio increases
as p_s is increased. This ratio is sometimes referred to simply as the dynamic range.
However, to avoid confusion with the dynamic range involving distortion – to be discussed below – it is better to use a more descriptive term, such as the signal-to-noise
dynamic range (*SNDR*). Since the noise power is dependent on bandwidth, so is the
SNDR. For each order of magnitude increase in bandwidth, the total noise power
increases by 10 dB, thereby decreasing the *SNDR* by 10 dB for a fixed signal level.
Thus it is common to quote the *SNDR* in a 1 Hz bandwidth, since this value can readily be scaled to any other bandwidth using the above scaling relation; this is listed

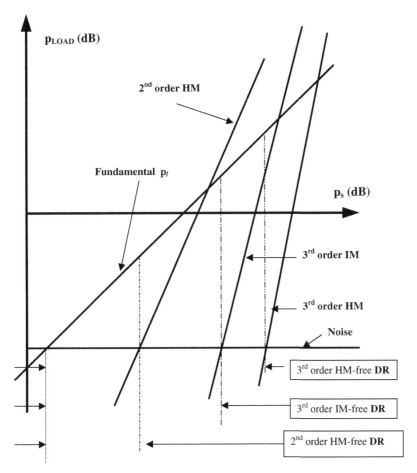

Figure 1.15. Graphical illustration of the spurious free dynamic range (SFDR) in a P_{LOAD} vs. p_s plot.

Table 1.1.

	Bandwidth scaling
SNDR	Hz
SFDR	
2nd	$Hz^{1/2}$
3rd	$Hz^{2/3}$

in the first line of Table 1.1. The maximum fundamental level – and hence the maximum *SNDR* – is limited by the gain reduction or compression. That inevitably occurs as the signal level is increased. A commonly used value is the 1 dB compression point; i.e., when the gain has been reticent by 1 dB below its low amplitude value.

As long as the harmonic output is smaller than the noise, the *SNDR* determines the system performance. When the fundamental is large enough to generate distortion terms that rise above the noise floor, such terms can often degrade system performance. Therefore when the effects of distortion are included, they set a limit on the maximum level of the fundamental, which is typically less than the compression level. The signal level that generates distortion terms equal to the noise floor is the maximum signal for which the link output is free from distortion. Hence this is the spur-free, or intermodulation-free, dynamic range. The bandwidth dependence of the *SFDR* depends on the order of the distortion, because of different signal power dependencies of the fundamental and distortion terms. Further, since each distortion order has a different power dependency, consequently there is a different *SFDR* for each distortion order. Table 1.1 also lists common distortion orders and their bandwidth depedence.

Let us consider first a link in which the 2nd harmonic is the dominant contribution to the nonlinear distortion. As p_s increases, the 2nd harmonic power increases. The signal-to-harmonic ratio becomes eventually less than the signal-to-noise ratio after the 2nd harmonic output surpasses the noise output. The signal-to-harmonic ratio that determines the system performance becomes smaller as p_s is increased further. Therefore, the 2nd order harmonic free dynamic range (DR) of the link is the range of p_s (or p_{LOAD}) in dB from the point where $p_{LOAD} = N$ to the point where the 2nd harmonic power $= N$. This is shown in Fig. 1.15. Similarly the 3rd harmonic free dynamic range is the range of p_s (or p_{LOAD}) in dB from $p_{LOAD} = N$ to $p_{3HM} = N$ as shown in Fig. 1.15. The 3IM free dynamic range is also shown in the same figure.

To find the *SFDR* of a given link, we will first determine all the orders of nonlinearity that are considered to be important for the desired application. The *SFDR* is the smallest range of p_s (or p_{LOAD}) within which $p_{LOAD} = N$ and the largest harmonic power of all those orders $= N$.

1.4.2 An alternative graphical representation of nth order distortion free DR

The preceding procedure of determining the nth order harmonic or IM free dynamic range can be more conveniently obtained by considering the N to p_{LOAD} ratio and the IM or harmonic to p_{LOAD} ratio. Let us define first the modulation index m as

$$m = \frac{i_{LOAD}}{I_{B,LOAD}}.$$

For a given $I_{B,LOAD}$, let us plot $10 \log_{10}(N/p_{LOAD})$ in dBc and $10 \log_{10}(p_{nHM}$ or $p_{nIM}/p_{LOAD})$ in dBc as a function of $10 \log_{10} m$ in dB as shown in Fig. 1.16. Since N

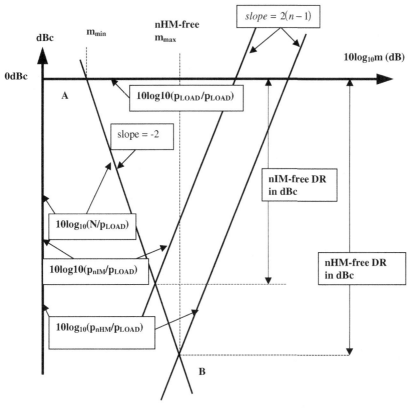

Figure 1.16. Graphical illustration of the spurious free dynamic range (SFDR) in a plot of dBc of "noise or nonlinear distortion power/p_{LOAD}" vs. modulation index m.

is independent of i_{LOAD} and since $p_{\text{LOAD}} \propto i_{\text{LOAD}}^2$, the line representing N/p_{LOAD} has a slope of -2. The vertical position of this noise line depends on the dominant noise mechanism and its dependence on $I_{\text{B,LOAD}}$, the NF, Δf and other parameters discussed in Section 1.3. The line representing $10\log_{10}(p_{n\text{HM}}/p_{\text{LOAD}})$ has a slope of $2(n-1)$. The line representing $10\log_{10}(p_{n\text{HM}}/p_{\text{LOAD}})$ has the same slope, but is higher in the vertical position. The definition of nth order harmonic (or IM) free DR in dBc is from point A, at $N = p_{\text{LOAD}}$ (i.e., 0 dBc), to point B which is the intersection of the $10\log_{10}(N/p_{\text{LOAD}})$ line with the $10\log_{10} p_{n\text{HM}}$ (or $p_{n\text{IM}}/p_{\text{LOAD}}$) line. The value of m from A to B varies from $m = m_{\text{min}}$ to $m = m_{\text{max}}$. In Fig. 1.16, only points A and B for the harmonic free DR are shown. As the vertical positions of the noise and the $p_{n\text{HM}}$ or $p_{n\text{IM}}$ line changes because of changes in the link design (e.g., change of NF, P_{L} or $T(V)$), the nth order distortion free DR is just the vertical coordinates in dBc of the intersection point B.

1.4.3 General comments on dynamic range

Characterization of the link distortion via the *SFDR* is commonly applied in the design of radar and communication systems. In such cases the strong, distortion-producing signals are typically few and randomly located in frequency.

However, there are applications where there are many, equally spaced carriers – among the common examples are cable TV and wireless systems. In principal it would be possible to use the *SFDR* to characterize the distortion in these systems too, but in practice other means are used. The most common measures for such multiple-tone systems are the composite second-order (CSO) and the composite triple beat (CTB). Unlike the *SFDR*, these later measures are functions of the number of carriers and the frequency within the band. Thus these measures are more application-specific and as a result there are no simple scaling laws as there were with *SFDR* and bandwidth. Further the conversion between 2nd order *SFDR* and CSO as well as between 3rd order *SFDR* and CTB are complex and approximate [9].

In some applications the linearity is insufficient to meet the application requirements. In such cases a wide variety of linearization techniques have been developed. The principal categories of electro-optical techniques are predistortion, feedback and feed-forward. Predistortion of a diode laser is widely used in CATV distribution applications for example. A number of optical linearization techniques have also been developed. To date these techniques have only been implemented in external modulation. An analytical comparison of a number of these techniques has been prepared by Bridges and Schaffner [10].

One of the important findings of their study, which has been confirmed experimentally by Betts [11], is that there is a trade-off between broadband linearization (i.e., 2nd and 3rd order) and noise figure. Narrowband linearization (i.e., 3rd order only) does not impose such a trade-off.

1.5 Summary and conclusion

In this chapter we have presented the basics for the design of photonic microwave links. This field has emerged over the last ten years as a discipline that is distinct from both device and system design. Although there are two modulation methods – direct and external – we have shown that there exists a single formalism for analysis of both that is based on slope efficiency.

However, the link design methodology also builds on the traditional measures – such as G_t, *NF*, *SFDR* – for characterizing the performance of RF devices. This facilitates combining links with other RF components to meet system requirements.

At present photonic microwave links have achieved impressive performance in all key parameters; bandwidth to 150 GHz, gain of 30 dB, noise figure of 4.2 dB and *SFDR* of 130 dB Hz.

References

1. C. Cox, *Analog Optical Links: Theory and Practice*, Cambridge, Cambridge University Press. To be published.
2. E. Ackerman, D. Kasemset, S. Wanuga, D. Hogue, and J. Komiak, "A high-gain directly modulated L-band microwave optical link," *IEEE MTT-S Int. Microwave Symp. Dig.*, Dallas, Texas, pp. 153–5, 1990.
3. R. Fano, "Theoretical limitations on the broadband matching of arbitrary impedances," *J. Franklin Inst.*, **249**, 57–84 and 139–54, 1950.
4. C. Cox, G. Betts, and L. Johnson, "An analytic and experimental comparison of direct and external modulation in analog fiber-optic links," *IEEE Trans. Microwave Theory Tech.*, **38**, 501–9, 1990.
5. K. Williams, L. Nichols, and R. Esman, "Photodetector nonlinearity limitations on a high-dynamic range 3 GHz fiber optic link," *J. Lightwave Technol.*, **16**, 192–9, 1998.
6. C. Cox, H. Roussell, R. Ram, and R. Helkey, "Broadband, directly modulated analog fiber optic link with positive intrinsic gain and reduced noise figure," *Proc. IEEE International Topical Meeting on Microwave Photonics*, Princeton, New Jersey, 1998.
7. C. Cox, E. Ackerman, and G. Betts, "Relationship between gain and noise figure of an optical analog link," *IEEE MTT-S Int. Microwave Symp. Dig.*, San Francisco, California, pp. 1551–4, 1996.
8. E. Ackerman, C. Cox, G. Betts, H. Roussell, K. Ray, and F. O'Donnell, "Input impedance conditions for minimizing the noise figure of an analog optical link," *IEEE Trans. Microwave Theory Tech.*, **46**, 2025–31, 1998.
9. T. Darcie and G. Bodeep, "Lightwave subcarrier CATV transmission systems," *IEEE Trans. Microwave Theory Tech.*, **38**, 524–33, 1990.
10. W. Bridges and J. Schaffner, "Distortion in linearized electro-optic modulators," *IEEE Trans. Microwave Theory Tech.*, **43**, 2184–97, 1995.
11. G. Betts, "Linearized modulator for suboctave-bandpass optical analog links," *IEEE Trans. Microwave Theory Tech.*, **42**, 2642–9, 1994.

2

RF subcarrier links in local access networks

XIAOLIN LU*

AT&T Labs

2.1 Introduction

Today's communication industries are facing tremendous challenges and new opportunities brought about by deregulation, competition, and emerging technologies. The opportunities in broadband subscriber access networks have been stimulating large scale business efforts and technology evolution. Network operators and services providers must continuously upgrade their "embedded infrastructure" in order to protect their current revenue while searching for new markets. Depending on the economic situation and projected service/revenue potentials, different network operators and service providers may chose different network upgrade paths and business strategies utilizing different technologies [1–3].

With the cost being the primary consideration, the embedded metallic last-mile drop may exist longer than people would hope in wired networks. This therefore makes it necessary to embrace RF transmission technology in those access networks. Even in all-fiber networks (e.g., PON: passive optical networks), the capability of broadcasting multichannel TV signals is critical, therefore making it desirable to carry certain types of RF signals. All these also become very attractive thanks to the innovations in the wireless industry that continually improve the bits/Hz and bits/dollar ratios of RF technologies.

In this chapter, we discuss technology and system issues associated with RF links in the access networks. Instead of discussing all the physical details with complicated equations, we try to provide an intuitive view of those issues, their impacts on system performance, and on-going technology innovations that continually drive architecture evolution and revolution. Section 2.2 describes the different access networks and the roles of RF subcarrier links in those networks. Section 2.3 discusses RF lightwave technology and its applications, including optical transmitters, receivers, and fiber amplifiers. Section 2.4 discusses certain system issues

---

* Xiaolin Lu is currently with Morning Forest, LLC.

and the different network migration strategy that will affect and be affected by the application of RF subcarrier link technology.

2.2 Overview of local access networks

Historically, communication networks have been established based on supporting specific services and related business models. Therefore, each type of network inherently has its own characteristics that fit special needs. With entertainment and communication being the two major and distinct services, at least historically, the access networks can be categorized as broadcast-based and switch-based. Examples include hybrid fiber/coax (HFC) based tree-and-branch networks used by the cable industry and twist-pair based star architecture used by the telephony industry.

2.2.1 Broadcast networks

Broadcasting multichannel video to customer homes was and is still the primary business of the cable industry. The notion of "pushing" information to users and letting them choose-and-pick leads to the tree-and-branch architecture and the use of coaxial cable (broadband capacity). Using RF subcarrier links with frequency division multiplexing (FDM) then becomes very natural for this type of network. Figure 2.1 shows three primary categories of wired access systems. An example of an HFC network [4–7] originating from the cable industry is shown in Fig. 2.1a. The headend broadcasts multichannel analog video (AM–VSB) to multiple remote fiber nodes (FNs) using intensity modulation of high-power linear lasers or external modulators such that one laser may serve one or more FNs. The FN then further

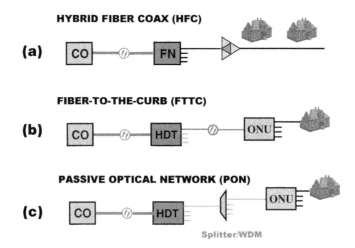

Figure 2.1. Three access network architectures: (a) HFC network, (b) FTTC network, (c) PON.

distributes the video signals over the coax distribution plants to all the users associated with the FN. For two-way services, each coax amplifier is upgraded with a return module and the upstream and downstream transmission is separated by diplexers that provide non-overlapping gain for the two bands [8]. Typically, upstream transmission uses the band of 5–40 MHz, and downstream transmission uses the band from 55 MHz to the highest frequency (i.e., 750 MHz) supported by the coax amplifiers. Because each coax bus is shared by many users, bandwidth capacity is allocated to each user by using a combination of FDM and TDM, in which users are provisioned with time slots within RF channels (typically 6 MHz downstream and 2 MHz upstream). To avoid collisions in upstream transmission, a scheduled or a dynamically assigned time division multiple access (TDMA) protocol is used for contention resolution (MAC: media access control) [9].

Most recently, to consolidate cable headends and to provide more flexible service provisioning, dense wavelength division multiplexing (DWDM) technology has moved from its high-end position in long-haul networks to HFC networks [10–13]. It provides cable operators with more network capacity and better trunking capability for emerging two-way services and simplifies operation by consolidating equipment, such as Cable Modem Termination System (CMTS), to a central location (e.g., super headend).

2.2.2 Switched networks

Traditionally, voice or telephone service was considered as point-to-point, on-demand communication. Local telephone networks were then established with star topology and central-control operation [2]. The relatively small bandwidth needed for voice also made twist-pair the cost-effective means for transportation.

Driven by competition and the need to provide more services for new revenue, local exchange carriers (LEC) have been defining or re-defining evolution strategy for providing broadband services. These include using digital subscriber line (xDSL) technology to explore more bandwidth over existing twist-pair [14,15], and bringing fiber to the network for upgrades. The latter leads to solutions like digital loop carrier (DLC) and fiber-to-the-curb (FTTC) architectures.

Figure 2.1b illustrates an FTTC system. The host digital terminal (HDT) connects multiple optical network units (ONUs) with either dedicated fiber or PON configurations. Each user then connects to a distinct port of a switch or mux/demux in the ONU. The ONU performs the function of demultiplexing downstream signals from the HDT and switching them to each user, and multiplexing upstream signals from each user and transmitting them to the HDT. The dedicated point-to-point architecture provides each user with potentially higher switched bandwidth and privacy. The difficulties of providing analog video could be compromised by overlaying an

HFC network. Keeping the existing coax or twist-pair as the drops, RF subcarrier link technology, such as a CAP-16 modem, capable of delivering 50 Mbps over a short distance twist-pair, could be used for the transmission over the metallic drop. Unlike the pure HFC approach however, FTTC systems typically maintain digital baseband signal transmission over optical fiber for non-broadcast services.

2.2.3 Evolution and revolution

The need to future-proof networks and to provide abundant bandwidth capacity for any foreseeable services motivates industries to explore more network alternatives. With the advent of fiber optics, replacing metallic links with optical fiber becomes feasible. This leads to varieties of PONs for the last-mile connectivity, or for certain types of FTTx architecture (fiber-to-the-curb, fiber-to-the building, etc.) [16,17].

The PONs currently under development use a point-to-multipoint configuration, as shown in Fig. 2.1c. Due to its broadcasting nature, TDM is used for downstream transmission where each ONU receives baseband broadcast signals from which it selects the appropriate information. Scheduled or dynamic TDMA is used for upstream transmission to avoid collisions over the shared fiber trunk. Upstream and downstream transmission could be separated by time division duplexing (TDD) or by different wavelengths (coarse WDM, 1.3 μm/1.5 μm) if a single fiber is used for the drop, or could be carried by two separate fibers. Replacing optical splitters with wavelength splitters or routers, DWDM technology could be used to establish virtual point-to-point connections over the physically point-to-multipoint network.

Even though the ultimate transmission scheme over optical fiber is digital baseband, to be compatible with tuner-based customer permise equipment (e.g., TV set, VCR, set-top box, etc.) and current service delivery mechanisms (broadcast video, etc.), deploying RF subcarrier link technology over even purely fiber based networks becomes very attractive.

2.3 RF subcarrier lightwave technology

Generally speaking, "RF subcarrier link" means to transmit signals over an electromagnetic wave. In that sense, the conventional lightwave system in which the digital signals are directly applied to the optical wave is already a "subcarrier link" except at higher frequency. As discussed in the previous section, adding another RF layer on top of the optical carrier is mainly driven by the hybrid fiber/metallic architecture and the notion of keeping a transparent link so that the signals, like analog TV signals, can be directly sent to the customer premises (e.g., TV set). This section will discuss some lightwave technologies that make such RF subcarrier lightwave systems feasible.

2.3.1 Linear lightwave technology

Delivering multichannel analog video was the primary business of the cable industry and continues to be the industry's bread-and-butter. The ability to transmit multiple analog video channels over optical fiber had a profound impact on the cable industry and enabled the HFC systems, which are also used by other access network operators.

Traditionally, the AM–VSB signals are frequency multiplexed and transmitted over coaxial cable with multiple coax amplifiers in cascade. To have a good signal quality at the TV set, the end-to-end link must satisfy strict noise and distortion requirements. Typically, this means 48–49 dB CNR and −53 dBc composite distortion, including composite second order (CSO) and composite triple beat (CTB), must be maintained at the TV receiver. This has been a big challenge to traditional coaxial networks due to the accumulation of noise and distortion over long cascaded coax amplifier chains. The advent of linear lightwave technology allows operators to replace the trunk part of the cable networks with optical fiber while only keeping the last-mile coax plants for distribution. This HFC architecture substantially improves system performance but certainly imposes challenges to lightwave technology which was traditionally only used for digital transmission systems.

The history of transmitting analog signals over fiber began in the early 1980s [18]. The commonly used systems were intensity modulated/direct detected systems. Limitations in laser output power and relative-intensity noise (RIN) restricted early efforts to just a few channels over short distances. Innovation in semiconductor devices improved laser structure that led to increased output power, and the single-frequency-distributed-feedback (DFB) lasers provided lower RIN and better linearity. The linear lightwave family further expanded to include 1.3 μm and 1.5 μm DFB lasers, and external modulators that were combined with erbium-doped fiber amplifiers (EDFAs) to further extend the reach. The system performance was also enhanced by sophisticated predistortion and noise reduction techniques. All these enabled analog links capable of delivering more than 80 channels of analog video plus a broad spectrum of digital-RF signals over distances in excess of 20 km with performance approaching the theoretical optimum.

2.3.1.1 Generic system characteristics

An RF lightwave system consists of transmitters, fiber link, and receivers. The performance of this kind of system is then determined by the performance of those active and also passive components, such as the effect of fiber link "stimulated" by the light. All these can be quantified by carrier-to-noise-ratio (CNR), and second and third order distortions (CSO and CTB), which directly determine the received

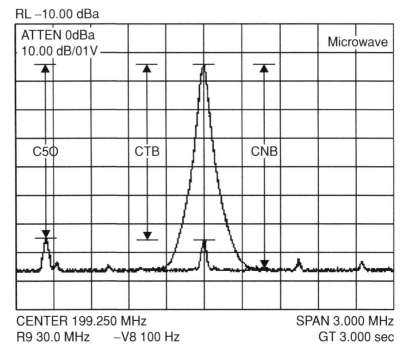

RL −10.00 dBa

CENTER 199.250 MHz   SPAN 3.000 MHz
R9 30.0 MHz   −V8 100 Hz   GT 3.000 sec

Figure 2.2. CNR, CTB, CSO in RF subcarrier system.

signal quality at the customer premises equipment (e.g., TV set) [18–22]. This is
shown in Fig. 2.2.

Excluding fiber effects, the CNR at the receiver is given by

$$CNR = \frac{\frac{1}{2}(m I_0)^2}{B_e \left(2 e I_0 + n_{th}^2 + RIN \times I_0^2\right)}. \tag{2.1}$$

The signal power per channel is $(m I_0)^2/2$, where m is the optical modulation
depth (ODM) per channel, and I_0 is the average received photocurrent. The denominator consists of several noise factors and B_e is the noise bandwidth (4 MHz
for NTSC channels). The thermal noise of the receiver is $B_e n_{th}^2$, and the shot noise
is $2 B_e e I_0$, where the e is the electron charge. The laser's relative intensity noise is
RIN (dB/Hz).

The noise and quantum efficiency of the receiver determine the necessary received optical power to satisfy the CNR requirement. The received optical power
is related to the photocurrent by:

$$P_0 = \frac{h \nu}{\eta e} I_0, \tag{2.2}$$

where $h \nu$ is the photon energy, and η is the quantum efficiency of the photodiode.
From Eqs. (2.1) and (2.2), one can determine the necessary optical power to achieve
the desired CNR.

Besides noise, when multiple RF signals are multiplexed (FDM), the second order and third order distortion products that are generated within the multichannel multioctave band will affect the signal quality. For a simple (memoryless) nonlinear characteristic, the optical power can be expressed as a Taylor-series expansion of the electrical signal:

$$P \propto X_0(1 + x + ax^2 + bx^3 + \cdots), \tag{2.3}$$

where X_0 is the DC bias and x is the modulation signal:

$$x(t) = \sum_i^{N_{ch}} m_i(t) \cos(2\pi f_i t + \phi_i), \tag{2.4}$$

where $m_i(t)$ is the normalized modulation signal for channel i, f_i is the channel i frequency, and N_{ch} is the number of channels. A general practice is to simulate the AM channels with RF tones of certain modulation depth $m = m_i(t)$. The CTB and CSO can then be given by:

$$CTB = 10 \log \left[N_{CTB} \left(\frac{3}{2} bm^2 \right)^2 \right], \tag{2.5}$$

$$CSO = 10 \log[N_{CSO}(am)^2], \tag{2.6}$$

where N_{CTB} and N_{CSO} are third order and second order product counts. Figure 2.3 shows the product distribution in the multichannel multioctave band.

In an RF lightwave system, transmitter, receiver and optical fiber will contribute to the CNR, CTB, and CSO performance in many different ways. The system (end-to-end) performance is then a balance among those variables.

2.3.1.2 Receiver

From Eq. (2.1), the shot noise and laser's RIN dominate the CNR performance with higher received power. With lower received power, the receiver thermal noise becomes significant. In that situation, the link performance can be improved by reducing this noise. This can be achieved by improving the impedance match between the photodiode, which is an infinite-impedance current source, and the low-noise amplifier. Nevertheless, this approach leads to two limitations. First, the RC circuit to accomplish the impedance match imposes a bandwidth limitation. Second, the increased impedance increases the signal level that requires much more linear preamplifier.

2.3.1.3 Transmitter

The stringent requirements of AM–VSB on lightwave systems makes the transmitter the most challenging component to implement. Two types of transmitters have been

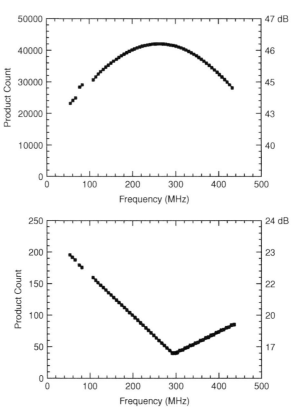

Figure 2.3. Third order (top) and second order (bottom) distortion product distribution.

developed: intensity modulated semiconductor laser and external modulator. The former has the advantages of being simple, compact, and low cost. The external modulator, on the other hand, offers high power and low chirp, therefore is widely used in optical amplified trunk systems that need higher power budgets and super performance [23,24].

As discussed in the previous sections, system performance is determined by the balance of CNR and distortion (CTB, CSO). Based on Eqs. (2.1), (2.5), and (2.6), CNR, CTB, and CSO are presented in Fig. 2.4. Two approaches could improve system performance. First, increasing optical power would improve CNR performance. Second, driving the laser harder would also improve the CNR performance but degrade the CTB and CSO. Further, when driving the laser too hard, the electrical signal may drive the laser below the threshold current, which makes the output optical signal clip, and creates broadband distortion [25,26]. The distortion is a function of the root-mean-square (RMS) modulation depth of the total signal, $\mu = m\sqrt{(N/2)}$, as show in Fig. 2.5. Optimizing system performance is then a question of balancing the OMD, optical power, and other related issues as will be discussed later.

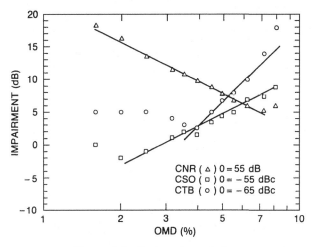

Figure 2.4. System degradation.

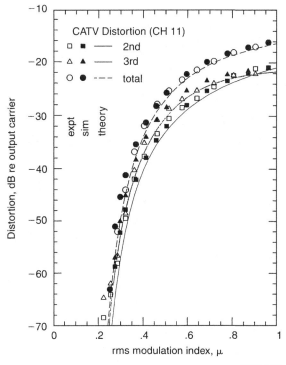

Figure 2.5. Distortion due to clipping. $\mu = \sqrt{(N_{ch}/2)}$.

Besides noise and linearity, the interaction between the linear and nonlinear effects of the fiber and the optical field spectrum can add extra intensity distortion or noise. Chirp, a phenomenon of dynamic shift of the lasing frequency due to modulation, is directly related to this effect. The laser's optical field spectrum is determined by FM efficiency β_{FM}. For transmitters with significant FM efficiency, a Gaussian distribution with standard deviation can approximate the time-average spectrum when the transmitter is driven by multiple subcarriers [27,28]:

$$\sigma_v = m\beta_{FM}I_b\sqrt{\frac{N}{2}}.\qquad(2.7)$$

This then directly relates to the half-width at half-maximum (HWHM) linewidth by:

$$HWHM = \sqrt{2\ln 2}\sigma_v.\qquad(2.8)$$

In the next subsection, we will discuss how this contributes to the intensity noise when interacting with the fiber's linear and nonlinear effects.

2.3.1.4 Fiber effects

When signals are transmitted over optical fiber, the received intensity spectrum can be modified by changes in the phase or amplitude of the optical field spectrum, which adds extra noise and distortion. These effects can be categorized as linear and nonlinear effects.

Linear effects *Multipath interference (MPI)* comes from the interaction between an optical signal and a doubly reflected version of itself. For a directly modulated laser in which the optical spectrum is approximately Gaussian distributed, the MPI induced intensity noise is given by [29,31].

$$RIN = \frac{2pR^2}{\sqrt{\pi}\sigma_v}\exp\left(-\frac{f^2}{4\sigma_v^2}\right),\qquad(2.9)$$

where p is a parameter determined by the polarization overlap of interfering fields and R is equivalent reflectivity due to double Rayleigh scattering. It can be seen that the MPI can be reduced by artificially broadening the optical spectrum using a dithering tone.

Dispersion consists of chromatic dispersion and polarization mode dispersion (PMD) [32–34]. The impact is the intermodulation distortion generated by the coupling between the dispersion and the highly chirped transmitter. For a single wavelength DFB laser, the chromatic dispersion only becomes the limiting factor over relatively long distance, and can be overcome with dispersion compensated fiber. The polarization mode dispersion of most current fiber is less than 0.04 ps/km for single mode fiber. Therefore, even for a 20 km fiber trunk, the maximum PMD is 0.18 ps, which will not be a significant performance degradation factor.

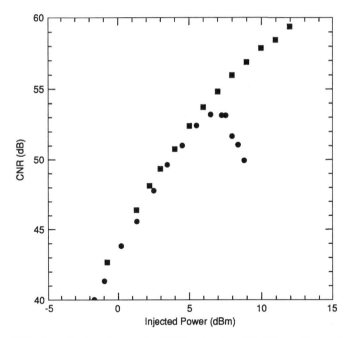

Figure 2.6. CNR degradation due to the SBS, stimulated Brillouin Scattering • w/SBS, ■ w/o SBS.

Nonlinear effects Unlike linear effects, nonlinear effects depend on the optical power. With the power budget being a critical factor in analog lightwave system design, the power distance limitation imposed by the fiber nonlinearities must be considered.

Stimulated Brillouin scattering comes from the interaction between the optical field and acoustic phonons in the fiber [35–38]. It scatters light into the backward direction and increases intensity noise in the forward direction, therefore degrading CNR performance. As shown in Fig. 2.6, the effect on CNR performance is insignificant until the injected optical power reaches the so-called SBS threshold. The threshold is determined by the Brillouin linewidth in comparison with the optical spectrum of the laser. Therefore, for a directly modulated DFB laser with a much broader optical spectrum than the Brillouin linewidth, the Brillouin threshold is relative high (e.g., 2 W). The SBS effect is not apparent. For an external modulator that has a much narrower spectrum, the Brillouin threshold can be increased by artificially broadening the spectrum with phase or frequency modulation.

Stimulated Raman scattering is similar to SBS except that optical phonons are interacting with the optical field. Compared with SBS, SRS has a higher threshold, and therefore will never be a problem in normal analog distribution systems. It may occur in longhaul DWDM/EDFA based analog systems [34].

2.3.1.5 Optical amplifier

An optical amplifier increases the system link budget, therefore allowing more optical splitting to cover wider areas. It also enables longer reach for headend consolidation favored by the cable industry. On the other hand, the increased power comes with additional noise and distortion [39,40].

One mechanism contributing to the distortion is the gain saturation. The upper bound of CSO is given by:

$$CSO \leq 20 \log \sqrt{N_{cso}} \frac{m P_{out}}{P_{sat}} \frac{1}{\sqrt{1 + (2\pi f_{ch1}\tau)^2}}, \tag{2.10}$$

where the τ is the lifetime of the upper state. A semiconductor amplifier has τ in the order of 300 ps, while an EDFA has τ in the order of 10 ms. This therefore makes an EDFA a good candidate for amplification of analog signals. Other alternatives include Raman fiber amplifiers and rare-earth-doped fiber amplifiers.

In addition to the distortion, intensity noise can also be generated from signal–spontaneous beat and spontaneous–spontaneous beat, especially in amplifier cascade scenarios:

$$i^2_{sig-sp} \propto \frac{B_e}{B_o}, \tag{2.11}$$

$$i^2_{sp-sp} \propto \frac{B_e}{B_o^2}(2B_o - 2f - B_e), \tag{2.12}$$

where B_o is optical bandwidth. For standard NTSC channels, $f, B_e \ll B_o$. With the input powers necessary to achieve the required CNR, the spontaneous–spontaneous beat noise is negligible compared with the signal–spontaneous beat noise.

2.3.2 Low-cost lightwave

The advent of RF modem technology enables the cost effective delivery of digital services using RF subcarriers. Digital RF subcarrier signals typically have much more relaxed requirements on CNR, CTB, and CSO than AM–VSB does. This therefore opens doors for relatively low-cost lightwave components being deployed for this application, such as uncooled DFB and FP lasers, and even LEDs.

Utilizing the low-cost lightwave components for downstream transmission is the first straightforward approach. Applications include broadcasting multichannel digital TV in HFC networks [41] and other fiber-based networks, such as PON, FTTC, etc., and narrowcasting, such as high speed data, telephony, and interactive

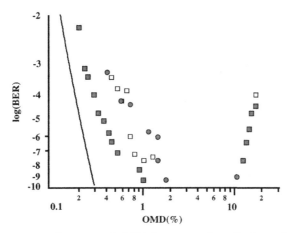

Figure 2.7. Performance of an uncooled FP laser carrying 60 channel QPSK signals at 2 Mbps/channel. – theory, ■ room temperature, ● $T > 80\,^{\circ}$C, □ reflection.

video services. Figure 2.7 shows the performance of an uncooled FP laser carrying 60 channel QPSK modulated video signals. Even though bit error rate (BER) is degraded due to thermal effects and clipping, a comfortable BER free range exists for good system performance.

Applications in the upstream direction are more challenging. The multipoint-to-point topology raises certain system issues such as:

1. Beat interference
2. Dynamic range

When passive optical combining is used, coherent interference among different light paths ends up as intensity noise at the receiver. The noise is a function of optical spectrum overlapping of the two interfering lasers. Therefore, it can be reduced by broadening the optical spectrum to reduce the combined energy in the same bandwidth.

In an HFC network, even though the users' upstream transmission level can be well defined by the HE and the balanced tap value (based on network design), the temperature effect and other effects may still vary the upstream transmission level. When those signals arrive at the optical node, the driving power of those RF signals to the upstream laser will vary. This therefore requires the laser (typically a low-cost laser) to have enough dynamic range to handle those variations while still maintaining good CNR performance. This is usually characterized by the noise power ratio (NPR), as shown in Fig. 2.8. A chunk of 5–40 MHz noise is applied to the laser, simulating the return spectrum in the HFC network [8]. A notch filter is used to create a dip in the spectrum for CNR measurement. The CNR is then measured as a function of the total "signal" level of the spectrum, as seen in Fig. 2.8.

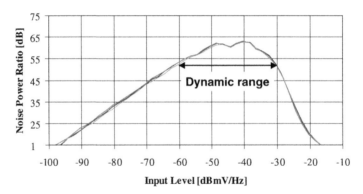

Figure 2.8. An example of an NPR curve.

On the left side of the curve, the laser's performance is determined by the noise level. On the right side, the performance is limited by the clipping effect.

2.4 System design and requirements

Each component contributes to system performance in many different ways. For a system operator, it is important to consider the end-to-end system performance and to balance the contributions from each subsystem and component.

On the other hand, technology innovations and new service opportunities enable architecture alternatives that can change the signal transportation mechanism, and potentially even totally change the application of RF subcarrier links (at least in wired networks).

This section will use a cable network as an example to illustrate how the balance can be achieved and how the RF subcarrier link technology is evolved to support architecture evolution.

2.4.1 End-to-end HFC system design

A modern HFC network [7,10,11] contains four major segments as shown in Fig. 2.9:

Primary ring This section of the network is placed in large service areas. Primary hubs are digitally linked to a regional headend with fiber optic ring architecture. This ring contains multiple fibers and may be configured for SONET, a proprietary digital transport system, other format of data networking, or all the above.

Secondary ring This section of the network transports analog and digital RF signals between the primary hub and secondary hubs. Route-diversed fiber optic links are used to transport the signals. The secondary hubs (SH) serve as signal concentration and distribution points using DWDM or other multiplexing/demultiplexing schemes to limit the number of

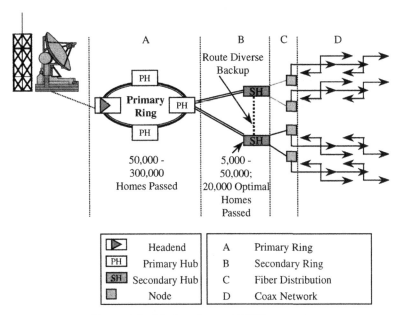

Figure 2.9. Modern layered HFC network.

fibers between the primary hub ring and secondary hubs to achieve cost reduction, for mean-time-to-repair (MTTR) improvement, and to allow for cost-effective backup switching (redundancy).

Fiber distribution section This section of the network transports analog and digital RF signals between the secondary hub and fiber nodes (FNs). Dedicated fiber optic links are used to transport these signals. On physical layer, the FN is the transition point at which the optical signals over fiber cable and RF electronic signals over coax cable are converted into each other.

Coaxial layer This section of the network transports analog and digital RF signals between the FN and subscribers using coaxial cable with coax amplifiers in cascade.

It can be seen that an end-to-end RF subcarrier link covers the entire PH–SH (secondary ring), SH–FN (fiber distribution), and FN–customers (coax plant) sections. Depending on the technology availability, the end-of-line (EOL) performance in both downstream and upstream directions will be determined by the cumulative performance of those three segments.

Downstream performance

For analog TV signals, the desired EOL performance at customer premises is 49 dB CNR, and −53 dB CSO and CTB. Table 2.1 illustrates one example of balancing the performance of different segments to meet these requirements.

Table 2.1. *Downstream performance of a
conventional HFC system*

	Performance		
Network section	CNR (dB)	CTB (dBc)	CSO (dBc)
HE/PH output	63	−80	−80
HE/PH–FN link	52	−65	−63
Coax plant	54	−56	−56
EOL	49	−53	−53

Table 2.2. *Upstream performance of a conventional
HFC system*

	Performance	
Network section	NPR (dB)	Dynamic range (dB)
Coax plant	46	13
FN–SH link	41	13
SH–HE/PH link	43	13
End-to-end	38	13

The additional digital RF signals, such as QPSK, QAM signals have much lower
CNR, CTB, and CSO requirements than AM–VSB signals (Table 2.2). They usually
are operated at 6 dB below the AM signals.

Upstream performance

Over the conventional 5–40 MHz upstream band, the CTB and CSO are negligible.
The major issue is to maintain good CNR with certain dynamic range. As discussed
in Section 2.3.2, temperature change and other effects may alter the upstream signal
level. When those RF signals go through RF amplifiers, are applied to the laser, and
are detected by the optical receiver, a certain operation range is needed for each of
the components to assure that the CNR over each segment, and therefore the accu-
mulated CNR, is above the operation threshold. Table 2.3 shows the performance
of individual segments and their accumulation over the entire RF link.

2.4.2 Architecture evolution and its impact

Cable networks have gone through several generations of upgrade, migrating from
a purely coax-based tree-and-branch architecture for broadcast services to an HFC-
based two-way broadband platform. Fiber optics has had profound impact on this

Table 2.3. *Downstream performance of a*
LightWire^{TM} system

Network section	Performance		
	CNR (dB)	CTB (dBc)	CSO (dBc)
HE/PH output	63	−80	−80
HE/PH–mFN link	49	−67	−63
mFN RF	63	−57	−57
EOL	48.7	−54	−53

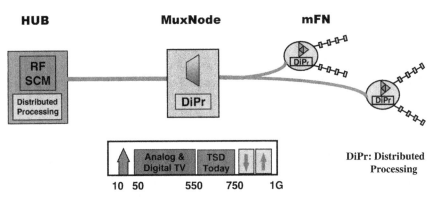

Figure 2.10 LightWire^{TM} system. MuxNodes replace FNs to perform certain "mux/demux" functions, and mFNs replace all coax amplifieres.

evolution. The challenge has been, and still is, how to utilize technology innovations to cost-effectively resolve issues like upstream bandwidth limitation and ingress-induced performance degradation, to further improve reliability and reduce operation costs, and to provide a migration path for the future [42].

One solution called *LightWire^{TM}* [43,44] is shown in Fig. 2.10. The mini fiber nodes (mFNs) eliminate all the coax amplifiers (in about a 1:3 ratio), and connect to customers over existing passive coax plant.

Fibers connecting multiple mFNs will be terminated at the MuxNode that resides either at the original fiber node location or at a location that "consolidates" multiple FNs. As its name implies, the MuxNode performs certain concentration and distribution functions. It "multiplexes" the upstream signals and sends them to the primary hub through the secondary hubs. It also "demultiplexes" the downstream signals received from the PH–SH fiber trunks and distributes them to mFNs.

One of the interesting features of this architecture is that it maintains the characteristics of conventional HFC networks of being transparent to different signal formats and protocols, therefore fully supporting the existing operation for current services. To future-proof the network with more capacity and simple operation, this

Table 2.4. *Upstream performance of a LightWire^{TM} system*

	Performance	
Network section	NPR (dB)	Dynamic Range (dB)
mFN–MuxNode Link	49	11
After MuxNode combining (12)	38	11
MuxNode–PH/HE Link	44	13
End-to-end	37	11

architecture can also support a distributed-processing strategy by moving the RF and MAC demarcation to the mFN, thereby enabling pure digital baseband transmission between PH and mFNs. This would substantially reduce the transport costs and simplify system operation. Moreover, it can open the higher frequency band (750 MHz–1 GHz) for emerging new services. The development can be partitioned into two phases.

***Phase 1*: Service delivery over RF subcarrier link** In *Phase 1*, the fiber-deep platform will support all current service delivery and operation over the end-to-end RF link. It should be noted that the previous coax link between FN and coax amplifier is replaced by a fiber link (MuxNode–mFN), and the amplifier chain is eliminated with only passive coax cable connecting customers to the mFN. Therefore, the requirements on each segment of the RF link will be different for the same (or even better) end-to-end performance. This can be seen in Table 2.4.

***Phase 2*: Add-on distributed processing platform** As shown in Fig. 2.10, a distributed processing platform can be transparently added when it is needed. Narrowcast services (data, voice, interactive video, etc.) could be delivered to the MuxNode in baseband format over a DWDM link separate from the one carrying broadcast services (analog and digital TV channels). At the MuxNode, the baseband signals are demultiplexed and further distributed to multiple mFNs.

At each mFN, the downstream digital baseband signals are received and modulated onto RF carriers. They are then combined with the received downstream broadcast RF signals and transmitted over the coax buses to customers. In the upstream direction, customers transmit upstream signals to the mFN, at which the signals are demodulated to baseband and transmitted to the MuxNode. At the mFN, the MAC function is also performed to resolve contention over the shared coax plant, and the transmission between each mFN and the MuxNode is in a point-to-point fashion.

With this *Phase 2* approach, the broadcast and narrowcast service delivery are totally decoupled over the lightwave path. The broadcast services are still carried by the RF link while the narrowcast services are carried over a digital baseband path. Even though we still need to maintain the same RF link performance for broadcast service delivery, the requirement on two-way service delivery is substantially relaxed. The only RF link for two-way services is the passive coax plant.

2.5 Summary

The advent of linear lightwave technology, in which the RF subcarrier link and fiber optics combine, allows access network providers to bring fiber deeper into the networks in a cost-effective way and stimulates tremendous technology innovations that further enable wide varieties of architecture alternatives. This opens doors to service providers with many different service delivery mechanisms and new service opportunities. The trend has been continuing for at least the past 15 years and is still in accelerating mode. It is interesting to notice that many technologies have been transforming themselves from high-end applications (e.g., high-speed DWDM long-haul networks) to applications in access networks (e.g., DWDM RF link in cable networks).

On the other hand, with fiber penetrating deeper into access networks, the advances of RF link technology, together with the power of DSP, motivates distributing certain control functions into the networks to simplify operation and improve network scalability. Terminating the RF subcarrier link at the fiber/metallic transition point therefore makes more sense, if not just from a cost reduction point of view. This will become more compelling when analog TV is fully replaced by digital TV. Perhaps we should pave a migration path to the new digital world?

References

1. T. E. Darcie, *Broadband Subscriber Access Architectures and Technologies*, OFC'96 tutorial, 1996.
2. N. J. Frigo, "A survey of fiber optics in local access architecture", in *Optical Fiber Telecommunications: IIIA*, Academic Press, 1997.
3. X. Lu, *Broadband Access: Technologies and Opportunities*, Globecom'99 tutorial, 1999.
4. T. E. Darcie, "Subscriber multiplexing for multiple-access lightwave networks", *J. Lightwave Technol.*, **LT-5**, 1103–10, 1987.
5. J. A. Chiddix, H. Laor, D. M. Pangrac, L. D. Williamson, and R. W. Wolfe, "AM video on fiber in CATV systems, need and implementation", *J. Select. Areas Commun.*, **8**, 1990.
6. J. A. Chiddix, J. A. Vaughan, and R. W. Wolfe, "The Use of Fiber Optics in Cable Communication Networks", *J. Lightwave Technol.*, **11** (1), 154–66, 1993.
7. A. E. Werner and O. J. Sniezko, "HFC optical technology: ten years of progress and today's status, choices and challenges", *CED*, September 1998.

8. O. J. Sniezko, "Reverse path for advanced services – architecture and technology", *NCTA Technical Papers*, 1999.

9. *Data Over Cable Service Interface Specifications*, Cable Labs, July 1998.

10. A. E. Werner, "Regional and metropolitan hub architecture consideration", *SCTE Conference on Emerging Technology*, 1995.

11. T. G. Elliot and O. J. Sniezko, "Transmission technologies in secondary hub rings – SONET versus FDM once more", *NCTA Technical Papers*, 1996.

12. O. J. Sniezko and A. E. Werner, "Invisible hub or end-to-end transparency", *NCTA Technical Papers*, 1998.

13. O. J. Sniezko, "Video and Data Transmission in Evolving HFC Network", *OFC'98*, 1998.

14. F. T. Andrews, "The heritage of telephony", *IEEE Commun. Mag.*, **27**, 8, 1989.

15. F. T. Andrews, "The evolution of digital loop carrier", *IEEE Commun. Mag.*, **29**, 3, 1991.

16. P. W. Shumate and R. K. Snelling, "Evolution of fiber in the residential loop plant", *IEEE Commun. Mag.*, **29**, 3, 1991.

17. P. W. Shumate, "What's happening with fiber to the home?", *Opt. Photon. News*, **7**, 1996.

18. M. R. Phillips and T. E. Darcie, "Lightwave analog video transmission", in *Optical Fiber Telecommunications: IIIA*, Academic Press, 1997.

19. K. Y. Lau and A. Yariv, "Intermodulation distortion in a directly modulated semiconductor injection laser", *Appl. Phys. Lett.*, **45**, 1984.

20. R. Olshasky, V. Lanzisera, and P. Hill, "Design and performance of wideband subcarrier multiplexed lightwave systems", *ECOC'88*, 1988.

21. T. E. Darcie, R. S. Tucker, and G. J. Sullivan, "Intermodulation and harmonic distortion in InGaAsP lasers", *Electron. Lett.*, **21**, 1984.

22. G. P. Agrawal and N. K. Dutta, *Long-Wavelength Semiconductor Lasers*, New York, Van Nostrand Reinhold, 1986.

23. M. Nazarathy, J. Berger, A. J. Ley, I. M. Levi, and Y. Kagan, "Progress in externally modulated AM CATV transmission systems", *J. Lightwave Technol.*, **11**, 1993.

24. A. Gnauck, T. Darcie, and G. Bodeep, "Comparison of direct and external modulation for CATV lightwave transmission at 1.55 μm wavelength", *Electron. Lett.*, **28**, 1992.

25. N. Frigo, M. Phillips, and G. Bodeep, "Clipping distortion in lightwave CATV systems: model, simulation and measurements", *J. Lightwave Technol.*, **11**, 1993.

26. A. Saleh, "Fundamental limit of number of channels in subcarrier multiplexed lightwave CATV system", *Electron. Lett.*, **25**, 1989.

27. U. Cebulla, J. Bouayad, H. Haisch, M. Klenk, G. Laube, H. Mayer, R. Weinmann, and E. Zielinski, "1.55 mm strained layer quantum well DFB lasers with low chirp and low distortions for optical analog CATV distribution systems", *Proceedings of Conference on Lasers and Electron-optics*, 1993.

28. F. Koyama and K. Iga, "Frequency chirping in external modulators", *J. Lightwave Technol.*, **6**, 1988.

29. A. Judy, "The generation of intensity noise from fiber Rayleigh back-scattering and discrete reflections", *Proceedings of European Conference on Optical Communications*, 1989.

30. T. Darcie, G. Bodeep, and A. Saleh, "Fiber-reflection induced impairments in lightwave AM–VSB CATV systems", *J. Lightwave Technol.*, **9**, 1991.

31. S. Woodward and T. Darcie, "A method for reducing multipath interference noise", *IEEE Photon. Technol. Lett.*, **6**, 1994.

32. M. Phillips, T. Darcie, D. Marcuse, G. Bodeep, and N. Frigo, "Nonlinear distortion generated by dispersive transmission of chirped intensity-modulated signals", *IEEE Photon. Technol. Lett.*, **3**, 1991.

33. C. Poole and T. Darcie, "Distortion related to polarization-mode dispersion in analog lightwave systems", *J. Lightwave Technol.*, **11**, 1993.

34. M. Phillips, G. Bodeep, X. Lu, and T. Darcie, "64QAM BER measurements in an analog lightwave link with large polarization-mode dispersion", *Proceedings of Conference on Optical Fiber Communication*, 1994.

35. A. Chraplyvy, "Limitation on lightwave communication imposed by optical-fiber nonlinearities", *J. Lightwave Technol.*, **8**, 1990.

36. X. Mao, G. Bodeep, R. Tkach, A. Chraplyvy, T. Darcie, and R. Derosier, "Brillouin scattering in externally modulated lightwave AM–VSB CATV transmission systems", *IEEE Photon. Technol. Lett.*, **4**, 1992.

37. X. Mao, G. Bodeep, R. Tkach, A. Chraplyvy, T. Darcie, and R. Derosier, "Suppression of Brillouin scattering in externally modulated lightwave AM–VSB CATV transmission systems", *Proceedings of Conference on Optical Fiber Communication*, 1993.

38. G. Agrawal, *Nonlinear Fiber Optics*, Academic Press, 1989.

39. A. Saleh, T. Darcie, and R. Jopson, "Nonlinear distortion due to optical amplifiers in subcarrier-multiplexed lightwave communications systems", *Electron. Lett.*, **25**, 1989.

40. E. Desurvire, *Erbium-doped Fiber Amplifiers: Principles and Applications*, Wiley, New York, 1994.

41. S. L. Woodward and G. E. Bodeep, "Uncooled Fabry–Perot lasers for QPSK transmission", *IEEE Photon. Technol. Lett.*, **7**, 558, 1995.

42. X. Lu, T. E. Darcie, A. H. Gnauck, S. L. Woodward, B. Desai, and Xiaoxin Qui, "Low-cost cable network upgrade for two-way broadband", *ET'98*, 1998.

43. O. Sniezko, T. Werner, D. Combs, E. Sandino, Xiaolin Lu, T. Darcie, A. Gnauck, S. Woodward, B. Desai, "HFC architecture in the making", *NCTA Technical Papers*, 1999.

44. O. Sniezko and Xiaolin Lu, "How much "F" and "C" in HFC", *ET2000*, 2000.

3

Analog modulation of semiconductor lasers

JOACHIM PIPREK AND JOHN E. BOWERS

University of California, Santa Barbara

3.1 Introduction

The laser is one of the most important elements in fiber optic links since it generates the coherent optical wave that carries the signal. Typical laser wavelengths are 1.3 μm and 1.55 μm corresponding to the dispersion and absorption minimum, respectively, of silica fibers. The laser frequency is about 200 THz and the RF (10 kHz–300 MHz) or microwave (300 MHz–300 GHz) signal can be modulated onto the laser beam either directly or externally. This chapter focuses on direct modulation. It is simpler to implement than external modulation but the usable bandwidth is limited to a few GHz. Applications of direct analog laser modulation include cable TV, base station links for mobile communication, and antenna remoting. Laser performance requirements include high slope efficiency to obtain high link gain, low laser noise to keep the link noise figure low, and low distortion to achieve a large spurious free dynamic range (SFDR) [1].

Section 3.2 outlines basic physical mechanisms of semiconductor lasers. We emphasize the quantum nature of electrons and photons which helps to understand efficiency and noise issues. The slope efficiency is discussed in detail. Section 3.3 presents the laser rate equations which are the common basis for the analysis of analog performance. Numerical solutions to the rate equations allow for an exploration of a wide spectrum of lasing effects. However, analytical formulas based on the small signal approximation are valid in many cases and they are given throughout this chapter. Section 3.3 includes the effects of gain saturation, gain compression, carrier density dependent lifetimes of electrons and photons, carrier transport delay, as well as frequency effects on the current injection efficiency (parasitics). Section 3.4 discusses intensity modulation (IM) including relative intensity noise (RIN), harmonic distortion (HD), intermodulation distortion (IMD), and the dynamic range. Section 3.5 describes frequency modulation (FM), frequency noise, and spectral line broadening. Throughout this chapter, we give typical laser parameters and we highlight performance advantages of specific laser types: Fabry–Perot (FP) lasers, distributed feedback (DFB) lasers, or vertical-cavity lasers (VCLs).

3.2 Laser diode fundamentals

3.2.1 Gain, loss, and recombination rates

Traveling through a semiconductor, a single photon is able to generate an identical second photon by stimulating the recombination of an electron–hole pair. This process is called stimulated recombination or stimulated emission and it is the key physical mechanism of lasing. The second photon exhibits the same wavelength and the same phase as the first photon, doubling the amplitude of their monochromatic wave. Subsequent repetition of this process leads to strong light amplification. However, the competing process is the absorption of photons by generation of new electron–hole pairs. Stimulated emission prevails when more electrons are present at the higher energy level (conduction band) than at the lower energy level (valence band). This situation is called inversion. In semiconductor lasers, inversion can be achieved at p-n junctions by providing conduction band electrons from the n-doped side and valence band holes from the p-doped side. The photon energy is given by the band gap which depends on the semiconductor material. Continuous current injection into the device leads to continuous stimulated emission of photons, but only if enough photons are constantly present in the device to trigger this process. Thus, only part of all photons can be allowed to leave the laser diode as the lasing beam, the rest must remain inside the diode to generate new photons. This optical feedback and confinement of photons in an optical resonator is the second basic requirement of lasing.

The rate of stimulated recombination events $R_{st} = v_g g S$ is calculated from the photon density S, the optical gain g, and the photon group velocity v_g ($\sim 10^{10}$ cm/s). The material parameter g depends on the density N of electron–hole pairs, the temperature T, the photon wavelength λ, and the photon density. The gain function $g(N, T, \lambda, S)$ is the heart of laser physics [2]. Typical curves for $g(N)$ are shown in Fig. 3.1 at different temperatures. Positive net gain occurs above the transparency density N_{tr} (inversion). The differential gain dg/dN is a key parameter for analog modulation of laser diodes and high values of dg/dN are desirable. The differential gain decreases with higher carrier density. This can be described by the logarithmic function (3.1). The room temperature gain curve $g(N)$ in Fig. 3.1 is reproduced by $g_0 = 1246$ cm^{-1}, $N_{tr} = 1.38 \times 10^{18}$ cm^{-3}, and $N_s = -0.83 \times 10^{18}$ cm^{-3}. The linear gain approximation $g = a(N - N_0)$ is commonly used for small signal analysis. However, the optical gain is reduced at higher photon densities which lead

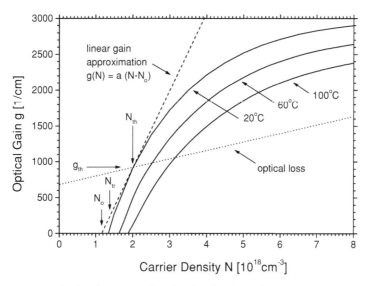

Figure 3.1. Peak optical gain vs. carrier density for an InGaAsP strained quantum well active layer (1.55 μm) at different temperatures (solid). The lasing threshold is given by the intersection with the optical loss (dotted). The linear gain approximation (dashed), its parameter N_0, the transparency density N_{tr}, the threshold density N_{th}, and the threshold gain g_{th} are indicated for $T = 20\,°C$.

to a depletion of electron–hole pairs. This spectral hole burning within the energy distribution of carriers restricts further stimulated recombination. The spectral hole is filled by electrons from other energy levels on the time scale of the intraband relaxation time (~ 1 ps). This effect is usually considered by introducing the gain compression factor ε ($\sim 10^{-17}$ cm^3) into the gain function

$$g(N, S) = \frac{g_0}{1 + \varepsilon S} \ln\left[\frac{N + N_s}{N_{tr} + N_s}\right]. \tag{3.1}$$

Photons are also generated by spontaneous electron–hole recombination, without correlation to other photons. The spontaneous emission rate can be written as $R_{sp} = BN^2$ using the spontaneous recombination parameter B ($\sim 10^{-10}$ cm^3 s^{-1}). Such photons are needed to trigger lasing initially but most spontaneously emitted photons do not contribute to the laser beam. From a quantum efficiency point of view, each injected electron that does not generate a stimulated photon constitutes a loss. Other carrier losses are caused by non-radiative recombination and by carrier leakage. Non-radiative recombination can be triggered by crystal lattice defects (Shockley–Read–Hall recombination, rate $R_{SRH} = AN$, $A \sim 10^8$ s^{-1}), with the lost electron energy being transferred into lattice heat (phonons). It can also be caused by Auger recombination (rate $R_{Aug} = CN^3$, $C \sim 10^{-28}$ cm^6 s^{-1}) which transfers the lost energy to another electron or hole. Leakage includes all processes that prevent carriers from recombining at the p-n junction at all.

Stimulated photons get lost from the laser cavity by emission (mirror loss α_m) or by internal absorption and scattering (internal loss α_i) giving the modal photon loss rate $v_g S(\alpha_i + \alpha_m)$. Since the optical gain needs to balance these modal optical losses, lasing does not start at the transparency level of carrier injection, but at a slightly higher threshold carrier density N_{th} (Fig. 3.1).

3.2.2 Basic laser structures

Modern laser diodes use a sandwich-like structure of different semiconductor materials to form the p-n junction. The center part of this structure has a lower bandgap and a higher refractive index than the cladding material, thus confining electron–hole pairs by energy barriers (Fig. 3.2) and photons by total reflection. For fiber optic applications at 1.3 µm or 1.55 µm wavelength, InP is the most common cladding and substrate material (bandgap 1.35 eV, refractive index 3.2). The center layers are made of InGaAsP or other III–V compound semiconductors whose composition and thickness are tailored to obtain optimum laser performance. In particular, stimulated recombination is confined to quantum wells (QWs) that are only a few nanometers thick to give maximum gain [2]. The QW bandgap is tailored to give the desired photon wavelength. Multiple quantum wells (MQWs) are often used to multiply the gain, thus reducing the required carrier density and increasing the differential gain (Fig. 3.2).

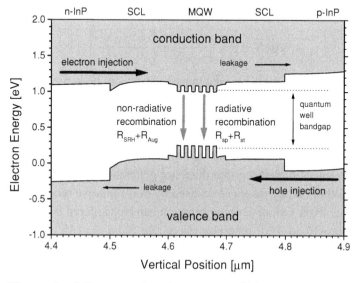

Figure 3.2. Electron band diagram and carrier transport within an InGaAsP/InP multiquantum well (MQW) active region sandwiched between separate confinement layers (SCLs).

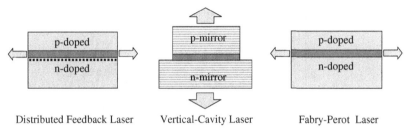

Distributed Feedback Laser   Vertical-Cavity Laser   Fabry-Perot Laser

Figure 3.3. Schematic pictures of basic laser types: DFB laser, VCL, and FP laser (arrows indicate light emission).

Carriers first need to cross the waveguiding separate confinement layer (SCL) before they can be captured by one of the quantum wells. This transport and capture process causes some delay in the carrier supply to the quantum wells and it is described by the transport time constant τ_t [3]. Within the quantum wells, electron–hole pairs recombine by one of the mechanisms described above. Carriers may also escape from the wells by thermionic emission generating vertical leakage current (escape time constant τ_e). These time constants can be translated into rates ($R = N/\tau$) and they control the balance between the quantum well density N and the barrier density N_b of carriers.

Stimulated photons establish an optical field inside the laser diode which is governed by the design of the optical cavity. Figure 3.3 shows three types of optical cavity designs. The most simple design uses the reflection at the two laser facets to form a Fabry–Perot (FP) resonator in the longitudinal direction. Constructive interference of forward and backward traveling optical waves is restricted to specific wavelengths. Those wavelengths constitute the longitudinal mode spectrum of the laser. The resonator length L is typically in the order of several hundred micrometers, much larger than the lasing wavelength, so that many longitudinal modes may exist. The actual lasing modes are given by the quantum well gain spectrum. Single mode lasing is hard to achieve in simple FP structures, especially under modulation. Dynamic single mode operation is required in many applications and it is achieved using optical cavities with selective reflection. The distributed feedback (DFB) laser is widely used for single mode analog applications [4]. Typical DFB lasers employ a periodic longitudinal variation of the refractive index within one layer of the edge-emitting waveguide structure as shown in Fig. 3.3. An emerging low-cost alternative to DFB lasers are vertical-cavity lasers (VCLs) which emit through the bottom and/or top surface of the layered structure. In VCLs, distance and layer thickness of the two distributed Bragg reflectors (DBRs) control the lasing wavelength (Fig. 3.3) [5].

In transverse directions, the optical wave is typically confined by the refractive index profile, using a ridge (FP laser, DFB laser) or a pillar (VCL) to form the

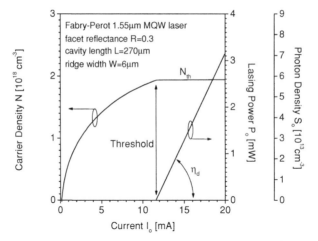

Figure 3.4. Steady state dependence of carrier density, lasing power and photon density on the injection current.

waveguide. Even with restriction to one longitudinal mode, multiple transverse optical modes occur in all three types of lasers. The first or fundamental mode has only one intensity maximum at the laser axis. Higher order modes have multiple maximums in the transverse direction. Their competition with the fundamental mode may cause serious problems for analog modulation applications.

3.2.3 Threshold current and slope efficiency

The most basic laser characteristic is the plot of light power vs. injected current (*PI* curve, Fig. 3.4). At small currents, low intensity light may be generated by spontaneous emission. Lasing starts when the QW carrier density reaches its threshold value N_{th} at which the optical gain compensates for all optical losses (cf. Fig. 3.1). The threshold current I_{th} compensates for all the carrier losses: $\eta_a I_{th} = q V_a (A N_{th} + B N_{th}^2 + C N_{th}^3)$ (q, electron charge; V_a, gain producing active volume). The injection efficiency η_a gives the fraction of electrons that recombine within the active volume. In other words, $(1 - \eta_a)$ is the fraction of current consumed by leakage. The effect of leakage is often neglected in modulation analysis but it can strongly degrade laser performance, especially at high temperatures. Leakage can be separated into lateral carrier spreading and vertical carrier escape. Typical threshold current values are in the lower mA range, depending on the size of the active volume (threshold current density ~ 1 kA/cm^2). The small active volume of VCLs allows for record low μA threshold currents.

$$P_o = \eta_s(I - I_{th}) = \eta_d \frac{h\nu}{q}(I - I_{th}) = \eta_i \frac{\alpha_m}{\alpha_i + \alpha_m} \frac{h\nu}{q}(I - I_{th}). \qquad (3.2)$$

Equation 3.2 gives the common expression for the *PI* curve above threshold ($h\nu$, photon energy; q, electron charge). The slope efficiency η_s [W/A] is related to the differential quantum efficiency η_d (%) which is the probability that an electron injected above threshold contributes a photon to the laser beam. It can be separated into limitations from carrier losses (η_i) and from photon losses (α_i). The differential internal efficiency η_i is the fraction of the total current increment above threshold that results in stimulated emission of photons [6]. It is commonly assumed to be identical to the total injection efficiency η_a (~0.8) as far as the DC carrier density remains constant above threshold. Recent findings suggest some deviation from this rule in MQW lasers [7]. Photon losses are governed by the internal loss parameter α_i (~10 cm^{-1}). The mirror loss parameter $\alpha_m = (1/L)\ln(1/R)$ accounts for the emitted laser beam (R, average power reflectance of both mirrors). Thus, the slope efficiency depends on the cavity length L and it is $\eta_s = 0.37$ W/A ($\eta_d = 47\%$) in our short FP laser example (Fig. 3.4). To achieve high slope efficiencies, internal losses of electrons and photons must be minimized. Better lateral and vertical carrier confinement in the active region reduces carrier leakage. Photon absorption can be lowered, e.g., by less doping of confinement layers. Low mirror reflectivity and low cavity length support higher slope efficiencies but at the expense of a higher threshold current density. At $\lambda = 1.3$ μm emission wavelength, the theoretical maximum of the combined slope efficiencies from both facets is $h\nu/q = 0.95$ W/A ($\eta_d = 100\%$). The maximum front facet slope efficiency measured on 1.3 μm DFB lasers is $\eta_d = 0.65$ W/A (front coating: 2% reflection, rear coating: 90% reflection) [8]. However, typical numbers are lower and coupling loss further reduces the fiber-coupled slope efficiency to 0.02–0.2 W/A for common FP and DFB lasers [9]. VCLs have the advantage of a circular beam leading to very small coupling loss. 850 nm VCLs with $\eta_s = 1.3$ W/A have been demonstrated, which translates into an equivalent value of 0.85 W/A at 1.3 μm emission wavelength [10]. However, long-wavelength VCLs are still under development and their current slope efficiencies are relatively low [11,12].

3.3 Rate equation analysis

3.3.1 Single mode rate equations

The dynamic performance of laser diodes is usually analyzed in terms of rate equations which add up all physical processes that change the densities of photons and carriers. In the following, we discuss a set of three rate equations for the single mode photon density S averaged over the optical volume V_o, the density of quantum well carriers N averaged over the active volume V_a, and the density of barrier carriers N_b averaged over the barrier volume V_b (including SCL) [4]. In Eq. (3.3a), the

change of the barrier density N_b is caused by current injection, carrier capture by
the active layers, and carrier escape from the active layers (recombination within
the barrier layers is neglected). The density of quantum well carriers N in (3.3b)
is increased by carrier capture from the barriers and it is reduced by carrier escape
as well as by all four quantum well recombination processes (cf. Fig. 3.2). The
photon density S in Eq. (3.3c) is increased by stimulated emission and by the fraction β of spontaneously emitted photons that enter the lasing mode ($\Gamma = V_a/V_o$ is
the confinement factor, $\Gamma \sim 0.1$). The factor β is usually very small ($\sim 10^{-4}$) and
spontaneous emission can often be neglected. The density S is reduced by pho-
ton loss mechanisms. The dependence $\alpha_i(N)$ reflects the influence of free carrier
absorption and intervalence band absorption which both rise proportional to the
carrier density ($d\alpha/dN \sim 35 \ldots 140 \times 10^{-18}$ cm^2 $\times \Gamma$) [13]. This results in a
reduced net differential gain.

$$\frac{dN_b}{dt} = \frac{\eta_a I}{q V_b} - \frac{N_b}{\tau_t} + \frac{V_a}{V_b} \frac{N}{\tau_e}, \tag{3.3a}$$

$$\frac{dN}{dt} = \frac{V_b}{V_a} \frac{N_b}{\tau_t} - \frac{N}{\tau_e} - [AN + BN^2 + CN^3] - v_g g(N, S)S, \tag{3.3b}$$

$$\frac{dS}{dt} = \Gamma v_g g(N, S)S - v_g [\alpha_i(N) + \alpha_m] S + \beta \Gamma BN^2. \tag{3.3c}$$

Under steady state conditions with vanishing β, Eq. (3.3c) gives the relation
$\Gamma g(N, S) = \alpha_i(N) + \alpha_m$ describing the balance of gain and optical loss. This lasing
condition implies that for any current above threshold the carrier density N remains
close to the threshold level N_{th} – a slight increase is caused by gain compression
or by MQW carrier non-uniformity [7]. At room temperature, we often can neglect
carrier escape ($\tau_t \ll \tau_e$) and the steady state rate equations yield the DC photon
density $S_o = \eta_i \tau_p (I - I_{th})(q V_o)^{-1}$. The photon lifetime τ_p (~ 1 ps) is defined by
$\tau_p^{-1} = v_g(\alpha_i + \alpha_m)$. The total optical power emitted through both mirrors is given
by $P_o = v_g \alpha_m h\nu V_o S_o$ ($h\nu$, photon energy) leading to the steady state result given
in Eq. (3.2). With analog modulation, sinusoidal variations are added to the steady
state input current I_0. In the simple case of just one angular frequency $\omega = 2\pi f$
and constant amplitude ΔI, the injection current in Eq. (3.3a) becomes

$$I(t) = I_0 + \Delta I \times \sin(\omega t) = I_0 + \Delta I \times \text{Re}\{\exp(i\omega t)\}. \tag{3.4}$$

Thus far, we neglected the non-uniformity of carrier and photon distributions
[14,15,16]. A non-uniform spatial distribution of photons can result in spatial hole
burning (SHB) into the carrier distribution in regions with high photon density
which reduces the gain and increases the refractive index (see Section 3.5.1). Carrier
diffusion into these spatial holes is relatively slow. Longitudinal hole burning can
be severe in DFB lasers since longitudinal variations in gain and refractive index

change the feedback from the grating layer [15,16]. This causes nonlinearities of the *PI* curve and variations of the lasing wavelength with large signal modulation. Lateral hole burning often leads to the appearance of higher order transversal modes that are better supported by the non-uniform profile of gain and refractive index. Multiple optical modes require the use of multiple rate equations (3.3c) [17]. However, lateral hole burning effects can be minimized by well designed lateral index guiding. Lateral diffusion of carriers out of the quantum well active region is part of the lateral carrier leakage which is already considered by the efficiency η_a. However, it also affects the AC response [18]. Lateral transport effects can be minimized by lateral carrier confinement.

3.3.2 Small signal analysis

Many applications use current variations ΔI which are much smaller than $(I_o - I_{th})$ leading to variations ΔN, ΔS, and ΔP which are much smaller than the steady state values N_{th}, S_o, and P_o, respectively. This small signal case allows the rate equations to be linearized and solved analytically [6]. Linearization includes the introduction of the differential carrier lifetime τ_c ($\sim$1 ns) using $\tau_c^{-1} = A + 2BN_{th} + 3CN_{th}^2$ and the use of the linear gain coefficient a representing the differential gain at threshold (Fig. 3.1). Assuming $\beta = 0$ and using $\chi = 1 + \tau_t/\tau_e$, the small signal solution to the rate equations (3.3) is [3]

$$\Delta P(\omega) = M(\omega) \times \Delta P(0) = \frac{1}{1 + i\omega\tau_t} \frac{\omega_r^2}{\omega_r^2 - \omega^2 + i\omega\gamma} \times \eta_i \frac{\alpha_m}{\alpha_i + \alpha_m} \frac{h\nu}{q} \Delta I,$$

$$(3.5a)$$

with the angular electron–photon resonance frequency $\omega_r = 2\pi f_r$ given by

$$\omega_r^2 = \frac{a}{\chi} \frac{v_g S_o}{\tau_p(1 + \varepsilon S_o)} \left(1 + \frac{\varepsilon}{v_g a \tau_c} \right),$$

$$(3.5b)$$

and the damping constant

$$\gamma = \frac{a}{\chi} \frac{v_g S_o}{(1 + \varepsilon S_o)} + \frac{\varepsilon S_o}{\tau_p(1 + \varepsilon S_o)} + \frac{1}{\chi \tau_c}.$$

$$(3.5c)$$

Equation (3.5a) shows that long transport delay times τ_t can reduce the AC signal ΔP substantially at high frequencies when carriers are too slow to reach the active layer within a modulation time period. With reduced carrier escape time τ_e (e.g., at higher temperatures), the transport factor χ lowers the effective differential gain to a/χ and it elongates the carrier lifetime to $\chi \tau_c$. As result, the resonance frequency (3.5b) is reduced to $\omega_r/\sqrt{\chi}$ and the damping constant (3.5c) becomes γ/χ. With negligible gain compression ($\varepsilon = 0$) and no carrier escape ($\chi = 1$), we obtain

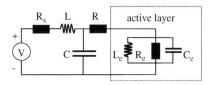

Figure 3.5. Equivalent laser circuit for the active layer (R_e, C_e, L_e) plus parasitic elements: laser resistance R, contact capacitance C, bond wire inductance L. ($R_s = 50\,\Omega$, resistance of ideal current source.)

more simple solutions for the angular resonance frequency

$$\omega_r = 2\pi f_r = \sqrt{\frac{v_g a S_o}{\tau_p}} = \sqrt{\frac{\Gamma a}{\tau_p \alpha_m h\nu V_a}}\sqrt{P_o} = \sqrt{\frac{\Gamma \eta_i v_g a}{q V_a}}\sqrt{I - I_{th}} \quad (3.6a)$$

and the damping constant

$$\gamma = \omega_r^2 \tau_p + \tau_c^{-1}. \quad (3.6b)$$

3.3.3 Equivalent circuits and parasitics

When the QW carrier density N is related to the junction voltage V, small signal analysis of the rate equations leads to an impedance function $Z(\omega) = \Delta V(\omega)/\Delta I(\omega)$ which is equivalent to that of an electronic circuit [19]. In the most simple case (no transport effects, $\varepsilon = 0$, $\beta = 0$), the impedance function corresponds to that of a parallel RLC circuit (Fig. 3.5) with resistance R_e, inductance L_e, and capacitance C_e given by ($\Gamma = 1$) [20]

$$R_e = R_d \frac{I_{th}}{I_o}, \quad L_e = \frac{I_{th} R_d \tau_p}{I_o - I_{th}}, \quad C_e = \frac{\tau_c}{R_d}, \quad \text{with } R_d = \frac{2kT\tau_c}{N_{th} V_a}.$$

$$(3.7)$$

More extended equivalent circuits are obtained when spontaneous emission [20], noise [21], gain compression [22], or transport effects [23] are considered. These equivalent circuits describe processes within the active layer and its intrinsic circuit elements should not be confused with parasitic circuit elements representing processes outside the active layer as discussed in the following.

The internal efficiency η_i of the small signal solution (3.5a) takes into account that only part of the current injected into the device actually arrives in the active region. In practical lasers, the injection efficiency depends on the modulation frequency and it can reduce the AC lasing power $\Delta P(\omega)$ considerably at higher frequencies. This behavior is caused by several physical mechanisms inside and outside the semiconductor device. Inside the laser, space charge regions at the p-n junction and at other heterojunctions establish internal capacitors. Outside the laser, metal contact pads form external capacitors and bond wires act as external inductors. The contribution of capacitors and inductors to the device impedance depends on the modulation frequency and so does the fraction η_i of the current that enters the active region. The function $\eta_i(\omega)$ is enhanced at LC resonance and it decreases with higher frequency as more current bypasses the active region through parallel capacitors. A common circuit model for laser parasitics is shown in Fig. 3.5 resulting in [24]

$$\frac{\eta_i(\omega)}{\eta_i(0)} = \left[\left(1 - \frac{\omega^2}{\omega_0^2} \right)^2 + \left(\frac{\omega}{\omega_0 Q} \right)^2 \right]^{-\frac{1}{2}}$$

$$\text{with} \quad \omega_0^2 = \frac{R_s + R}{LRC}, \qquad Q = \frac{\sqrt{LRC(R_s + R)}}{L + RR_sC}. \tag{3.8}$$

This equation may be introduced into (3.5a) to account for parasitics in small signal analysis. However, more sophisticated circuits are required in some cases, e.g., to separate external and internal parasitic capacitors [25]. External impedance matching networks are often used to reduce the RF signal insertion loss [26].

3.4 Intensity modulation

3.4.1 Fundamental response characteristics

Analog intensity modulation (IM) applies sinusoidal variations to the injection current (Eq. 3.4). The electrical modulation depth is given by $m = \Delta I/(I_o - I_{th})$. Small signal analysis can be applied for $m < 0.1$. Figure 3.6 shows plots of the normalized AC output power $|M(\omega)|$ as function of modulation frequency (Eq. 3.5).

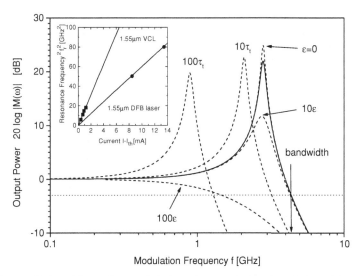

Figure 3.6. Small signal intensity modulation characteristics (Eq. 3.5a) with variations of the gain compression parameter $\varepsilon = 10^{-17}$ cm^3, and the transport time $\tau_t = 1$ ps ($\tau_e = 10$ ps, $P_o = 3$ mW). The insert shows best results of modulation current efficiency measurements on long wavelength VCLs and DFB lasers.

This IM response characteristic reaches a maximum at the peak frequency ω_p which is (slightly) smaller than the resonance frequency ω_r. Without transport or parasitic effects, the peak frequency is given by $\omega_p^2 = \omega_r^2 - \gamma^2/2$. With increasing current, plots of ω_p^2 vs. $(I_o - I_{th})$ or ω_p^2 vs. P_o give a straight line as long as restricting mechanisms are negligible. The modulation bandwidth is usually defined as the frequency f_b at which the output signal drops to $|M(2\pi f_b)| = 2^{-1/2} = -3\,\text{dB}$ (Fig. 3.6). With low damping at low injection current, we obtain $f_b \approx 1.55 f_r$ and the linear plot $f_b = \text{MCEF} \times (I_o - I_{th})^{1/2}$ provides the modulation current efficiency factor (MCEF). High-speed in-plane lasers [27] give $\text{MCEF} = 3\,\text{GHz}\,\text{mA}^{-1/2}$ whereas the small active region of vertical-cavity lasers has led to a five times higher MCEF [28]. However, the bandwidth reaches a maximum at high currents due to increased damping, device heating, gain compression, transport effects, or parasitics [29]. A record bandwidth of 30 GHz has recently been achieved using 1.55 μm DFB lasers [27]. VCLs show high resonance frequencies [30] and bandwidths up to 20 GHz [28].

The effects of carrier transport delays and gain compression are illustrated in Fig. 3.6. Gain compression may cause a sub-linear increase or even a decrease of the squared peak frequency with higher power P_o [31]. Transport effects can result in a low-frequency cut-off of the response function. However, such a cut-off may also be related to device parasitics.

Large signal analog modulation ($m > 0.1$) causes some deviation of the fundamental response characteristic from the small signal case. The peak frequency ω_p decreases with increasing modulation depth. Reductions up to about 50% of the small signal value have been observed with $m = 1$ [32]. This is related to photon density oscillations at frequencies other than the fundamental input frequency (see Section 3.4.3).

3.4.2 Intensity noise

So far we have only considered light intensity variations due to deliberate current modulation. Steady state carrier and photon densities were assumed to be constant. However, the quantum nature of electrons and photons and the random nature of physical processes produce random fluctuations (noise) of the output power, even without current modulation. For analog applications, intensity noise is quantified using the signal-to-noise ratio (SNR) which is linked to the relative intensity noise (RIN) commonly used with laser diodes: $\text{SNR} = m^2/2\,\text{RIN}$ [6,17]. For example, for a high quality TV transmission with 50 dB SNR (electrical dB after conversion by the photodetector, i.e., $\text{SNR} = 10^5$) with $m = 0.5$ modulation depth, the laser must have a $\text{RIN} < -59\,\text{dB}$. Since noise is proportional to the measurement bandwidth Δf, RIN is often divided by Δf and given in dB/Hz. In single mode operation, RIN is almost constant at lower RF frequencies and it peaks at the laser resonance

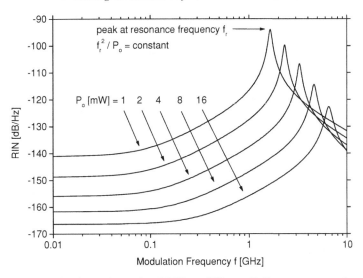

Figure 3.7. Relative intensity noise (RIN) at different DC output power levels P_o.

frequency. Higher output power gives lower RIN (Fig. 3.7). For theoretical analysis, Langevin noise terms are added to the rate equations [6]. This results in the low-frequency (RF) noise formula (noise floor in Fig. 3.7, $f = 100$ kHz to $f_r/10$)

$$\frac{\text{RIN}_{RF}}{\Delta f} = \frac{\langle \delta P \rangle^2}{\Delta f P_o^2} = \frac{4h\nu V_a \alpha_m \nu_g \beta R_{sp}}{\omega_r^4 \tau_c^2 P_o} + \frac{2h\nu}{P_o}\left[\eta_o \frac{I + I_{th}}{I - I_{th}} + 1 - \eta_o\right]. \quad (3.9)$$

The first term in (3.9) is due to spontaneous emission and it decreases as $1/P_o^3$ ($\omega_r^4 \propto P_o^4$). The second term dominates with high output power P_o and it is caused by the shot noise of the injected current. This shot noise can be reduced by using a high impedance source to drive the laser [33]. At very low frequencies ($f < 100$ kHz), RIN increases considerably showing a $1/f$ dependence which is typical for electronic devices (flicker noise) and which is assumed to be caused by defect recombination [17].

With multiple laser modes, the RF noise is still low if the total power of all modes is detected. However, if a single mode is filtered out, it typically exhibits a much higher RIN due to mode partition noise. Spontaneous emission into one of several modes exhibits stronger fluctuations than the total spontaneous emission into all laser modes. Mode partition noise can usually be neglected for sidemode suppression ratios above 30 dB [17]. The coupling of laser diodes to fiber systems causes additional laser noise. Optical feedback from external interfaces such as connectors or photodiodes may cause periodic spikes in the noise spectrum. Such excess noise spikes are strongest near the resonance frequency. The average RF RIN starts to increase with about 10^{-4} feedback rising by 30 dB/decade with higher feedback [34].

High-power DFB lasers have shown RIN $< -165\,\text{dB/Hz}$ at $f = 100\,\text{MHz}$ [35]. The noise increases when the modulation frequency f comes closer to the resonance frequency (Fig. 3.7). Typical RIN values for FP lasers are from -120 to $-140\,\text{dB/Hz}$ [36]. With reduced external feedback, FP lasers can give a link noise figure similar to DFB lasers [37]. Lower output power and unstable polarization of VCLs causes higher noise than with in-plane lasers, especially with polarization sensitive applications. However, polarization controlled VCLs have shown RIN below $-140\,\text{dB/Hz}$ [38]. VCL noise was found to be less sensitive to external feedback than with in-plane lasers [39].

3.4.3 Harmonic and intermodulation distortion

A sinusoidal current modulation (3.4) generates light power modulations at the input frequency ω and also at the harmonics 2ω, 3ω, $\ldots$, $n\omega$. The amplitude of the nth order harmonic is proportional to the nth power of the optical modulation depth $|\Delta P(\omega)|/P_\text{o} = m|M(\omega)|$ and it decreases rapidly with higher order. Harmonic distortion (HD) depends on the nonlinearity of a transmission system. In the case of laser diodes, it is often related to the nonlinearity of the PI characteristic which can be expressed by the Taylor series

$$P(I) = P(I_\text{o}) + (I - I_\text{o})\frac{\partial P}{\partial I} + \frac{1}{2}(I - I_\text{o})^2\frac{\partial^2 P}{\partial I^2} + \frac{1}{6}(I - I_\text{o})^3\frac{\partial^3 P}{\partial I^3} + \cdots \quad (3.10)$$

With analog modulation, the last two terms of (3.10) yield second and third order distortions, respectively $[(I - I_\text{o})^n = m^n(I - I_\text{th})^n e^{in\omega t}]$. With low frequencies well below resonance, the steady state formula (3.2) can be used to evaluate (3.10). Any slight current dependence of the parameters in (3.2) causes higher order distortions: (a) Carrier leakage from the active region affects the differential internal efficiency η_i depending on the injection current [7,40]. (b) Gain compression at high photon densities causes the effective threshold current to increase and it may dominate the low-frequency distortion in Fabry–Perot lasers [41]. (c) The internal loss parameter α_i increases with higher carrier density leading to an amplification of distortion effects [42,43]. (d) Spatial hole-burning (SHB) dominates the low-frequency distortion in DFB lasers [16]. The DFB coupling constant and the facet reflectivity are crucial design parameters to achieve low distortion [44,45]. In DFB lasers, the phase difference between distortions of different origin is often utilized to obtain distortion dips at a specific frequency or bias [41]. For example, a PI super-linearity caused by SHB can be used to cancel a PI sub-linearity due to gain compression, leading to a more linear PI curve [45]. This cancellation is bias dependent since SHB decreases and gain compression increases with higher bias.

Even with perfectly linear PI characteristics, distortions are generated by the intrinsic interaction of electrons and photons during stimulated recombination. This can easily be recognized from the $N \times S$ terms in the linearized rate equations which generate time dependences $\exp(i2\omega t)$, $\exp(i3\omega t)$, etc., in addition to the $\exp(i\omega t)$ dependence described earlier. Intrinsic distortion dominates at frequencies near the resonance frequency f_r and often limits the usable bandwidth to low frequencies $f \ll f_r$. At low frequencies, the distortion is governed by the nonlinearity of the PI curve discussed above (static distortion).

Intermodulation distortion arises when two or more signals at different modulation frequencies are transmitted. Two signals at ω_1 and ω_2, for example, are accompanied by second order distortions at frequencies $2\omega_{1,2}$ and $\omega_1 \pm \omega_2$, third order distortions at frequencies $3\omega_{1,2}$, $2\omega_1 \pm \omega_2$, and $2\omega_2 \pm \omega_1$, etc. The third order intermodulation distortions (IMDs) at $2\omega_1 - \omega_2$ and $2\omega_2 - \omega_1$ are of special interest since they are close to the original signals and they might interfere with other signals in multichannel applications (Fig. 3.8). The third order IMD increases as the cube of the modulation depth [46]. The amplitudes of intermodulation distortions can be related to the amplitudes of harmonic distortions (those relations depend on the dominating cause of the distortion: intrinsic or static distortion) [16]. In multichannel applications, distortions from several channels add up and they are described by composite second order (CSO) and composite triple beat (CTB) quantities. Additional distortions from clipping occur when the combined modulation depth of all channels is larger than one, i.e., when the total modulation current drops below the threshold current [47].

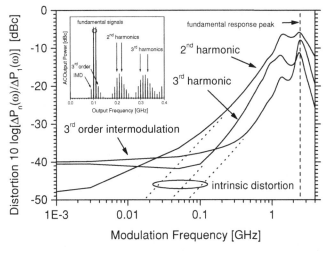

Figure 3.8. Harmonic and intermodulation distortions vs. modulation frequency f_1 (Eq. 3.3) with two AC input signals ($f_2 = f_1 + 10$ MHz, $P_o = 20$ mW, $m = 0.25$ each). The insert illustrates the variety of distortions of the AC output spectrum. The dotted lines give the decay of intrinsic distortions with lower frequencies.

Analytical formulas can been derived for magnitude and phase of specific distortion mechanisms [41,43,48]. However, the variety of distortion mechanisms in the RF range and their complex interaction often requires a numerical evaluation of rate equations [15,16]. For a simple Fabry–Perot laser without spatial non-uniformities, Fig. 3.8 plots solutions to the rate equations (3.3) with two input signals at $f_1 = 100$ MHz and $f_2 = 110$ MHz, respectively, and $m = 0.25$ electrical modulation depth each. Second order HD arises at 200 MHz and 220 MHz, third order HD at 300 MHz and 330 MHz, as well as third order IMD at 90 MHz and 120 MHz, respectively. Intrinsic distortion dominates at higher frequencies where multiple peaks occur in all three distortion spectra. HD signals of nth order have a peak at f_r/n input frequency when their output frequency is equal to the large-signal resonance frequency $f_r = 2.6$ GHz (small signal resonance at 2.8 GHz). Another distortion peak occurs at input frequency f_r when the photons of the fundamental signal are at resonance. Additional maximums can occur at peak frequencies of other distortion signals. Intrinsic distortions decline steadily with lower frequencies (dotted lines in Fig. 3.8) [48] and distortions from the nonlinear PI curve dominate for frequencies below a few hundred MHz [41]. Figure 3.8 illustrates that the usable low-distortion bandwidth is mainly limited by intrinsic relaxation oscillation effects. A higher DC lasing power gives a larger resonance frequency and it expands the low-distortion bandwidth.

Besides internal mechanisms, external feedback can affect the strength of laser distortion substantially. This type of distortion is proportional to the amount of feedback entering the laser and it exhibits periodic changes with increasing modulation frequency ω (amplitude $\propto \omega^{-1}$) [32]. Optical isolators are commonly used to reduce feedback effects.

3.4.4 Dynamic range

The dynamic range of linearity is of paramount importance for many analog modulation applications. Even if the total lasing power changes linearly with injection current, intrinsic distortions draw power from the fundamental signal. At low modulation depth, distortions are still below the noise floor but the signal-to-noise ratio is also small. With increasing modulation depth, distortions rise above the noise floor and grow faster than the fundamental signal. Thus, the largest distortion free signal-to-noise ratio (dynamic range) is reached when the amplitude of the distortion is equal to the noise floor (Fig. 3.9). Since the noise floor depends on the measurement bandwidth Δf, the spurious free dynamic range SFDR refers to $\Delta f = 1$ Hz. SFDR is the same for input and output. External optical feedback can cause the SFDR to vary periodically with changing modulation frequency [49]. Predistortion of the input signal or specific feedback schemes are used to increase the intermodulation-free dynamic range of DFB laser diodes [36].

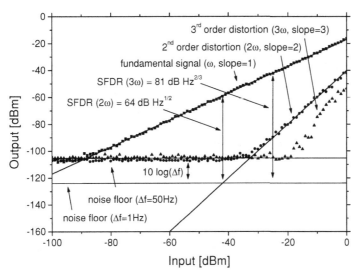

Figure 3.9. Measured 1.55 μm VCL output power (dots) at different frequencies vs. input power (input frequency $f = 1$ GHz). Lines illustrate the extraction of the spurious free dynamic range (SFDR) for second and third order harmonic distortion.

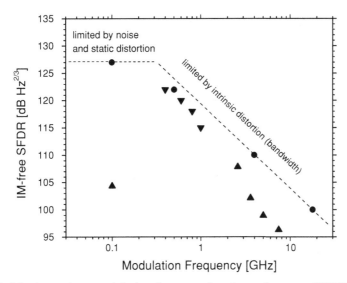

Figure 3.10. Maximum intermodulation-free spur-free dynamic range (SFDR) vs. modulation frequency as published for different laser types. ● DFB lasers ▲ VC lasers ▼ FP lasers.

Figure 3.9 gives the first results on the SFDR for HD-free operation of 1.55 μm VCLs at 1 GHz modulation frequency [50]. Figure 3.10 reviews the maximum intermodulation-free SFDR as measured at different modulation frequencies. Results are shown for DFB lasers [51], Fabry–Perot lasers [37], and vertical-cavity lasers [49,52]. The decline with higher frequency reflects the enhancement of intrinsic distortion close to the resonance frequency (Fig. 3.8). However, there is no further increase of SFDR with lower frequencies (cf. Figs. 3.7, 3.8). To achieve a large dynamic range, it is advantageous to use laser designs with high differential gain (strained quantum wells) and low photon lifetime (low mirror reflectivity). In DFB lasers, an increase of the facet loss usually reduces the distortion from spatial hole burning. However, SHB effects in DFB lasers are often utilized to cancel other nonlinearities leading to maximum SFDR within a narrow frequency range [41]. Since SBH is less important with FP lasers and VCLs, these laser types have a potential advantage if low distortion is required within a wide frequency band. With reduced external feedback, FP laser performance in optical links has been shown to be comparable to that of high-linearity DFB lasers [37]. Promising first measurements of the intermodulation-free dynamic range of long-wavelength VCLs have been reported [53].

3.5 Frequency modulation

3.5.1 Modulation characteristics

The carrier density N affects the optical gain g, but it also affects the refractive index n of the active region ($dn/dN \sim -2 \times 10^{-20}$ cm^3). The refractive index describes the magnitude of semiconductor polarization in response to the electromagnetic wave. Active region carriers take part in this polarization. The complex refractive index $n + i(gc/4\pi\nu)$ combines both aspects of the interaction between light and matter (c, speed of light; ν, lasing frequency). The spectra $n(\nu)$ and $g(\nu)$ are closely connected by the Kramers–Kroenig formulas [15]. With variation ΔN of the carrier density, the corresponding variations Δn and Δg are related by the linewidth enhancement factor $\alpha = -4\pi\nu\Delta n(c\Delta g)^{-1}$ ($\alpha \sim 4$–6) [54]. The lasing frequency ν depends on the optical length nL of the laser cavity and the change ΔN causes the frequency shift (chirp) $\Delta\nu = (\alpha/4\pi)\Gamma v_{\mathrm{g}}a\Delta N$.

Analog modulation of the laser injection current leads to the variation $\Delta N(\omega)$ of the active region carrier density which results in a variation $\Delta\nu(\omega)$ of the laser frequency. Small signal analysis including gain compression gives the frequency modulation (FM) characteristic [54]

$$\Delta\nu(\omega) = \frac{\alpha}{4\pi}\left[\frac{\Gamma\beta R_{\mathrm{sp}}}{S_{\mathrm{o}}} + \frac{\varepsilon\Gamma v_{\mathrm{g}}aN_{\mathrm{th}}S_{\mathrm{o}}}{1 + \varepsilon S_{\mathrm{o}}} + i\omega\right]\frac{\Delta P(\omega)}{P_{\mathrm{o}}}. \qquad (3.11)$$

Thus, frequency chirping increases linearly with the optical modulation depth $\Delta P(\omega)/P_{\mathrm{o}}$. The normalized FM characteristic $|\Delta\nu(\omega)|/\Delta I$ is plotted in Fig. 3.11. The FM response has its maximum at ω_{r}, slightly above the IM peak frequency ω_{p}. At high frequencies $\omega \gg \omega_{\mathrm{r}}$ the FM response drops twice as fast (40 dB/decade)

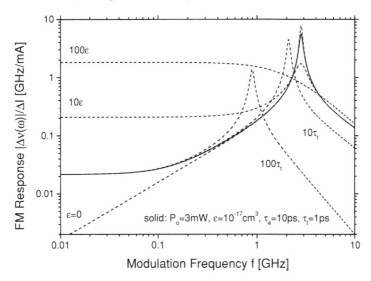

Figure 3.11. Frequency modulations response (Eq. 3.11) with variations of the gain compression parameter ε and the transport time τ_t.

as the IM response. At low frequencies, laser modulation also causes the active region temperature to change periodically. This leads to additional variations of the refractive index $(dn/dT = 2 \ldots 5 \times 10^{-4} \text{ K}^{-1})$ translating into $dv/dT \sim -20$ GHz/K. Temperature modulation can dominate the FM response up to 50 MHz [55]. The effects of transport delay and gain compression are illustrated in Fig. 3.11. In DFB lasers, FM characteristics can be strongly affected by spatial hole burning [16]. External frequency or phase modulation is preferred over direct laser modulation in coherent analog fiber optic links [56]. However, experiments with external cavity lasers show high signal-to-noise ratios at low optical power [57].

3.5.2 Frequency noise and linewidth

Random fluctuations of the lasing frequency are caused by the carrier density noise and by the random phase of spontaneously emitted photons (phase noise). Carrier noise is mainly due to spontaneous emission and it is related to the intensity noise discussed above. The frequency noise spectral density $W_\nu(\omega)$ is directly related to the full-width half-maximum linewidth $\Delta \nu_{FW}$ of the lasing spectrum [6]

$$\Delta \nu_{FW}(\omega) = 2\pi W_\nu(\omega) = \frac{\Gamma \beta R_{sp}}{4\pi S_o} \left[1 + \frac{\alpha^2 \omega_r^4}{\left(\omega_r^2 - \omega^2\right)^2 + \omega^2 \gamma^2} \right], \qquad (3.12)$$

with the frequency independent part representing phase noise (Schawlow–Townes linewidth). The linewidth is inversely proportional to the lasing power which is one of the unique features of lasing. The steady state linewidth is typically in the low

MHz range and a lasing power well above 10 mW is required to obtain sub-MHz linewidths.

3.6 Conclusion

At the present stage of development, DFB lasers clearly outperform FP lasers and VCLs in analog optical fiber links. At 1.3 μm or 1.55 μm wavelength, DFB lasers exhibit record values of main performance parameters: 30 GHz bandwidth, [27] 127 dB Hz$^{2/3}$ intermodulation-free SFDR (100 MHz) [58], relative intensity noise of less than −165 dB/Hz [35], and 0.65 W/A single facet slope efficiency [8]. However, these record parameters are hard to combine in a single device. Fabry–Perot lasers are low-cost alternatives for less demanding applications which do not require dynamic single mode lasing. Long-wavelength vertical-cavity lasers are currently under development and they promise to combine the low cost advantage with a high performance potential.

References

1. E. I. Ackerman, C. H. Cox III, and N. A. Riza (eds.), *Selected Papers on Analog Fiber-Optic Links*, SPIE Milestone Series, vol. 149, Bellingham, 1998.
2. S. L. Chuang, *Physics of Optoelectronic Devices*, Wiley & Sons, New York, 1995.
3. R. Nagarajan, T. Fukushima, M. Ishikawa, J. E. Bowers, R. S. Geels, and L. A. Coldren, "Transport limits in high-speed quantum-well lasers: experiment and theory," *IEEE Photon. Technol. Lett.*, **4** (2), 121–3, 1992.
4. G. Morthier, "Design and optimization of strained-layer-multiquantum-well lasers for high-speed analog communication," *IEEE J. Quantum Electron.*, **30**, 1520–7, 1994.
5. C. Wilmsen, H. Temkin, and L. A Coldren (eds.), *Vertical Cavity Surface Emitting Lasers*, Cambridge University Press, 1999.
6. L. A. Coldren and S. W. Corzine, *Diode Lasers and Photonic Integrated Circuits*, John Wiley & Sons, New York, 1995.
7. J. Piprek, P. Abraham, and J. E. Bowers, "Carrier non-uniformity effects on the internal efficiency of multi-quantum-well laser diodes," *Appl. Phys. Lett.*, **74** (4), 489–91, 1999.
8. T. R. Chen, P. C. Chen, J. Ungar, J. Paslaski, S. Oh, H. Luong, and N. Bar-Chaim, "Wide temperature range linear DFB lasers with very low threshold current," *Electron. Lett.*, **33**, 963–4, 1997.
9. T. Olson, "An RF and microwave fiber-optic design guide," *Microwave J.*, **39** (8), 54–78, 1996.
10. R. Jaeger, M. Grabherr, C. Jung, R. Michalzik, G. Reiner, B. Weigl, and K. J. Ebeling, "57% wallplug efficiency oxide-confined 850nm wavelength GaAs VCSELs," *Electron. Lett.*, **33**, 330–1, 1997.
11. N. M. Margalit, J. Piprek, S. Zhang, D. I. Babic, K. Streubel, R. P. Mirin, J. R. Wesselmann, J. E. Bowers, and E. L. Hu, "64 °C continuous-wave operation of 1.5 μm vertical-cavity laser," *IEEE J. Sel. Top. Quantum Electron.*, **3**, 359–65, 1997.

12. V. Jayaraman, J. C. Geske, M. H. MacDougal, F. H. Peters, T. D. Lowes, and T. T. Char, "Uniform threshold current, continuous-wave, singlemode 1300nm vertical cavity lasers from 0 to 70 °C," *Electron. Lett.*, **34**, 1405–6, 1998.

13. I. Joindot and J. L. Beylat, "Intervalence band absorption coefficient measurements in bulk layer, strain and unstrained multiquantum well 1.55 μm semiconductor lasers," *Electron. Lett.*, **29** (7), 604–6, 1993.

14. N. Tessler and G. Eisenstein, "On carrier injection and gain dynamics in quantum well lasers," *IEEE J. Quantum Electron.*, **29** (6), 1586–93, 1993.

15. J. Carroll, J. Whiteaway, and D. Plumb, *Distributed Feedback Semiconductor Lasers*, IEE/SPIE Press, London/Washington, 1998.

16. G. Morthier and P. Vankwikelberge, *Handbook of Distributed Feedback Laser Diodes*, Artech House, Norwood, 1997.

17. K. Petermann, *Laser Diode Modulation and Noise*, Kluwer Academic Publishers, Dordrecht, 1988.

18. G. H. B. Thompson, *Physics of Semiconductor Laser Devices*, John Wiley & Sons, New York, 1980.

19. K. Y. Lau and A. Yariv, "High-frequency current modulation of semiconductor injection lasers," Ch. 2 in *Semiconductors and Semimetals*, vol. 22, Academic Press, Orlando, 1985.

20. J. Katz, S. Margalit, C. Harder, D. Wilt, and A. Yariv, "The intrinsic equivalent circuit of a laser diode," *IEEE J. Quantum Electron.*, **17** (1), 4–7, 1981.

21. C. Harder, J. Katz, S. Margalit, J. Shacham, and A. Yariv, "Noise equivalent circuit of a semiconductor laser diode," *IEEE J. Quantum Electron.*, **18** (3), 333–7, 1982.

22. E. J. Flynn, "A note on the semiconductor laser equivalent circuit," *J. Appl. Phys.*, **85** (4), 2041–5, 1999.

23. K. Y. Lau, "Dynamics in quantum well lasers," Ch. 5 in *Quantum Well Lasers*, ed. P. S. Zory, Academic Press, San Diego, 1993.

24. J. E. Bowers, "Modulation properties of semiconductor lasers," in *Optoelectronics for the Information Age*, ed. C. Lin, Van Nostrand, New York, 1989.

25. D. A. Atlas and A. Rosiewicz, "A 20 GHz bandwidth InGaAsP/InP MTBH laser module," *IEEE Photon. Technol. Lett.*, **5** (2), 123–6, 1993.

26. C. L. Goldsmith and B. Kanack, "Broad-band reactive matching of high-speed directly modulated laser diodes," *IEEE Microwave & Guided Wave Lett.*, **3** (9), 336–8, 1993.

27. O. Kjebon, R. Schatz, S. Lourdudoss, S. Nilsson, B. Stalnacke, and L. Backbom, "30 GHz direct modulation bandwidth in detuned loaded InGaAsP DBR lasers at 1.55 μm wavelength," *Electron. Lett.*, **33**, 488–9, 1997.

28. K. L. Lear, A. Mar, K. D. Choquette, S. P. Kilcoyne, R. P. Schneider, and K. M. Geib, "High-frequency modulation of oxide-confined vertical-cavity surface emitting lasers," *Electron. Lett.*, **32**, 457–8, 1996.

29. O. Kjebon, R. Schatz, S. Lourdudoss, S. Nilsson, and B. Stalnacke, "Modulation response measurement and evaluation of MQW InGaAsP lasers of various design," *Proc., SPIE*, **2687**, 138–52, 1996.

30. D. Tauber, G. Wang, R. S. Geels, J. E. Bowers, and L. A. Coldren, "Large and small dynamics of vertical cavity surface emitting lasers," *Appl. Phys. Lett.*, **62**, 325–7, 1993.

31. J. E. Bowers, "High speed semiconductor laser design and performance," *Solid State Electron.*, **30**, 1–11, 1987.

32. K. Kishino, "Direct modulation of semiconductor lasers," in *Handbook of Semiconductor Lasers and Photonic Integrated Circuits*, eds. Y. Suematsu and A. R. Adams, Chapman & Hall, London, 1994.

33. Y. Yamamoto and S. Machida, "High-impedance suppression of pump fluctuation and amplitude squeezing in semiconductor lasers," *Phys. Rev. A*, **35** (12), 5114–30, 1987.
34. N. Schunk and K. Petermann, "Numerical analysis of feedback regimes for a single-mode semiconductor laser with external feedback," *IEEE J. Quantum Electron.*, **24** (7), 1242–7, 1988.
35. T. R. Chen, W. Hsin, and N. Bar-Chaim, "Very high-power InGaAsP/InP distributed feedback lasers at 1.55 μm wavelength," *Appl. Phys. Lett.*, **72**, 1269–71, 1998.
36. C. H. Cox, "Optical transmitters," in *The Electronics Handbook*, ed. J. C. Whitaker, CRC Press & IEEE Press, 1996.
37. H. Roussell, R. Helkey, G. Betts, and C. Cox, "Effect of optical feedback on high-dynamic-range Fabry–Perot laser optical links," *IEEE Photon Technol. Lett.*, **9**, 106–8, 1997.
38. T. Yoshikawa, T. Kawakami, H. Saito, H. Kosaka, M. Kajita, K. Kurihara, Y. Sugimoto, and K. Kasahara, "Polarization-controlled single-mode VCSEL," *IEEE J. Quantum Electron.*, **34** (6), 1009–15, 1998.
39. J. W. Bae, H. Temkin, S. E. Swirhun, W. E. Quinn, P. Brusenbach, C. Parson, M. Kim, and T. Uchida, "Reflection noise in vertical cavity surface emitting lasers," *Appl. Phys. Lett.*, **63**, 1480–2, 1993.
40. M. S. Lin, S. J. Wang, and N. Dutta, "Measurement and modeling of the harmonic distortion in InGaAsP distributed feedback lasers," *IEEE J. Quantum Electron.*, **26**, 998–1004, 1990.
41. G. Morthier, F. Libbrecht, K. David, P. Vankwikelberge, and R. G. Baets, "Theoretical investigation of the second-order harmonic distortion in the AM response of 1.55 μm FP and DFB lasers," *IEEE J. Quantum Electron.*, **27**, 1990–2002, 1991.
42. G. Morthier, "Influence of the carrier density dependence of the absorption on the harmonic distortion in semiconductor lasers," *J. Lightwave Technol.*, **11**, 16–19, 1993.
43. V. B. Gorfinkel and S. Luryi, "Fundamental limits for linearity of CATV lasers," *J. Lightwave Technol.*, **13**, 252–60, 1995.
44. A. Takemoto, H. Watanabe, Y. Nakajima, Y. Sakakibara, S. Kakimoto, J. Yamashita, T. Hatta, and Y. Miyake, "Distributed feedback laser diode and module for CATV systems," *J. Select. Areas Commun.*, **8**, 1359–64, 1990.
45. J. Chen, R. Ram, and R. Helkey, "Linearity and third order intermodulation distortion in DFB semiconductor lasers," *IEEE J. Quantum Electron.*, 1999.
46. K. Y. Lau and A. Yariv, "Intermodulation distortion in a directly modulated semiconductor injection laser," *Appl. Phys. Lett.*, **45** (10), 1034–6, 1984.
47. Q. Shi, R. S. Burroughs, and D. Lewis, "An alternative model for laser clipping-induced nonlinear distortion for analog lightwave CATV systems," *IEEE Photon Technol. Lett.*, **4** (7), 784–7, 1992.
48. T. E. Darcie, R. S. Tucker, and G. J. Sullivan, "Intermodulation and harmonic distortion in InGaAsP lasers," *Electron. Lett.*, **21** (16), 665–6, 1985 (correction in *Electron. Lett.*, **22** (11), 619, 1986).
49. H. Lee, R. V. Dalal, R. J. Ram, and K. D. Choquette, "Resonant distortion in vertical cavity surface emitting lasers for RF communication," *Optical Fiber Conference*, Paper FE1, Digest F, pp. 84–6, Optical Society of America, Washington, 1999.
50. J. Piprek, K. Takiguchi, A. Black, P. Abraham, A. Keating, V. Kaman, S. Zhang, and J. E. Bowers, "Analog modulation of 1.55 μm vertical cavity lasers," *Proc. SPIE*, **3627**, *Vertical-Cavity Surface-Emitting Lasers III*, eds. K. D. Choquette and C. Lei, 1999.
51. C. Cox, E. Ackerman, R. Helkey, and G. E. Betts, "Techniques and performance of intensity-modulation direct-detection analog optical links," *IEEE Trans. Microwave Theory Tech.*, **45** (8), 1375–83, 1997.

52. R. V. Dalal, R. J. Ram, R. Helkey, H. Roussell, and K. D. Choquette, "Low distortion analogue signal transmission using vertical cavity lasers," *Electron. Lett.*, **34**, 1590–1, 1998.

53. J. R. Wesselmann, N. M. Margalit, and J. E. Bowers, "Analog measurements of long wavelength vertical-cavity lasers," *Appl. Phys. Lett.*, **72**, 2084–6, 1998.

54. T. L. Koch and J. E. Bowers, "Nature of wavelength chirping in directly modulated semiconductor lasers," *Electron. Lett.*, **20** (25–26), 1038–40, 1984.

55. S. Kobayashi, Y. Yamamoto, M. Ito, and T. Kimura, "Direct frequency modulation in AlGaAs semiconductor lasers," *IEEE J. Quantum Electron.*, **18** (4), 582–95, 1982.

56. T. Kimura, "Coherent optical fiber transmission," *J. Lightwave Technol.*, **5** (4) 414–28, 1987.

57. B. Chai and A. J. Seeds, "Optical frequency modulation links: theory and experiments," *IEEE Trans. Microwave Theory Tech.*, **45** (4), 505–11, 1997.

58. Fujitsu 1.3 μm DFB laser with bulk active region, R. Ram, personal communication.

4

LiNbO$_3$ external modulators and their use in high performance analog links

GARY E. BETTS

MIT Lincoln Laboratory

4.1 Introduction

This chapter will cover the design of modulators for high performance externally modulated analog links. This includes development of a model of the externally modulated analog link so that the connection between modulator design parameters and link performance is clear. The only type of link considered here is one using amplitude modulation of the light and direct detection.

Section 4.2 will go through the basic modulator designs in detail. Section 4.3 will relate the modulator performance to link performance, including an explanation of modulator linearization and an overview of several types of linearized modulator.

The link analysis in Section 4.3 is applicable to any externally modulated link using a modulator that can be characterized by a transfer function that depends on voltage (i.e., the optical transmission depends on voltage). It applies to modulators of any type in any material that meet this criterion. The modulator designs given in Section 4.2, and the linearized modulators described in Section 4.3.2, can be built on a variety of materials.

Lithium niobate (LiNbO$_3$) presently is the material in widest use as a modulator substrate for modulators used in analog links. It is an insulating crystal whose refractive index changes with voltage. It has a high electro-optic coefficient, it is stable at normal electronic operating temperatures, low-loss (0.1 dB/cm) optical waveguides can be made in it, there are manufacturable ways to attach optical fibers with low coupling loss, and its dielectric constant is low enough that high speed modulators can be made without too much trouble. Further details specific to lithium niobate will be discussed in Section 4.2.5. There are other materials that can exceed the performance of lithium niobate in just about every one of

This work was sponsored by the Department of the Air Force under AF Contract No. F19628-95-C-0002. The opinions, interpretations, conclusions and recommendations are those of the authors and are not necessarily endorsed by the United States Air Force.

the above areas individually; however, at the present time each of those materials comes with an unacceptable penalty in one of the other necessary areas. For this reason, all of the modulators discussed in this chapter have been built in lithium niobate (except for the electroabsorption modulator, which can only be built in a semiconductor). Several of these modulators have also been built in other materials, notably polymers and semiconductors, and improvement in these materials is a continuing process. Therefore, even though lithium niobate is the present material of choice, the generality of this chapter to the other materials should be kept in mind.

As is the case with materials, there is also a dominant modulator design at the present time: the Mach–Zehnder interferometric modulator (MZ), described in Section 4.2.1, is the design in widest use for both analog and long-distance digital applications at the time of this writing. The electroabsorption modulator is the modulator of choice for medium speed and distance digital applications, and it is emerging as a viable analog modulator also, but it is the subject of another chapter and will not be treated in detail here.

4.2 Basic modulator designs

A tremendous variety of modulator designs have been published over the last 30 years. In this section, only a few of these, the ones with the most current applications, will be described. Since these modulators must be used in systems based on single mode fiber, the modulators described here are all based on single mode dielectric optical waveguides [1,2]. The modulator is a combination of an underlying mechanism by which an electric field or current affects some optical property of a material, and an arrangement of waveguides that converts this change into an optical intensity change. (Phase and frequency modulators are also possible, but we are only considering amplitude modulation here.)

Examples of the underlying mechanism are the electro-optic effect (an applied electric field changes the optical refractive index), the acousto-optic effect (a sound wave, generally produced by an electric field in a piezoelectric material, changes the optical refractive index), and electroabsorption (where an applied electric field changes the optical absorption). The acousto-optic effect is limited to a few GHz or less and has generally not been applied to intensity modulators for fiber optic communication. The electro-optic effect is found in a wide variety of materials, including lithium niobate, other inorganic crystals such as strontium barium niobate, polymers, and III–V semiconductors. The electro-optic effect may be linear (index change depends on voltage, as in lithium niobate) or quadratic (index change depends on voltage squared). The electroabsorption effect occurs in semiconductors. Modulators in semiconductors can use electroabsorption directly, or they can use refractive index changes associated with the electroabsorption effect. For more detail on the underlying operating mechanisms, the reader is referred to other chapters of this book, and to textbooks such as those by Yariv [1] and Pankove [3].

The particular characteristics of each type of modulator lead to commonly used figures of merit that are specific to a given modulator type. Examples of this are V_π (the voltage for π phase shift between arms) for the MZ modulator, or V_{10} (the voltage for 10 dB extinction) in an electroabsorption modulator. These are the most convenient and physically meaningful ways to describe specific modulators, but they do not allow easy comparison between modulators of different types. Section 4.3 of this chapter will focus on the simplest overall description of a modulator's performance in analog applications, using a Taylor series expansion of the modulator's transfer function. This method produces a general small-signal link analysis in which the performance of different modulators is easily compared, and which includes all the modulator properties (voltage sensitivity, optical loss, impedance, and nonlinearity) that affect link performance. The MZ modulator will be used occasionally as an example of the relation of the modulator measures (such as V_π) and the Taylor coefficients.

Modulators based on the electro-optic effect or the electroabsorption effect (or a combination of these) have their transmission dependent on voltage, and thus may be characterized by a transfer function

$$p_{\text{out}} = p_{\text{in}} t_{\text{m}} T(V). \tag{4.1}$$

The input and output optical power are p_{in} and p_{out} respectively. t_{m} represents the voltage-independent transmission, from sources such as fiber-to-waveguide coupling, material absorption or scattering, waveguide bend loss, and so forth. $T(V)$ is the voltage-dependent transmission and depends on the modulator design and underlying mechanism. $T(V)$ is bounded by 0 and 1, although in some modulators it may not reach either extreme. (The split of losses between t_{m} and a choice of $T(V)$ that does not reach 1 can be made to achieve the simplest mathematical form for $T(V)$. As long as the treatment is self-consistent, no results here will change.)

Modulators of a given design (e.g., the MZ) using a given mechanism (e.g., the linear electro-optic effect) will have the same transfer function independent of the material in which they are built. The difference between materials in this case will be in the magnitude of constants describing the transfer function, not in its functional form.

No frequency dependence will be explicitly shown here. The basic electro-optic and electroabsorption effects in most materials of interest are invariant from low frequencies (where acoustic effects can occur) up to several hundred GHz (where molecular absorption peaks start to occur). The change with frequency over the microwave frequency range is determined by the electrical characteristics of the modulator electrodes, which is covered in detail in other chapters of this book.

4.2.1 Mach–Zehnder interferometric modulator

The MZ modulator [4,5,6] has found wide application because it is relatively tolerant to fabrication errors, it can be optimized for low or high frequency, and, when fabricated on lithium niobate, it provides very repeatable performance that can be predicted very accurately by a simple trigonometric transfer function. These features will become more apparent as the modulator is described in this section.

The simplest form of the MZ is shown in Fig. 4.1. Optical power p_{in} enters in a single mode dielectric optical waveguide which splits into two single mode waveguides at a waveguide Y branch. The Y branch splits the power into upper and lower arms. The optical phase of the lower arm is shifted with respect to the phase in the upper arm by an amount $\Phi(V)$. This is represented in Fig. 4.1 by the schematic phase modulator in the lower arm. (In this chapter we will only consider intensity modulation/direct detection links with no dispersion or nonlinear effects in the optical fiber, so only the total phase difference between arms matters, not how the absolute phase is distributed between the two arms. In cases where the optical phase matters, such as long distance fiber transmission, it is important to balance the phase modulation properly [7].) In actual devices, the phase modulation is accomplished by placing electrodes around the waveguides. The physical details depend on the material being used; for lithium niobate, the electrode arrangements are shown in Section 4.2.5. The two waveguides have their powers recombined in the output Y branch. The relative phase of the light in the two arms is determined by the phase modulator. If the light from the two arms is in phase, it adds to form the fundamental mode of the output waveguide, and the output power p_{out} is at a maximum. If the light from the two arms is out of phase, it adds in the output Y to form the second order mode of the output waveguide; this mode is not guided by the single mode output waveguide, so the power is radiated into the substrate and the output power is at a minimum (this radiation port makes the Y branch a four-port device which obeys the principle of reciprocity [8]). In the idealized case where there is no loss and both Y branches provide exactly equal splits, the phase

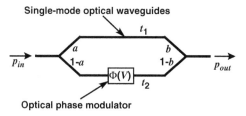

Figure 4.1. Mach–Zehnder (MZ) interferometric modulator built using two Y branches. The phase modulator typically is a pair of electrodes placed on either side of the waveguide. The optical power splitting ratios at the Y branches are represented by a and b, and the optical transmission of the arms by t_1 and t_2.

modulation is turned into intensity modulation according to the equation

$$p_{\text{out}} = p_{\text{in}} \cdot \frac{1}{2}\{1 + \cos[\Phi(V)]\}. \tag{4.2}$$

In practical devices, there will be losses at the Y branches and in the arms, and the Y branches will have unequal power splits. Losses at the Y branches or in the input and output waveguides will contribute to the total loss but will not otherwise affect the modulator's operation. We will represent these losses outside the interferometer by a transmission t_o. If the input Y branch has a power fraction of a going to the upper branch and $(1 - a)$ going to the lower branch, the arms have transmissions t_1 and t_2, and the output Y branch has the power split b to the upper branch and $(1 - b)$ to the lower branch (refer to Fig. 4.1 for illustration), then the output power becomes

$$p_{\text{out}} = p_{\text{in}}t_o\{abt_1 + (1-a)(1-b)t_2 + 2\sqrt{a(1-a)b(1-b)t_1t_2}\cos[\Phi(V)]\}. \tag{4.3}$$

If some of the losses or imbalances are voltage-dependent, the transfer function can become complicated [9]. It still can be put in the form of Eq. (4.1), and the general analysis based on Taylor coefficients can still be used, but the simple transfer functions for the MZ described in this section have to be modified to show the extra voltage dependence.

One or both Y branches may be replaced by directional couplers. If the output Y branch is replaced by a directional coupler, as shown in Fig. 4.2, the device becomes a 1×2 switch or modulator often referred to as a "Y-fed balanced bridge modulator (YBBM)" [10]. The optical power output of each waveguide depends on the amplitude and phase of the light entering the two inputs of the coupler (see Section 4.2.2 for a discussion of how the directional coupler works):

$$p_{\text{out}} = p_{\text{in}}t_o\left[\frac{1}{2} + \frac{1}{2}(2a - 1)\cos(2\kappa l) + \sqrt{a(1-a)}\sin(2\kappa l)\sin[\Phi(V)]\right], \tag{4.4}$$

$$p_{\text{outc}} = p_{\text{in}}t_o\left[\frac{1}{2} - \frac{1}{2}(2a - 1)\cos(2\kappa l) - \sqrt{a(1-a)}\sin(2\kappa l)\sin[\Phi(V)]\right]. \tag{4.5}$$

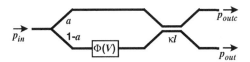

Figure 4.2. Mach–Zehnder modulator built as a 1×2 switch by replacing the output Y branch with a directional coupler (YBBM). The optical power transfer in the coupler is determined by the product of coupling coefficient κ and length l.

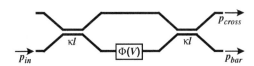

Figure 4.3. Mach–Zehnder modulator built as a 2×2 switch.

The performance of the coupler is characterized by a coupling angle κl, where κ is the coupling strength and l is the coupler length. When $\kappa l = \pi/4$, the coupler is an exact 3-dB coupler that splits a single input into two equal outputs. The ideal values for the YBBM, which allow full on–off modulation in each output, are $a = 0.5$ and $\kappa l = \pi/4$. The two output waveguides always have the same total power output, and the effect of the phase modulation is to move part or all of the light from one output to the other. (An exception to this can occur when the transitions in and out of the output coupler are poorly designed so that its loss is higher for the symmetric mode than for the antisymmetric mode. This causes reduction in total power output for the state with equal split between the two outputs, compared to the state with all light in either output.) This device behaves just like the double-Y version of the MZ if only one output is used; in broadcast applications like cable TV, the second output can be sent to a second receiver. Alternatively, both outputs can be used together in a push–pull detection scheme to double the signal and cancel common-mode intensity noise [11].

When both Y branches are replaced by couplers, the device becomes a 2×2 switch [12], as shown in Fig. 4.3. For light incident from one input, the power at the two outputs is given by

$$P_{\text{cross}} = P_{\text{in}} t_\text{o} \cdot \frac{1}{2} \sin^2(2\kappa l)[1 + \cos[\Phi(V)]] \tag{4.6}$$

$$\text{and} \quad P_{\text{bar}} = P_{\text{in}} t_\text{o} \left\{ 1 - \frac{1}{2} \sin^2(2\kappa l)[1 + \cos[\Phi(V)]] \right\}. \tag{4.7}$$

Here we have assumed that the coupling angles for the input and output couplers are the same. As for the YBBM, this modulator moves the light from one output to the other. This form of the MZ has not found wide application because the YBBM has similar modulation properties in a simpler structure, and the directional coupler modulator provides a more compact 2×2 switch. Note that the choice of equal coupling angles means that the output P_{cross} can always reach 0 and P_{bar} can always reach 1, but each individual output can reach both 0 and 1 only if $\kappa l = \pi/4$.

All of these forms of the MZ (considering just one output of the YBBM and 2×2) can be described by a single transfer function:

$$P_{\text{out}} = P_{\text{in}} t_\text{m} T(V) = \frac{1}{2} P_{\text{in}} t_\text{m}[(1 + E) + (1 - E)\cos[\Phi(V)]]. \tag{4.8}$$

This is diagrammed in Fig. 4.4. The extinction E is the ratio of the minimum to the maximum transmission. The maximum transmission is t_m, which includes all

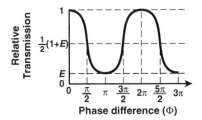

Figure 4.4. Transfer function of the MZ modulator: optical transmission as a function of the optical phase difference between the two arms. The extinction is E.

voltage-independent losses. (In common usage t_m is called the insertion loss, and t_m and E are usually expressed in dB without bothering with a negative sign.) This form of the transfer function is very convenient for use in link modeling, and it also directly incorporates the measurable quantities t_m and E.

The quantities t_m and E have a complicated relationship to the design parameters of the modulator. The equations relating t_m and E to the modulator design parameters can be found by using the definitions of t_m and E, along with the maximum and minimum transmissions found by substituting the appropriate values of $\Phi(V)$ into one of the transfer functions given in Eqs. (4.3)–(4.7). For the double-Y case (Eq. 4.3),

$$t_m = t_0\left[\sqrt{abt_1} + \sqrt{(1-a)(1-b)t_2}\right]^2 \tag{4.9}$$

$$\text{and} \quad E = \left[\frac{\sqrt{abt_1} - \sqrt{(1-a)(1-b)t_2}}{\sqrt{abt_1} + \sqrt{(1-a)(1-b)t_2}}\right]^2. \tag{4.10}$$

For the ideal case where $a = b = 0.5$ and $t_1 = t_2 = 1$, this gives $E = 0$ and $t_m = t_0$. For other cases, the extinction will generally not be zero. Some values of a and b can cause t_m to be less than t_0 even if $t_1 = t_2 = 1$. It is important to be sure to include this second effect in the transmission when one is looking at the effect of different modulator designs or fabrication errors, and not simply equate t_m to the losses physically outside the modulator (t_0). If $a = b$ and $t_1 = t_2 = 1$, the extinction is $E = (2a - 1)^2$ so a value of $a = 0.55$ gives -20 dB extinction. The procedure described above can be followed to get equations similar to Eqs. (4.9) and (4.10) for each output channel of the YBBM or 2×2 forms of the MZ. These devices can have different values of t_m and E in the two different output channels if the input Y or coupler does not provide a perfectly equal split or if the two arms have different losses. To give an example of the tolerance required on the coupler, if a YBBM has a perfect input Y the coupler angle κl must be in the range $40° \leq \kappa l \leq 50°$ for an extinction < -20 dB. For MZs built in lithium niobate, it is common to achieve an extinction ratio of -20 dB or better and a fiber-to-fiber transmission of -3 dB for single modulators. Other materials are not as well developed and generally do not achieve these numbers at the time of this writing, but improvement can be expected.

Now we will look at the details of the phase shift $\Phi(V)$. Assuming a linear electro-optic effect (which is the case for lithium niobate), $\Phi(V)$ can be expressed as

$$\Phi(V) = \phi_o + \frac{\pi}{V_\pi} V, \tag{4.11}$$

where V is the voltage on the phase modulating electrodes and ϕ_o is a constant phase shift (ϕ_o is referred to as the bias point). V_π is the voltage change required to cause a phase change of π between the arms, so V_π is the voltage change required to move the modulator from maximum to minimum transmission. If the electro-optic effect includes a quadratic term, the voltage dependence would change to $AV + BV^2$ and the modulation could not be represented by the single switching voltage V_π. We will just consider the case in Eq. (4.11) in the remainder of this section to keep the analysis as brief as possible.

With the two simplifying assumptions of (1) all modulator parameters independent of voltage except for the phase difference between arms and (2) linear electro-optic effect, the transfer function for any form of the MZ can be stated as

$$T(V) = \frac{1}{2}\left[(1 + E) + (1 - E)\cos\left(\phi_o + \frac{\pi}{V_\pi} V\right)\right]. \tag{4.12}$$

The bias point ϕ_o represents any built-in phase difference as well as any phase difference applied by a DC bias voltage. In the majority of applications, some DC voltage will be used to precisely keep ϕ_o at a specified value. ϕ_o will also include the phase difference caused by length differences between the arms, variations in the refractive index of the material, and differences in the waveguide parameters between the arms. In lithium niobate, several techniques have been tried to physically build in the necessary value of ϕ_o [13,14]. To grossly summarize the present situation, built-in bias techniques can hold ϕ_o to a tolerance of 0.1π ($18°$) or so. This is acceptable for some applications but not for most high-performance analog links where spurious signals can depend critically on ϕ_o, so a bias voltage is generally used.

The parameter V_π completely determines the sensitivity of the MZ to electrical input. V_π is frequency dependent, but it is a measurable quantity that is all that is required for characterization of the modulator sensitivity in analog links. Like the extinction, though, there is a wealth of detail relating physical parameters of the modulator to the single result V_π. Ignoring the possible variation of voltage along the electrode due to frequency and microwave loss, V_π is given by

$$V_\pi = \frac{\lambda_o}{n^3 r_i} \frac{d}{L}\left[\frac{\iint dx\, dy\, f(x, y)\, I(x, y)}{\iint dx\, dy\, I(x, y)}\right]^{-1}, \tag{4.13}$$

where λ_o is the vacuum wavelength, n is the refractive index of the electro-optic material, r_i is the linear electro-optic coefficient, L is the length of the modulating electrodes along the interferometer arm, and d is the gap between electrodes across which the modulating voltage is applied. The quantity in brackets is a dimensionless factor that arises from computing the average electric field over the mode volume, weighting the average by the optical mode intensity $I(x, y)$; this factor is referred to as the overlap of the electric field and optical mode. The function $f(x, y)$ is the spatial variation of the appropriate component of the electric field:

$$E_i(x, y) = \frac{V}{d} f(x, y). \tag{4.14}$$

It is easily seen that V_π varies inversely with L, so often the "$V_\pi L$ product" is referred to as a figure of merit. Lithium niobate MZs typically have a VL product of 50 to 100 V mm at 1550 nm, with the higher value associated with special high-frequency electrodes. Because L can be made as large as 50 mm (effectively double this if reflection is used), V_π can be made very small. The minimum V_π demonstrated to date ranges from a few tenths of a volt for low-frequency electrodes in a resonant circuit [15] to 4 V at 20 GHz in a broadband design [16]; commercially available broadband (≥ 20 GHz) devices presently have V_π between 5 and 10 volts. This is an active area of research and lower V_π values should appear soon, although this may occur using materials other than lithium niobate. For a fixed modulator design, V_π varies linearly with wavelength as seen explicitly in Eq. (4.13). When considering the effect of wavelength on modulator design, though, the dependence is closer to λ_o^2 because the electrode gap d also scales approximately with wavelength. (This is based on the assumption that the electrodes are placed just outside the optical mode, which scales in size with wavelength; this assumption may be violated, making the wavelength dependence approximate.) The final point to note is that a material or waveguide technology that allows smaller optical mode size will give lower V_π because d can be made smaller while still maintaining a large overlap between optical and electric fields.

For low-frequency, lumped element electrodes [15,17], the above analysis is completely relevant and the frequency response is given by the frequency dependence of the voltage across the electrode capacitance. At high frequencies, the phase of the applied electric field changes along the length of the electrode. In this case the electrode is built as an electrical transmission line (referred to as a "traveling wave" modulator) and the electric and optical fields must be integrated along the length of the device to obtain V_π as a function of frequency [18,19] (high-frequency design considerations are discussed in another chapter of this book). MZs on lithium niobate have been demonstrated up to 105 GHz [19]. Except for the variation in V_π, there is no change in the transfer function with frequency. This is because the optical phase difference between arms is converted to intensity at a single point (the interferometer output Y or coupler), so only the total phase difference matters and not the details of how the phase was changed along the device. This makes the distortion independent of frequency, in contrast to modulators such as the directional coupler [20].

After going through all this detail, it can now be seen that the practical utility of the MZ modulator on lithium niobate is based on a combination of fundamental advantages of the modulator and on fortunate properties of lithium niobate. The basic MZ design allows length scaling, where the V_π can be reduced without fundamental limit just by lengthening the modulator (of course there are practical limits to this). The phase is converted to intensity at a single point, so the details of how the voltage was applied along the device do not cause changes in the transfer function. The effects of imperfect fabrication change the extinction and the modulation efficiency, but they do not change the functional form of the transfer function. Lithium niobate allows fabrication of very long (125 mm wafers are available) low-loss waveguides and easy fiber coupling. It also provides an extremely linear electro-optic effect, so the assumptions stated before Eq. (4.12) are met to a very high degree of accuracy. This gives links using the MZ very predictable performance, and allows a number of linearization schemes that are based on compensating the known cosine nonlinearity.

4.2.2 Directional coupler

The directional coupler modulator, shown in Fig. 4.5, was one of the first modulators considered for integrated optic use [21]. In the early years of integrated optics, this modulator and the MZ were both considered as probable choices for widespread use. As noted above, the MZ has emerged as the primary modulator design today, but the directional coupler still is the preferred structure for 2×2 switches because of its compactness. Also, it is still being investigated as a linearized modulator, as will be discussed in Section 4.3.2.

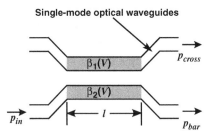

Figure 4.5. Directional coupler modulator. The waveguides' propagation constants β_1 and β_2 are functions of the applied voltage V in the shaded regions. Electrodes may be over the waveguides so the shaded regions represent metal, or they may be placed beside the waveguides so the shaded regions represent the gaps between electrodes.

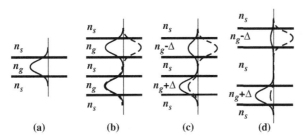

Figure 4.6. Normal modes of various waveguide structures. The substrate index is n_s, the waveguide material index is n_g, $n_g > n_s$, and Δ is a small ($\Delta < n_g - n_s$) index change. The solid curve is the lowest order mode and the dotted curve is the second order mode. (a) Single mode waveguide. (b) Identical single mode waveguides spaced close together. (c) Single mode waveguides with slightly different indices spaced closely together. (d) Same as (c) but spaced further apart.

To understand the operation of the directional coupler modulator, one must look at the normal modes of the five-layer structure formed by the two waveguides and also consider what happens at the input and output transitions. In Fig. 4.6b, the two normal modes of a five-layer structure consisting of two identical waveguide layers are shown. The lowest order mode is the symmetric mode, shown as the solid curve. The second order mode is the antisymmetric mode, shown as the dotted curve. Both modes have equal powers in the two waveguide layers, but the sign of the optical field is reversed in one layer for the antisymmetric mode. Now consider what happens to light entering the modulator from one waveguide. The input (and output) tapering regions, where the two waveguides are brought from a wide separation to a narrow separation, must be non-adiabatic so that mode conversion can occur. Thus the mode of the isolated input waveguide (Fig. 4.6a) converts to a combination of the two normal modes that gives the same optical power distribution as that of the isolated mode. This combination has equal powers in both normal modes, with the relative phases as shown in Fig. 4.6b so the modes add in one waveguide and cancel in the other. The lowest order (symmetric) mode has the higher effective index, and therefore travels more slowly along the waveguides than the second order (antisymmetric) mode. After some distance (referred to as the "coupling length"), the second order mode has exactly reversed its phase compared to the lowest order mode because of this velocity mismatch. If the waveguides are separated just at this point, the optical power will leave the coupler in the opposite waveguide to the one in which it entered because the two normal modes now add to form the mode of the upper waveguide. The further the waveguides are separated in the five-layer region, the closer are the effective indices of the two modes and the longer the coupling length.

To achieve modulation, electrodes can be placed near the waveguides so that when voltage is applied the refractive index of one waveguide is lowered and the other is raised due to the electro-optic effect. As one might expect, this increases the velocity difference between the modes so that the coupling length decreases. When enough voltage is applied, the modulator that was exactly one coupling length long with no voltage applied now becomes exactly two coupling lengths long and the modes add to produce output in the same waveguide in which the power entered (the "bar" state).

There is a second effect of applied voltage, however: the field distributions of the two normal modes change, with the lowest order mode having more of its power in the higher-index waveguide layer and the second order mode having more of its power in the lower-index layer, as shown in Fig. 4.6c. Again, consider what happens at the input transition. To match the mode of the isolated input waveguide, unequal powers must occur in the two normal modes. If the input guide is the lower guide in Fig. 4.6c, the lowest order mode will have the larger power so that the two modes add to give zero power in the upper guide. As the modes propagate, the change in their relative phases causes the power to move from the lower (input) guide to the upper waveguide (and back) as described above. In this case, though, all of the input optical power can never be transferred to the upper waveguide because the small amount of power in the higher order mode is not enough to cancel all the power in the lower order mode in the lower waveguide.

This latter effect of applied voltage means the transfer function of this modulator is not periodic. In fact, the only way to build a directional coupler modulator with a single electrode that has the ability to fully transfer power from one waveguide to the other is to build it out of identical waveguides and build it exactly one coupling length long at zero volts. Different waveguide widths or different material indices in the two waveguides will act just like the applied voltage described above – they will distort the modes and prevent complete power transfer. The modulator can be made so that an applied voltage can adjust for these fabrication errors, but only by nearly doubling its length and adding a second electrode [22].

An alternative method of modulation is to change the index between the waveguides instead of the index of the waveguides themselves. This can produce a more periodic transfer function because the shapes of the modes are hardly affected, but it is only applicable to devices with very small (or no) separation between waveguides. This will be discussed further in Section 4.2.3; it has not often been used with the traditional directional coupler configuration shown in Fig. 4.5.

The most accurate form of modeling this modulator is to numerically compute the properties of the normal modes directly; however, for simple quantitative calculations in the case where the waveguides are separated far enough that almost all the power of the two lowest order normal modes is concentrated in one guide or the other, the "coupled mode" approximation provides good accuracy [23]. The coupled mode approximation leads to simple equations for this modulator [22] that will be used here to obtain the transfer function.

To simplify the equations, we will define

$$\delta(V) = \frac{1}{2}[\beta_2(V) - \beta_1(V)], \tag{4.15a}$$

$$A = \cos[l\sqrt{\kappa^2 + \delta^2(V)}] + \frac{j\delta(V)}{\sqrt{\kappa^2 + \delta^2(V)}} \sin[l\sqrt{\kappa^2 + \delta^2(V)}], \tag{4.15b}$$

and $\quad B = \frac{\kappa}{\sqrt{\kappa^2 + \delta^2(V)}} \sin[l\sqrt{\kappa^2 + \delta^2(V)}]. \tag{4.15c}$

$\delta(V)$ is related to the difference in propagation constants (β_1 and β_2) between the waveguides, and is assumed to be the only place that voltage dependence appears. The strength of coupling between the waveguides is represented by κ, the coupling coefficient, and l is the length over which κ is effective, so κl characterizes the total coupling between the waveguides. (κl has the units of angle, and could physically be thought of as half the phase difference between the normal modes propagating along the five-layer structure as shown in Fig. 4.6. This picture is just for illustration, though, and is not the basis for this analysis.) The power out of the two output waveguides can be related to the optical fields in the two input waveguides (refer to Fig. 4.5; $p_{\text{in}} = |\varepsilon_{\text{in}}|^2$) by the matrix equation

$$\begin{pmatrix} p_{\text{cross}} \\ p_{\text{bar}} \end{pmatrix} = t_{\text{m}} \left| \begin{pmatrix} A & -jB \\ -jB & A \end{pmatrix} \begin{pmatrix} 0 \\ \varepsilon_{\text{in}} \end{pmatrix} \right|^2. \tag{4.16}$$

Equation (4.16) applies to the general case of input power from both input waveguides, but here the input field in the top waveguide is taken to be zero. This is normally the case when this device is used as a simple intensity modulator. Using the top waveguide as the output, so $p_{\text{out}} = p_{\text{cross}}$, the optical output can be written

$$p_{\text{out}} = \frac{p_{\text{in}} t_{\text{m}}}{1 + \delta^2(V)/\kappa^2} \sin^2[\kappa l\sqrt{1 + \delta^2(V)/\kappa^2}]. \tag{4.17}$$

The only condition that gives full transmission is $\kappa l = \pi/2$ and $\delta(V) = 0$. This is as expected from the discussion above of the physical basis of this modulator. To obtain a simple example of the transfer function, we will assume $\kappa l = \pi/2$ and $\delta(0) = 0$. As in the interferometric modulator above, we will also assume a linear electro-optic effect. Using these assumptions and defining V_{s} as the switching voltage (voltage required to go from fully on to fully off), then

$$\delta(V) = \sqrt{3} \frac{\pi}{2l V_{\text{s}}} V \tag{4.18}$$

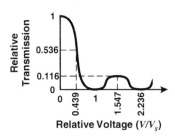

Figure 4.7. Directional coupler modulator transfer function in ideal case, where $\kappa l = \pi/2$ and $\delta = 0$ at $V = 0$ ($\delta = \beta_2(V) - \beta_1(V)$). The switching voltage (fully on to fully off) is V_s.

and the transfer function is

$$T(V) = \frac{1}{1 + 3(V/V_s)^2} \sin^2\left[\frac{\pi}{2}\sqrt{1 + 3(V/V_s)^2}\right]. \tag{4.19}$$

This is plotted in Fig. 4.7. The first half-cycle is the one that would be used for analog modulation. When biased near the half-power point (the slope is maximized at $V_{bias}/V_s = 0.439$), the performance is fairly similar to the MZ [24]. The more complex transfer function does give this device some interesting properties that will be discussed further in Section 4.3.2.

The switching voltage V_s characterizes the voltage response in much the same way as did V_π for the MZ. For the case defined above V_s is given by

$$V_s = \sqrt{3}\frac{\lambda_o}{n^3 r_i}\frac{d}{l}\left[\frac{\int\int dx\, dy\, f(x, y)\, I(x, y)}{\int\int dx\, dy\, I(x, y)}\right]^{-1}, \tag{4.20}$$

where l is the coupling length and all the other quantities are the same as in Eq. (4.13) for the MZ. The modulation has the same physical basis as the MZ, a change in the waveguide propagation constant β via the electro-optic effect, so it results in a similar expression for the switching voltage. For a given physical situation, the VL product for the directional coupler modulator is $\sqrt{3}$ higher than for the MZ. The more significant practical difference is that the directional coupler modulator's electrode length l is related to the coupling coefficient κ, so it can only be increased by decreasing κ, typically by increasing the waveguide separation. When the waveguide separation is increased, the coupler becomes more sensitive to wavelength and to fabrication errors. As illustrated in Fig. 4.6d, the shape of the normal modes is much more radically affected by a given index difference between waveguides when the separation is large. These problems prevent scaling the directional coupler to the long lengths needed for low V_s in a material such as lithium niobate. In cases where a short length is all that is necessary, such as in a switch array [25], the directional coupler modulator can be fabricated accurately enough to be useful.

4.2.3 Other designs based on refractive index change

There are many more configurations possible for modulators using electro-optic index change than the two described above. So far, none have been found that are as sensitive to applied voltage as the MZ. Some designs have properties such as a saturated switching characteristic (where the transmission changes from on to off once for the entire range of possible voltages), compact and simple structure, or improved linearity (modulators with improved linearity are generally compound structures based on the MZ or directional coupler modulators, and will be covered in Section 4.3.2). The desire for some of these other properties has kept the search for new modulator structures alive. Here a few alternative modulator designs will be described to give an idea of the variety of structures possible.

A modulator that gives a saturated switching characteristic is the mode evolution modulator [26,27,28], shown in Fig. 4.8. This modulator is an adiabatic structure, meaning that no mode conversion occurs in it. Its operation is based on this principle, and if mode conversion occurs the modulator will not work properly. This requirement results in very shallow branching angles (a few tenths of a degree in lithium niobate), so these modulators are long. (The input and output Ys are fundamentally different in this respect from the input and output tapers on the directional coupler described above, where mode conversion is required for proper operation.) The angle required for the structure to be adiabatic is a function of the branching angle, the two waveguide mode propagation constants β_1 and β_2, and the substrate refractive index [29].

Once it is understood that no mode conversion occurs, the operation mechanism of this modulator can be described by following the modes through the modulator. The two input waveguides are made with different propagation constants, most easily by making one waveguide a bit wider. This makes the two normal modes of the five-layer structure formed by the two input waveguides asymmetric, as shown in Figs. 4.8 and 4.6d. Power coming in a particular input waveguide will thus be in a single, particular mode of the structure. As long as the structure is adiabatic, the power will stay in that mode all the way through to the output waveguides. At the center of the modulator, the device is a single waveguide that supports two modes. The power is still in only one of these two modes. The output waveguides are made

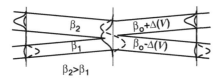

Figure 4.8. Mode evolution modulator. Electrodes are placed around the output waveguides in such a way as to produce the voltage-dependent asymmetry in propagation constants shown; with no voltage applied, both output guides have the same propagation constant β_o.

to be identical, except for the index change due to the applied voltage. The applied voltage raises the index of one waveguide and lowers the index of the other. The modes of this structure, like those of the input side, have their power identified with one of the two output waveguides, and the power will end up in the waveguide that has the non-zero amplitude for the mode that contained the optical power at the start of the output Y branch. In Fig. 4.8, one can follow the power through the device simply by following the progress of either the solid (lowest order) mode or the dotted (second order) mode. The polarity of the applied voltage determines which output guide has the higher index, and thus determines which output waveguide gets the power for the lowest order mode, which is the mode that contains the power that came in the wider input waveguide (and vice-versa for the second order mode).

This device has a saturated or "digital" switching characteristic. As more voltage is applied beyond the minimum necessary to steer the light to the desired output waveguide, the light stays in that guide, it does not start to switch back as it does in the MZ or directional coupler modulators. This is a very desirable characteristic for a switch. The drawback of this design is the much longer length required for the total device (including input and output waveguide separating regions) compared to a directional coupler with the same switching voltage.

A second example of an alternative modulator design is the total internal reflection (TIR) modulator. A typical design is shown in Fig. 4.9. Many variations of this design were tried; one of the earliest analyses was published by Sheem [30], and perhaps the best experimental result on lithium niobate was that of Neyer [31]. The initial idea behind this switch was simple: the applied field in the electrode gap produces a low index region that reflects the light by total internal reflection. Referring to Fig. 4.9, light would enter from one input waveguide and with no voltage applied would cross the intersection to exit on the output waveguide corresponding to the straight-thru path. When voltage was applied, the light would be reflected by the low index region and exit from the other waveguide. The big problem with this simple picture was that it required multimode waveguides with careful mode

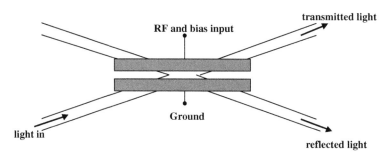

Figure 4.9. "Total internal reflection" (TIR) modulator.

control and/or very high voltages to work. When the devices are built with single mode waveguides, they operate most efficiently by mode interference and not by total internal reflection at all [31,32].

To understand the mode interference method of operation, one must return to consideration of the lowest two normal modes of the structure, as discussed above. The antisymmetric mode always has a null along the centerline of the device, while the symmetric mode has a maximum on the centerline of a two mode waveguide. If an electric field is applied along the device centerline, the index change due to it will affect the propagation constant of the symmetric mode much more strongly than that of the antisymmetric mode. The relative velocities of the modes can thus be controlled just as in the directional coupler. The optical power distribution at the output can then be controlled by controlling the relative phase of the two modes at the output. The input and output waveguide separation regions must be non-adiabatic so the output power distribution between the output waveguides depends upon the optical power distribution at the end of the intersecting region. This all is very similar to the directional coupler modulator described above. The one significant difference is that the shapes of the modes are not nearly as distorted by the electric field as they are in the directional coupler, so the modulator's transfer function is periodic over several cycles of switching.

The TIR modulator has the potential of being a compact 2×2 switch, but the switching voltage on lithium niobate is large (13 V at 633 nm in Ref. 31). In the III–V semiconductor, where devices can be engineered to give somewhat more effective modulation, 10 V at 1550 nm has been demonstrated [33].

The cut-off modulator, shown schematically in Fig. 4.10, will be the final example in this tour off the main modulator path. The cut-off modulator works by using the applied electric field to lower the index of a waveguide so that it no longer guides light. Structures literally like Fig. 4.10 have been built in lithium niobate [34] and other materials. The problem here is that the material index difference between waveguide and substrate for typical single mode waveguides is much larger (roughly an order of magnitude or more) than the index change typically achieved by the electro-optic effect. For the device as shown in Fig. 4.10 to work thus requires large

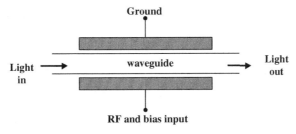

Figure 4.10. Cut-off modulator.

voltages and waveguides fabricated near cut-off. There are other ways to build a cut-off modulator, though. One interesting example is a device that works by pressing a structure like that of Fig. 4.10 against a fiber whose cladding has been polished back to expose the core along several mm [35]. The amount of light lost by the fiber is a function of the voltage applied to the electrode on the electro-optic material. The demonstrated device still has a very large switching voltage, but it has the advantage that light does not have to be fully coupled out of the fiber into a separate integrated optical chip to do modulation.

No transfer functions have been given for the modulators in this section because they are not described accurately by analytic functions. To model one of these structures, one must either do an accurate numerical calculation, or pick an analytic function that approximates the key features of the modulator's transfer function. Some caution must be used, though, because correct prediction of nonlinearities requires a much more accurate knowledge of the transfer function than simply predicting the sensitivity of the modulator to voltage. Once a transfer function of the form of Eq. (4.1) has been obtained, the analog link performance using any of these modulators can be calculated with the methods to be described in Section 4.3.

4.2.4 Electroabsorption

The electroabsorption modulator is one of the most widely investigated modulators, and is now in commercial use. It should be considered along with the MZ and directional coupler as one of the primary modulator designs. This device is described in another chapter of this book, and so will not be discussed in detail here. The electroabsorption modulator works by using an applied voltage to change the optical absorption of a semiconductor. This is fundamentally different than all of the other modulators described in Sections 4.2.1 to 4.2.3 in that the electroabsorption modulator does not depend on the electro-optic effect or refractive index change for its operation. However, it is still a device whose optical transmission depends on voltage, and it can be described by a transfer function of the form of Eq. (4.1), so its performance in a link can be described by the analysis below.

Because it depends on an effect specific to semiconductors, the electroabsorption modulator can only be built in a semiconductor material. The modulators described in Sections 4.2.1 to 4.2.3 can be built on any material that has an electro-optic effect, though, and this includes semiconductors. The electroabsorption effect is also accompanied by a refractive index change [36] (this is a quadratic electro-optic effect, different from the linear electro-optic effect that occurs in semiconductors independent of electroabsorption). Analysis of modulators on semiconductors that operate at wavelengths near the band gap must consider the effects of both absorption and refractive index, even though their intended function may depend on only one of these [9,37].

The sensitivity of the electroabsorption modulator to voltage cannot be improved by simply increasing its length as is the case for modulators based in refractive index change. This is not a severe drawback, though, because the electroabsorption effect is strong enough that devices a few hundred micrometers long have demonstrated switching voltages just over a volt along with bandwidths over 25 GHz [38].

4.2.5 Specific details of lithium niobate material

The modulator structures in Sections 4.2.1–4.2.3 are not limited to lithium niobate; however, given that lithium niobate is presently the most commonly used material for these devices, it is appropriate to cover some details specific to that material.

Lithium niobate is an insulating crystal, transparent to wavelengths from approximately 0.4 to 4.2 μm. Modulators have been demonstrated over virtually this entire range, but the most common use at present is in the fiber optic communication bands at 1.3 and 1.5 μm. Several of the key material properties for modulator fabrication are summarized in Table 4.1. Table 4.1 is only a partial list. A more complete list can be found in Refs. 1 and 39. There are several conventions used in labeling the crystal axes, and different constants use different conventions, so the equivalence is given in the table. Lithium niobate is birefringent, and its physical properties are different along the extraordinary axis than along the two ordinary axes. Many of the material properties are actually tensors and only a few key elements are shown.

Table 4.1. *Some properties of lithium niobate (at 300 K)*

Property	Element	Value
Axis identification	Ordinary	x,y; 1,2; a,b; o
	Extraordinary	z; 3; c; e
Refractive index at 1550 nm	n_o	2.223
	n_e	2.143
Electro-optic coefficient	r_{33}	30.8×10^{-12} m/v
	r_{13}	8.6×10^{-12} m/v
	$n_e^3 r_{33}$	303×10^{-12} m/v
Dielectric constant	ε_x and ε_y	43
	ε_z	28
Dielectric loss tangent		0.004
Thermal expansion coefficient	Along x or y	14×10^{-6} K^{-1}
	Along z	4.1×10^{-6} K^{-1}

The values for electro-optic coefficients and dielectric constant apply to frequencies above 10 MHz; at low frequencies some of these constants can change substantially due to mechanical effects.

The strongest electro-optic coefficient is r_{33}, which applies to light polarized along the z-axis and an applied electric field on the z-axis. There are several other non-zero coefficients as well; the only other one listed here is r_{13}, which applies to light polarized along the x-axis and an applied electric field on the z-axis. The effect of the different r values for different polarizations can be seen if some light is launched in the x-polarization through a modulator designed for z-polarized light: there is a small modulation with roughly triple the V_π superimposed on the desired modulation characteristic. Thus, lithium niobate is a polarization-sensitive material.

The index change produced by an applied field is a function of several factors in addition to the electro-optic coefficient. These are discussed in conjunction with Eq. (4.13) above. As far as basic material constants, though, it is the product $n^3 r$ that determines the index change so it is this product that should be used when comparing the basic electro-optic effect in different materials and not the r value alone.

The other constants in the table are relevant to modulator fabrication in particular circumstances. The dielectric constant is important when frequency response above a few MHz is needed. In lower frequency, lumped element electrode designs, it determines the capacitance. In high frequency traveling wave designs, it determines the microwave velocity. The dielectric constant of lithium niobate is high compared to materials such as optical polymers, III–V semiconductors, and common microwave substrates such as alumina. However, it is low compared to some electro-optic materials such as PLZT or strontium barium niobate. The dielectric loss tangent is important for very high frequency modulators. The measured value for lithium niobate is small enough that it is a minor source of loss compared to conductor loss for common modulator electrode structures [40]. The thermal expansion coefficients are important in packaging the modulator if it will be used over a range of temperatures. The stress level in the lithium niobate must be kept low enough not only to avoid physical damage, but also to avoid changes in the modulator bias point through the acousto-optic effect (stress couples to refractive index).

The two common orientations for waveguides and electrode structures are shown in Fig. 4.11. Both show a pair of waveguides, as is appropriate for the MZ or the directional coupler, but a single waveguide could also be used. Figure 4.11a shows the "x-cut" configuration, where the device surface is perpendicular to the crystal x-axis. Figure 4.11b shows the "z-cut" configuration. The buffer layer is a material such as SiO_2 that has a lower refractive index than lithium niobate.

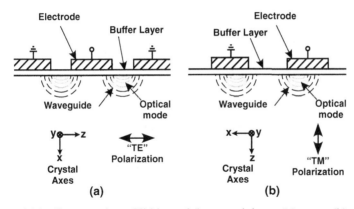

Figure 4.11. Cross-section of lithium niobate modulator: (a) *x*-cut, (b) *z*-cut.

This layer prevents attenuation of the light by the electrodes. It is essential for the *z*-cut configuration but is sometimes omitted in the *x*-cut configuration where the electrodes are not directly over the waveguides. In both *x*- and *z*-cut configurations the optical polarization is along the *z*-axis and the electrode pattern and placement are chosen to put a strong component of the applied electric field along this axis also. The terms "TE" and "TM" are often used to refer to the relation of the polarization to the device surface because the modes in the channel waveguides have TE- and TM-like properties; however, these are just labels and should not be interpreted too literally.

Waveguides are fabricated primarily by one of two methods. The titanium-indiffusion method [41] involves depositing a stripe of titanium metal, typically several micrometers wide and several hundred angstroms thick, along the path of the waveguide, then diffusing it into the substrate at a high temperature (e.g., 1050 °C) for a few hours. The annealed proton exchange method (APE) [42,43] involves exchanging most of the lithium ions in a very thin surface layer for hydrogen ions, usually in a hot benzoic acid bath, then annealing the device (typically a few hours at 350 °C) to redistribute the hydrogen ions over a deeper layer. The exchange is done through a mask with stripes several micrometers wide to define the channel waveguides. Both of these techniques result in refractive index changes of about 0.01 at the surface with a smooth drop-off with depth and with width away from the edge of the stripes. The size of the fundamental mode in these waveguides can be varied to some degree, but a typical mode at 1300 nm would be 7 μm wide by 4 μm deep (at the $1/e^2$ intensity points) centered about 2 μm below the device surface. Both these techniques can produce waveguides with loss on the order of 0.1 dB/cm at 1.3 to 1.5 μm wavelength, although the Ti-indiffused process seems to be slightly lower loss. The primary difference is in polarization: APE waveguides are strictly single polarization (they only guide light polarized on the

z-axis), while Ti-indiffused waveguides guide both polarizations (although the index change is greater for the z-axis). This gives the APE waveguides an advantage in analog modulators that require very high signal quality because they block light of the wrong polarization which would be modulated with a different bias point and therefore different distortion than the desired polarization. On the other hand, APE waveguides are more prone to drift in the optical bias point under an applied DC bias voltage than Ti-indiffused waveguides, so Ti-indiffused waveguides are preferred where high stability is required [44].

There are three mechanisms related to lithium niobate material that can cause problems in some circumstances: optical damage, the pyroelectric effect, and the piezoelectric effect. Optical damage refers to a refractive index change caused by high optical power density. It is reversible, but it is an operational problem if it occurs because it can cause changes in the bias point and increased loss in the waveguides. The mechanism is wavelength dependent, with the largest problem occurring at the shortest wavelengths. Less than a milliwatt at 0.632 μm wavelength can cause significant optical damage, while at 1320 nm waveguides can be stable at 400 mW [45]. At wavelengths below 1.0 μm, where optical damage can be a significant limit to operation, APE waveguides show a greater resistance to damage than Ti-indiffused waveguides.

The pyroelectric effect refers to the generation of an electric field due to a temperature change. Lithium niobate shows this effect along its z-axis. When the temperature of the crystal is changed, there is a large potential difference generated between the z-faces of the crystal. This causes rapid changes in the bias point of a modulator, especially in the z-cut configuration. (The effect is so strong that electric arcs can damage photoresist patterns during baking steps in fabrication on z-cut material!) The solution to this problem is to coat the z-faces of the modulator with conductive material and connect them together [46]. This is easy on x-cut devices, but is difficult on z-cut where a high resistance coating must be used so as to accomplish the z-face connection without interfering with the application of the modulating electric field to the waveguides [47,48].

Piezoelectric and acousto-optic effects are useful in many types of lithium niobate device, but they can be a problem for electro-optic modulators. At frequencies below about 1 GHz, substrate acoustic resonances and acoustic-related impedance changes can cause undesirable ripples in the frequency response of a modulator. The effect ranges from a few tenths dB on a traveling wave modulator (which is insignificant for many applications), to several dB on a lumped element modulator with a resonant match at <100 MHz [15]. If acoustic effects are a problem, they can be suppressed by roughening, angling, and/or grooving the back side of the modulator substrate, and by applying an absorbing coating over the top side including the modulator electrodes [49].

4.3 Modulator effects on link performance

We will now look at how modulator performance determines analog link performance. One particular improvement, linearization of the modulator, will be dealt with in detail.

When designing modulators for use in analog links, or when evaluating the utility of a new design such as a linearized modulator, the total effect on link performance must be evaluated. This is necessary because an improvement in one modulator parameter could come at the expense of another, and in the end the link performance could even be worsened. This is especially true when linearization of the modulator is being considered. Therefore, we will briefly look at link performance equations here.

Analysis of the externally modulated analog link can become quite involved, and it is dealt with in detail in Chapter 2 of this book. Here we will give the key equations without full derivation so that the important considerations relative to modulators may be discussed. The reader should refer to Chapter 2 for more background. (The reader is cautioned that notation details are not at all standard. The analysis in this chapter is self-consistent, as are the analyses in Chapter 2 and in earlier works; however, many similar symbols are defined differently in each of these different treatments.)

4.3.1 Link transfer function (gain and distortion)

A block diagram of the externally modulated optical link is shown in Fig. 4.12. This diagram shows several of the key variables that will be defined in the analysis below. In this first section the link gain and the various distortion signals produced by the link will be discussed. This will be followed by Section 4.3.2 on linearized modulators, which minimize these distortion signals. In Section 4.3.3, we will return

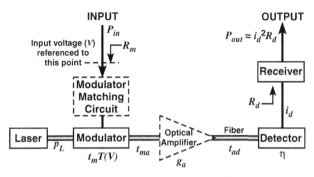

Figure 4.12. Block diagram of link. Components shown dotted may or may not appear in link. Variables are defined in the text.

to the link analysis and include the noise terms and distortion so that the overall link performance can be related to the modulator parameters.

In the analysis here, we assume the modulator is the only source of distortion (nonlinear dependence on input voltage). This is actually the case in many practical situations, such as a short (<100 m) link with a detector operated well below its saturation power. In links involving long fibers (several km) and optical amplifiers, nonlinearities in these components can cause nonlinearities in the link electrical performance [50]. The detector can also generate distortion [51]. This is generally a function of the bias voltage and the average intensity (optical power per unit area on the detector surface), so high frequency detectors are more problematic because of their small active areas. This detector consideration is one of the primary limitations on the optical power in a link. The reader is referred to the chapter on detectors for more detail.

The electrical input signal is applied to the modulator, and the electrical output signal is generated by the photodetector. (It is important to remember that the electrical power output of the detector is supplied primarily by the detector's DC bias supply, not shown in Fig. 4.12, and not by the optical power itself. The optical power merely controls the detector current.) The laser produces a constant optical power level, which is an important input to the link and gives the externally modulated link the characteristics of a three-terminal electrical device (somewhat analogous to a field-effect transistor). If used, an optical amplifier after the modulator increases the optical power level reaching the detector.

The modulator responds to the applied voltage. The detector is a current source, with the current determined by the optical input. The link transfer function thus relates the detector output current i_d to the modulator input voltage

$$i_d = p_L t_m T(V) t_{ma} g_a t_{ad} \eta = i_o T(V), \qquad (4.21)$$

where p_L is the laser power at the modulator input, t_m is the modulator optical transmission when $T(V) = 1$ (see Eq. (4.1) above), t_{ma} is the transmission from modulator to optical amplifier input, g_a is the optical amplifier gain, t_{ad} is the transmission from optical amplifier output to the photodetector, and η is the detector responsivity (A/W). If there is no optical amplifier, $g_a = 1$ and the transmission from modulator to detector is $t_{ma} t_{ad}$ (real systems often contain more than one optical amplifier in cascade, but a single amplifier will serve to illustrate the basic effects of optical amplification). We define a current

$$i_o \equiv p_L t_m t_{ma} g_a t_{ad} \eta, \qquad (4.22)$$

which is the detector current when $T(V) = 1$. This single quantity is all that enters the noise and gain equations as far as optical power and losses are concerned (the one exception to this is the noise term from the optical amplifier which will be discussed below). The impact of laser power, modulator insertion loss, fiber or connector losses, and detector responsivity are all equivalent in their impact on a non-optically-amplified system.

To examine the electrical performance of the link, especially as regards distortion, it is convenient to assume an input signal

$$V = V_B + V_a \sin \omega_a t + V_b \sin \omega_b t, \qquad (4.23)$$

where V_B is a DC bias voltage and V_a and V_b are the amplitudes of the signals at frequencies ω_a and ω_b, and then look at what output signals the link produces. (A two-tone input is the simplest input that will demonstrate mixing products, although the nonlinearities may be determined from the harmonics of a single signal. A phase term on the second frequency does not add any insight so it is omitted.)

One way to analyze this is to substitute the input signal (4.23) directly into the analytic form of $T(V)$. This is useful in the case of the MZ modulator (Eq. (4.12)) where the result is also an analytic function, namely a sum of Bessel functions, and some insights into the large signal behavior can be obtained directly. For other modulators, the resulting expression for the link output current will be complicated and may not really be helpful. Numerical modeling is another case where direct substitution may be used. Even though the expression for the output current is complicated, it can be evaluated numerically at several time samples and then a Fourier transform (FFT) can be done to get the frequency spectrum at the output. This is a very general and effective method for computer modeling of the link (provided one is careful about choosing the sampling frequency relative to the input frequencies, and other digital oddities!).

The best way to get insight into the small signal behavior of the link without slavishly plugging numbers into a computer is to look at the Taylor series expansion of the modulator transfer function (since by our assumptions above, this is the main source of nonlinearity in the link):

$$T(V) = \sum_{n=0}^{\infty} \frac{(V - V_B)^n}{n!} \frac{d^n T}{dV^n}\bigg|_{V=V_B} = c_0 + \sum_{n=1}^{\infty} c_n (V - V_B)^n. \qquad (4.24)$$

The expansion coefficients are thus

$$c_0 = T(V_B) \quad \text{and} \quad c_n = \frac{1}{n!} \frac{d^n T}{dV^n}\bigg|_{V=V_B}. \qquad (4.25)$$

The units of the c_n are (volts)$^{-n}$.

For a specific example, consider the MZ modulator with perfect extinction, so $T(V)$ is given by (4.12) with $E = 0$. The coefficients are then

$$c_0[MZ] = \frac{1}{2}[1 + \cos\phi_0], \tag{4.26a}$$

$$c_n[MZ] = \frac{1}{2}\frac{1}{n!}\left[\frac{\pi}{V_\pi}\right]^n (-1)^{\frac{n+1}{2}} \sin\phi_0 \quad (n \text{ odd}), \tag{4.26b}$$

$$\text{and} \quad c_n[MZ] = \frac{1}{2}\frac{1}{n!}\left[\frac{\pi}{V_\pi}\right]^n (-1)^{\frac{n}{2}} \cos\phi_0 \quad (n \text{ even}). \tag{4.26c}$$

To analyze the output signals, we start by combining Eqs. (4.21) through (4.24) to get the output current in terms of the input voltage as

$$i_d = i_0 c_0 + i_0 \sum_{n=1}^{\infty} c_n [V_a \sin\omega_a t + V_b \sin\omega_b t]^n. \tag{4.27}$$

We want to look at the power in each frequency component at the output. Physically, this corresponds to filtering the current and then measuring the power in each frequency component. Mathematically, this means expanding the expression in Eq. (4.27) and using trigonometric identities to convert all products of trigonometric functions into the sines or cosines of a sum of frequencies. The individual frequency components are then converted from voltage (input) or current (output) to rms electrical power by using the relations

$$P_{\text{out}}(\omega) = \langle i^2(\omega) \rangle R_d \quad \text{and} \quad P_{\text{in}}(\omega) = \frac{\langle V^2(\omega) \rangle}{R_m}, \tag{4.28}$$

where R_d and R_m will be discussed below (the $\langle \rangle$ brackets indicate time-averaging over a time long compared to $2\pi/\omega$). Although these resistances may appear trivial or arbitrary, they are actually important factors in determining link performance and must be considered carefully when evaluating one design compared to another (for example, two traveling wave modulators with the same V_π but different impedance can have quite different performance in a link).

To evaluate electrical performance measures such as gain and noise figure, the relation between input and output electrical power is required. Thus we must define some impedance for the input and output of the link. RF and microwave systems generally operate in an environment with a defined, resistive impedance such as 50 Ω (connection of an antenna directly to the modulator or detector is an important exception to this). However, because of the performance advantages that can be gained by presenting the modulator and detector with impedances different than the system impedance, situations other than simply connecting these components to the system impedance must be considered. This will be done as simply as possible here by considering the link's input impedance to be purely resistive (R_m), and by assuming the detector is connected to a receiver that presents it with an equivalent load resistance R_d. (In this analysis, the source impedance is always assumed to be ideally matched to the link input resistance R_m. The receiver following the detector is assumed to have an output impedance matched to that of any components following it.)

The modulator consists physically of an arrangement of metal electrodes around a dielectric material with electro-optic properties. The electrodes can be designed to have a particular impedance as a transmission line, but the electrical characteristics of the modulator are more determined by how the electrodes are connected to the signal source.

The modulator transfer function is generally reported relative to the voltage actually on the modulator electrodes, at low frequency. For some electrode configurations, this is useful directly; however, there are other cases where the actual voltage on the electrodes is quite different than the voltage applied to the input impedance R_m. As noted above, the transfer function must be related to this input voltage, not just to the actual voltage on the electrodes, in order to model the link performance.

The simplest design electrically (although the physical design can become quite involved) is the traveling wave configuration shown in Fig. 4.13a [18]. The electrodes form a transmission line of characteristic impedance $Z_o = R_m$, are driven from the signal source at one end, and are terminated with a matched load resistor R_m at the other end. At low frequencies, the voltage on the modulator is the same as the voltage at the input; as the frequency increases the voltage along the line drops due to electrical losses, and also the integrated total phase modulation along the electrode can decrease from its low frequency value if the microwave and optical velocities are not the same. For the MZ, these effects can be completely represented by an increase in V_π with frequency. For other modulators such as the directional coupler, the effect on gain and noise figure can be represented by an increase in switching voltage, but the effect on distortion may be more complicated [20].

The simplest design physically is the lumped element capacitor, shown in Fig. 4.13b [17]. The electrodes can be fabricated and connected to in any convenient configuration and they will look like a capacitor electrically. A parallel resistor is often connected, as shown in Fig. 4.13b. The modulator voltage is the same as the input voltage. As the frequency increases, the voltage will decrease for a given input electrical power. (It is especially important in this case to remember the assumption that the source impedance is R_m because the modulator's input impedance is resistive only at zero frequency. The modulator in Fig. 4.13b is not matched to a resistive source except at DC.)

An alternative connection to capacitive electrodes is the resonant match shown in Fig. 4.13c. This can lead to very high modulator sensitivity at low frequencies [15]. In this case, the voltage on the modulator can be much larger at resonance

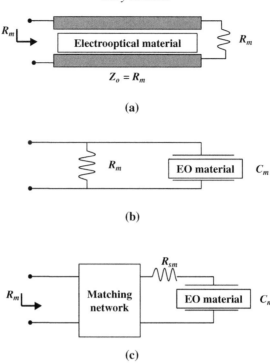

Figure 4.13. Modulator electrical configurations. (a) Traveling wave: electrodes form a transmission line with impedance R_m. (b) Lumped element: electrodes form a capacitor with capacitance C_m. (c) Resonant match to lumped element: an impedance-matching circuit causes all input power to be dissipated in R_{sm}.

than the input voltage. The modulator transfer function as seen from the input at R_m will be a much more sensitive function of voltage than when it is measured relative to the voltage on the actual modulator electrodes, but its other characteristics will not change.

Detector impedance-matching considerations and receiver design are more complex than for the modulator, and could fill a book on their own [52]. For this analysis, we will characterize the receiver by the parameter R_d, the equivalent input load resistance. The unity-gain receiver block shown in Fig. 4.12 thus has an output power of $i_d^2 R_d$, where i_d is the detector's output current. Note that R_d is not the physical resistance connected to the detector, but a parameter with units of ohms that describes the relation between detector current and receiver output power into the system impedance. (Some examples: (1) If the detector is directly connected to the input of an ideal 50-Ω amplifier, then R_d is 50 Ω. (2) If the detector is connected to a 50-Ω resistor inside its package, a 50-Ω output connector is also connected to the detector in parallel with the resistor, and the "receiver" output port is taken to be this 50-Ω connector, then R_d is 12.5 Ω because the true detector load is 25 Ω and only one-half the total power is transferred to the 50-Ω output port.)

Returning now to the calculation of the small signal performance of the link, we can apply the procedure discussed after Eq. (4.27) to obtain the frequency components present at the link output, and their power levels. Keeping terms up to fifth degree gives the following list of signals:

$$i_{dc} = i_0 \left[c_0 + c_2 R_m (P_a + P_b) + \frac{3}{2} c_4 R_m^2 (P_a^2 + 4 P_a P_b + P_b^2) \right] \quad (4.29a)$$

$$P(\omega_a) = i_0^2 R_d P_a \left[R_m c_1^2 + 3 R_m^2 c_1 c_3 (P_a + 2 P_b) + \frac{9}{4} c_3^2 R_m^3 (P_a + 2 P_b)^2 \right.$$
$$\left. + 5 c_1 c_5 R_m^3 (P_a^2 + 6 P_a P_b + 3 P_b^2) + O\left(P_{in}^3\right) \right] \quad (4.29b)$$

$$P(\omega_a \pm \omega_b) = 2 i_0^2 R_d P_a P_b \left[R_m^2 c_2^2 + 6 c_2 c_4 R_m^3 (P_a + P_b) \right.$$
$$\left. + 9 c_4^2 R_m^4 (P_a + P_b)^2 \right] \quad (4.29c)$$

$$P(2\omega_a) = \frac{1}{2} i_0^2 R_d P_a^2 \left[R_m^2 c_2^2 + 4 c_2 c_4 R_m^3 (P_a + 3 P_b) \right.$$
$$\left. + 4 c_4^2 R_m^4 (P_a + 3 P_b)^2 \right] \quad (4.29d)$$

$$P(2\omega_a \pm \omega_b) = \frac{9}{4} i_0^2 R_d P_a^2 P_b \left[R_m^3 c_3^2 + \frac{20}{3} c_3 c_5 R_m^4 \left(P_a + \frac{3}{2} P_b \right) \right.$$
$$\left. + \frac{100}{9} c_5^2 R_m^5 \left(P_a + \frac{3}{2} P_b \right)^2 \right] \quad (4.29e)$$

$$P(3\omega_a) = \frac{1}{4} i_0^2 R_d P_a^3 \left[R_m^3 c_3^2 + 5 c_3 c_5 R_m^4 (P_a + 4 P_b) \right.$$
$$\left. + \frac{25}{4} c_5^2 R_m^5 (P_a + 4 P_b)^2 \right] \quad (4.29f)$$

$$P(2\omega_a \pm 2\omega_b) = \frac{36}{8} i_0^2 R_d P_a^2 P_b^2 R_m^4 c_4^2 \quad (4.29g)$$

$$P(3\omega_a \pm \omega_b) = \frac{16}{8} i_0^2 R_d P_a^3 P_b R_m^4 c_4^2 \quad (4.29h)$$

$$P(4\omega_a) = \frac{1}{8} i_0^2 R_d P_a^4 R_m^4 c_4^2 \quad (4.29i)$$

$$P(3\omega_a \pm 2\omega_b) = \frac{100}{16} i_0^2 R_d P_a^3 P_b^2 R_m^5 c_5^2 \quad (4.29j)$$

$$P(4\omega_a \pm \omega_b) = \frac{25}{16} i_0^2 R_d P_a^4 P_b R_m^5 c_5^2 \quad (4.29k)$$

$$P(5\omega_a) = \frac{1}{16} i_0^2 R_d P_a^5 R_m^5 c_5^2 \quad (4.29l)$$

where P_a and P_b are the input powers at frequencies ω_a and ω_b. (Several more distinct frequencies also are present that can be obtained by reversing the subscripts *a* and *b*.) The reader should not be discouraged by the large number of terms in Eqs. (4.29); there are actually several useful observations that can be made by considering these signals and we will refer back to them several times. One observation that is immediately obvious is that linearization of the modulator consists of reducing the magnitudes of one or more of the c_n with $n \geq 2$; this will be discussed at length in Section 4.3.2.

The dc term is dependent not only on the dc coefficient of the transfer function c_0, but also on the even order coefficients. Modulators with $c_n \neq 0$ for any even n will show a dc shift under RF modulation. When c_1 can be determined from knowledge of c_2 and c_0, as it can for the MZ modulator (see Eq. (4.26)), this can be used to measure the RF modulation efficiency without actually having to use a high-speed detector [53]. This effect can cause problems in a link using erbium-doped fiber amplifiers (EDFAs) and a modulator with $c_2 \neq 0$ because even though the amplifiers respond too slowly to cause direct generation of intermodulation at RF frequencies, the EDFA can respond to the changes in the modulator's DC output power caused by low frequency amplitude modulation of the RF carriers so that there is a cross-modulation problem.

The small signal gain G of the link is the output power at the fundamental frequency $(P(\omega_a))$ divided by the input power at that frequency (P_a):

$$G = i_o^2 R_d R_m c_1^2, \tag{4.30}$$

where only the linear term is used because it is the largest for small signals. The higher order terms lead to compression for large signals. An interesting side effect of linearization that reduces the magnitude of c_3 is that the 1-dB compression point of the link is also raised.

The quantity $R_m c_1^2$ is the modulator response, with units of inverse power. For the MZ modulator with perfect extinction, we can substitute the c_1 value from Eq. (4.26b) to get

$$\left(R_m c_1^2 \right)_{MZ} = \frac{\pi^2 R_m}{4 V_\pi^2} \sin^2 \phi_o. \tag{4.31}$$

This serves to illustrate that the c_n are not determined solely by the modulator design, but also by its operating point (the bias point is ϕ_o in this example). When one is working exclusively with a single type of modulator, it is usually more useful to substitute expressions involving that modulator's particular design parameters (e.g., V_π, ϕ_o, and E for the MZ modulator) for the c_n. The MZ modulator also illustrates another property of the c_n: they can all be expressed in terms of a few basic modulator parameters. However, when comparing different types of modulators it is easier to keep the general formulation to derive the basic performance formulae and then substitute values for the particular modulators at the end.

A parameter that is often used in discussions of modulator performance is the modulation depth. This is a strictly internal parameter to the link, and it does not appear in the equations relating the input and output ports (although the distortion signals in Eq. (4.29) could be reformulated to express them as functions of a modulation depth instead of as functions of the input RF power). It is sometimes useful in that it can help in understanding why things happen, but it is often misleading when used to compare modulators because of the different ways in which the concept is applied. The optical modulation depth at the modulator output is

$$m_{\text{out}} = \frac{p_{\max} - p_{\min}}{p_{\max} + p_{\min}} = \frac{|c_1|}{c_0} V_a, \tag{4.32}$$

where $p_{\max}$ and $p_{\min}$ are the maximum and minimum values of the output optical power and V_a is the (peak) input voltage of a single signal applied to the modulator. For the MZ modulator with $E = 0$ this is

$$m_{\text{out,MZ}} = \frac{\pi V_a}{V_\pi} \frac{|\sin \phi_o|}{1 + \cos \phi_o} = m_p \frac{|\sin \phi_o|}{1 + \cos \phi_o}, \tag{4.33}$$

where $m_p = \pi V_a / V_\pi$ is the phase modulation depth in radians of phase difference between arms. For $\phi_o = \pi/2$, $m_{\text{out}} = m_p$ and the modulation depth is unambiguous. For $\phi_o \neq \pi/2$, $m_{\text{out}} \neq m_p$ and one must be more specific when discussing the modulation depth (consider the interesting and useful property that the output modulation depth can become much larger than m_p by operating with ϕ_o approaching π). Although a parameter like m_p is very useful for discussing a particular modulator, it becomes confusing in the general case, especially with compound modulators as found in various linearization schemes.

In the interest of keeping this analysis as general as possible, then, the modulator response will be left as $R_m c_1^2$ and the input RF powers will be used directly instead of combining these into a modulation depth of some kind.

4.3.2 Linearization

Of all the signals listed in Eq. (4.29), only the fundamental (4.29b) is the desired signal (unless one is purposely constructing a mixer). All the others must be minimized. These other signals are generally organized by "orders", depending on how many fundamentals (using the same one twice counts as two) are mixed to get them; thus $2\omega_a$ and $\omega_a \pm \omega_b$ are second order, $3\omega_a$ and $2\omega_a \pm \omega_b$ are third order, and so on. Linearization refers to modifying the performance available from a given modulator in some way so as to reduce one or more of these distortion signals.

There are three types of linearization in use today. The first type is analog electronic correction of the distorted electro-optic device so that the complete transmitter is linear. One example of this is predistortion [54], where an electronic circuit ahead of the modulator has equal and opposite nonlinearity to that of the modulator. Another example is feed-forward [55]. In feed-forward, part of the optical output of the modulator is detected and compared with the RF input to generate an error signal. The error is inverted and feed to a second electro-optic device whose output is added to that of the modulator to produce a linear total output. The second type of linearization is correction by digital signal processing after the link output has been detected and fed to an analog-to-digital converter [56]. The third type, sometimes called "optical linearization" is the modification of the modulator itself so that it produces smaller distortion signals. This is the only type of linearization we will discuss here.

All of these methods have their uses, and their relative attractiveness will depend on the system application and on technological advances. Predistortion is perhaps the simplest, but it usually does not give as big an advantage as the other techniques. Digital correction requires high speed signal processing; the extra processing load is not very significant in a system such as a digital radar that already has some heavy duty computational power, but it is beyond the capability of many simpler systems. Optical linearization can deliver significant improvements in performance by simply modifying the modulator, but this modification often proves to have difficult fabrication tolerances and/or difficult control problems.

Design of a linearized modulator involves trying to reduce the c_n for $n \geq 2$. In practice, the battle is actually just to minimize c_2 and c_3 because this gives a significant performance improvement and minimization of the coefficients with $n \geq 4$ leads to impractically tight control tolerances.

All modulators have a bias point where $c_2 = 0$, since this just corresponds to the inflection point of the transfer function. This is generally used as the normal operating point and therefore is not considered linearization. In the effort to reduce c_3, however, situations arise where the $c_2 = 0$ bias point does not correspond to the minimum c_3 so minimization of c_2 by means other than simply biasing at the $c_2 = 0$ point must be considered.

There are many systems that have a narrow bandwidth, much less than an octave, so that second order and even fourth order signals are outside the system bandwidth. In these cases linearization can be done simply by minimizing c_3, so that the third order signals are minimized. (Odd order distortion signals can always appear within the system bandwidth, no matter how narrow, as long as there are two or more input signals. This can be seen by examination of Eq. (4.29)) Therefore, linearization schemes that minimize c_3 but leave $c_2 \neq 0$ are useful for this type of system. When this method can be used, it leads to simpler linearized modulators and better performance from the link compared to true broadband linearized modulators.

When linearization is performed the relationship of the distortion signals to the input power can become complex. We will look at the third order signals as an

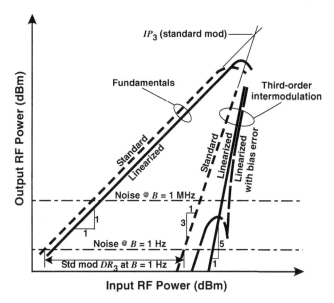

Figure 4.14. Fundamental and third order distortion signal power levels at the link output, as a function of the input power to the link. The output noise levels for two different noise bandwidths are also shown. The third order intercept point (IP_3) and third order intermodulation-free dynamic range (DR_3) are shown for the standard modulator.

example to explain some of the generic features. The second order signals also behave in a similar manner. Figure 4.14 shows a plot of the fundamentals and third order signals from a modulator with and without linearization. For both modulators, the fundamentals have a slope of 1 for small signals. The linearized modulator is shown with slightly reduced gain because the process of linearization often causes a reduction in c_1.

The third order intermodulation signals of the standard modulator have a slope of 3 for small signals, because the $c_3^2 P_{in}^3$ term is the largest term in the output power for this signal until compression occurs (see Eq. (4.29e); in this section we will assume two equal input signals so $P_a = P_b = P_{in}$). This behavior is typical of most electronic amplifiers as well, and the nonlinear behavior can be completely characterized by the output intercept point IP_3, extrapolated from small signals.

A linearized modulator can be designed so that $c_3 = 0$ (this is actually a design method – one expresses c_3 in terms of modulator parameters such as bias point and RF power split (in a compound modulator), then solves the equation $c_3 = 0$ to find the parameter values to use). By examining Eq. (4.29e), one can see that the third order intermodulation signals will have a slope of 5 because now the $c_5^2 P_{in}^5$ term is the only non-zero term remaining. The third order signal is still the largest odd order distortion signal. By comparing Eqs. (4.29e) and (4.29j), using $P_a = P_b$ and $c_3 = 0$, one can see that the third order signal is 25 times larger than the fifth order signal (an exception to this comment can occur if a system has a large number of evenly spaced carriers present, so that the larger number of fifth order intermodulation combinations contributing to distortion at a given frequency outweighs the larger magnitude of the individual third order contributions). The nonlinear behavior can still be characterized by IP_3, but the slope of 5 must be used instead of a slope of 3. The link can still be characterized in a simple way, but now the cascade of the link with electronic amplifiers that have the normal slope of 3 for third order intermodulation signals cannot be done with the usual simple formulae.

In practice, there is often an error in the operation of the linearized modulator, typically a small error in the bias point. This leads to the situation where c_3 is reduced but not equal to zero. c_3 may have the same sign as c_5 or the opposite sign; in the case of the opposite sign, the intermodulation signal behaves as shown in Fig. 4.14 for a bias error. At large input power, the slope of the third order signals is 5, because $c_5^2 P_{in}^5$ is larger than $c_3^2 P_{in}^3$. As the input power drops, $c_5^2 P_{in}^5$ decreases faster than $c_3^2 P_{in}^3$ and eventually the two become equal. When the signs are opposite, the two contributions cancel (this is clearer if one looks at the current instead of the power as in Eq. (4.29)) and there is a minimum in the third order signal. At very small input powers, the $c_3^2 P_{in}^3$ term dominates and the third order signal has a slope of 3, although with a much higher IP_3 than for the standard modulator (an easy way to solve the problem of cascading the linearized link with normal electronic amplifiers is to assume a bias error and then use the IP_3 from the smallest signal portion of the curve where the slope is 3, provided that the actual signals of interest fall within the smallest signal power range).

There are several ways that this behavior makes the design of links using linearized modulators more difficult. The dip in the third-order distortion referred to in the preceding paragraph is almost never a usable optimization because it depends very specifically on the power levels of the various input signals present. When using a bias controller that depends on pilot tones to generate a third order error signal, the pilot tones and feedback method must be thought out carefully because the tendency of the control loop will be to adjust the bias point so that the third order minimum for the pilot tones is achieved. Finally, both the amount of improvement from linearization and the sensitivity to error depend on the signal and noise power levels in the system. If a very narrow noise bandwidth is used (or equivalently, if a very high performance link is used with moderate bandwidth), the improvement from linearization is large but the degradation due to a small bias error is also large. If a large noise bandwidth is used, the amount of improvement is smaller but so is the sensitivity to a bias error. These effects can be seen in Fig. 4.14 by comparing the dynamic range (range of input power between the point where the fundamentals cross the noise level and the point where the intermodulation signals cross it) of the standard, linearized, and linearized-with-error modulators for the two noise bandwidths shown.

A large variety of linearized modulators have been demonstrated over the years. Not all can be listed here, but a sampling that will hopefully be representative will be presented. One way to classify the linearized modulators is by whether they are capable of true broadband ($c_2 = 0$ and $c_3 = 0$) linearization which minimizes both second and third order distortion, or whether they only perform narrowband ($c_3 = 0$) linearization which minimizes only the third order distortion. This is often determined by a choice of operating point, though, and the same modulator can be used in more than one way! So here the modulators will be divided into three classes by their physical design.

The first class of modulators are standard modulators that have a bias point where $c_3 = 0$ and $c_1 \neq 0$. These have $c_2 \neq 0$ so they are only applicable to narrowband systems where second order distortion is not a problem. The directional coupler modulator discussed above (Fig. 4.5) is one such modulator [57]. It has $c_3 = 0$ at $V_B/V_s = 0.7954$, which is a lower-transmission bias point than the $c_2 = 0$ point at $V_B/V_s = 0.439$. Another example is the electroabsorption modulator. The minimum third order behavior of the electroabsorption modulator was demonstrated in Ref. 58. The simple MZ modulator does not have such a bias point – by examining Eq. (4.26) one can see that $c_1 = 0$ when $c_3 = 0$ for the MZ modulator.

The second class of modulators are those made by combining two otherwise standard modulators. This can be done by combining the modulators in parallel or cascading them in series. Also, some versions of this technique use a single physical modulator but run two polarizations through it or make the light pass through it twice, effectively making it into a compound modulator. This class of modulators includes some that only achieve $c_3 = 0$ and some that achieve true broadband linearization.

The dual parallel MZ is shown in Fig. 4.15. There have been several variations of this device published [57,59,60]. This device was one of the earliest ways in which the MZ modulator was linearized because the concept behind it is simple: tap off some light and feed it to a second MZ with a different response from the primary MZ, and adjust the parameters so the third order distortion from the second MZ cancels that from the first. Figure 4.15 shows one way of doing this with two identical MZs. The optical power is split with a fraction a going to MZ #2 by using a passive directional coupler. The RF input power is also split, with a fraction s going to MZ #2 and the remainder $(1 - s)$ going to MZ #1. Each of the two MZs has a DC bias point that is controlled electrically. The values of these four variables can be found by solving the equations $c_2 = 0$ and $c_3 = 0$ subject to the additional constraint of maximizing c_1, as outlined above for the general linearization problem. The relative phase of the light coming from MZ #2 must be

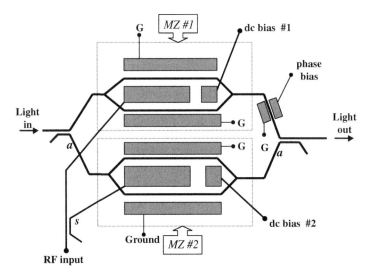

Figure 4.15. Dual parallel MZ. The fraction of optical power sent to MZ #2 is a, and the fraction of input RF power sent to MZ #2 is s. The extra phase bias control is necessary to ensure that the output light from the two MZs adds in phase optically at the output.

set at a particular value for proper operation, which in a practical device almost certainly requires an electrically adjustable phase control. Solutions for this do exist, so that true broadband linearization can be achieved by this device. There are significant practical problems with this device, however. The linearized solution requires that most of the light be directed to MZ #1, while most of the RF power is directed to MZ #2, so there is a substantial penalty in link gain and noise figure relative to a single MZ. (The RF power could be split equally between the MZs and a longer electrode used on MZ #2, but the basic problem is the same. If the longest practical electrode is used for MZ #2, then MZ #1, where most of the light is going, has a much lower sensitivity than a single MZ could have.) The penalty is generally shown in Ref. 60, although that reference does not directly calculate gain and noise figure. The control problem is also difficult. In common with all compound modulators that do broadband linearization, two independent DC biases must be controlled. In addition, the parallel arrangement as shown in Fig. 4.15 also requires control of the optical phase (this can be avoided by using two mutually incoherent lasers to drive the two modulators [59]). While the control itself is easily accomplished for all of these by means of DC voltages, the error detection and appropriate feedback loops are difficult. Also, the relation between a and s must be set precisely, which may require control of one of these in some situations as well.

One interesting simplification of this device is to use a single MZ but with both optical polarizations present [61]. This only works in a material system like

Ti-indiffused LiNbO$_3$ that guides both polarizations but has different electro-optic coefficients for the two polarization directions. The relative power in the two polarizations serves as the optical splitting factor a, and the difference in electro-optic coefficients serves as the RF power splitting s. There is no optical phase control required because the two polarizations do not interfere with each other (the polarizations must both arrive at the polarization-insensitive detector at the same time to within a small fraction of the period of the highest RF frequency present, though). The two DC biases are hard to control separately, but it can be done by using two sets of DC electrodes that produce electric fields primarily along two different crystal axes. The feedback loop and control of the polarization is difficult, and in LiNbO$_3$ there are rapid variations in the relative electro-optic coefficients for the two (ordinary and extraordinary) polarizations below about 500 MHz, so this technique also presents the user with some serious engineering challenges.

The series connection of two modulators is the other way to make a compound modulator. The linearization mechanism is a bit more difficult to visualize because the transfer functions of the two individual modulators are multiplied together instead of added together. Mathematically, however, the analysis can be done just as described earlier and it is no more difficult than for the parallel modulator case. Two different modulator types can be used, such as the MZ and the directional coupler [62], but connections of two MZs have produced the most effective practical devices. The series connection of two standard MZ modulators, which can provide third order linearization ($c_3 = 0$), will be described first, followed by more complex devices that can achieve broadband linearization. This is the reverse of the actual chronological sequence. Broadband linearized modulators were developed first, mainly in response to the requirements for cable TV transmission over analog fiber optic links. The third-order-only linearized modulators were developed later for a multiplicity of narrowband "niche" applications such as antenna remoting, after it was realized that there were advantages to be gained by dropping the unnecessary second order linearization when it was not needed.

The series connection of two standard MZ modulators [63] is shown in Fig. 4.16. There are three variables that describe the operation of this modulator: the two DC bias points and the RF power splitting ratio. The time delay from the RF input to the modulator output must be the same for both paths, through MZ #1 or through MZ #2; this requirement is the same on all compound modulators. When this condition is met, and both MZs have perfect extinction ($E = 0$), the transfer function of this modulator is

$$T_{\text{smz}}(V) = \frac{1}{4}\left[1 + \cos\left(\phi_1 + \frac{\pi}{V_\pi}V\sqrt{1-s}\right)\right]\left[1 + \cos\left(\phi_2 + \frac{\pi}{V_\pi}V\sqrt{s}\right)\right],$$

(4.34)

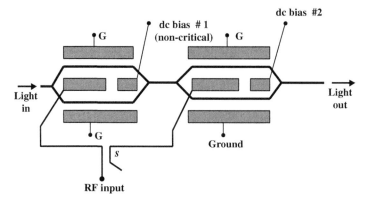

Figure 4.16. Series connection of two standard MZs. The fraction of input RF power sent to MZ #2 is s.

where both modulators have the same switching voltage V_π, ϕ_1 and ϕ_2 are the dc bias points of the two modulators (the result of the dc bias voltage, in the example of Fig. 4.16), s is the fraction of the input RF power sent to the second MZ, and V is the RF input voltage. This is just the product of the two MZ transfer functions, as can be seen by comparing with Eq. (4.12). (Transfer functions for the other compound modulators shown here can also be generated by the appropriate combination of individual modulator transfer functions.) To find the condition for third order linearization, one solves the equation $c_3 = 0$, where c_3 is calculated from $T_{smz}(V)$ by substituting $T_{smz}(V)$ into Eq. (4.25) with $n = 3$. Even though it would appear that there are enough degrees of freedom to do simultaneous second and third order linearization, this cannot be done with this device. However, third order linearization can be achieved for any value of the RF splitting ratio and DC bias #1, by setting DC bias #2 to the correct value. In practice, the RF splitting ratio and the DC bias point of MZ #1 are chosen to minimize the link noise figure (a detailed example is in Ref. 63). The RF power splitter is a passive device and small variations over temperature are easily compensated by small changes in the bias point of MZ #2. The bias point of MZ #1 can be set passively [13,14], or it can be controlled to a particular point using a dither signal on it to generate an error signal. The only critical control is the bias point of MZ #2, but even this is straightforward in that a single control loop with a single error signal (the third order distortion from a pilot tone) can be used. This device can be built using two separate standard MZs, with the output fiber of the first MZ fusion spliced to the input fiber of the second MZ (polarization must be controlled when both are LiNbO$_3$ devices), or by integrating both MZs on the same substrate.

A simplification of the series MZ is the reflective MZ [64] shown in Fig. 4.17. This is now a single physical modulator, but because of the second pass of the light through the device it acts as a series MZ with equal power to both modulators (but

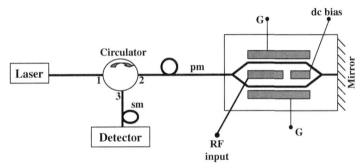

Figure 4.17. Reflective linearized MZ link. The modulator is a standard MZ but with a mirror attached to the output end.

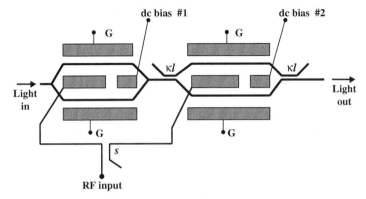

Figure 4.18. Series MZ capable of broadband linearization. The fraction of input RF power sent to MZ #2 is s. The second MZ is made with reduced extinction by designing the input and output couplers with $\kappa l < \pi/4$ so the fraction of optical power sent through the upper arm is <0.5.

no penalty due to a power split!) and equal DC bias points on the two modulators. This has only one bias point that gives $c_3 = 0$ (about 104.5 degrees of phase difference between arms). There is a delay between the first and second pass through the modulator, so this device has a frequency limitation because it cannot meet the equal-time-delay requirement noted above (a traveling wave electrode with an electrical reflection near the optical mirror [65] can remove this limitation). For a polarization-sensitive modulator, polarization must be maintained between the circulator and the modulator.

   The series connection of two MZs can achieve broadband linearization if one of the two MZs has non-zero extinction ($E \neq 0$). This can be done by adding loss to one arm of one MZ [66] ($t_1 < 1$ in Eq. (4.10)), or by using an unequal power split in one MZ [67] ($a \neq 0.5$ in Eq. (4.10)). Figure 4.18 shows an implementation of this using directional couplers. When equal coupling ratios are used at input and output, this method has an advantage over adding loss to one arm in that there is no

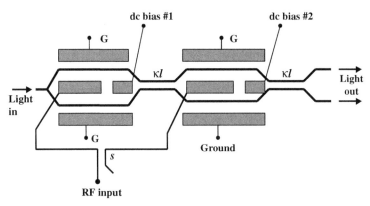

Figure 4.19. Series MZ for broadband linearization with balanced outputs [67]. This structure is different than the series MZs in Figs. 4.16 and 4.18 in that both input ports of the
second MZ have optical inputs.

extra optical insertion loss introduced, as can be seen by examining Eq. (4.7). The
transfer function is the product of two individual MZ transfer functions, similar
to Eq. (4.34) but now the second MZ has $E \neq 0$. The solution of the broadband
linearization problem requires simultaneous solution of $c_2 = 0$ and $c_3 = 0$ while
minimizing noise figure. The most practical implementation is to pick the extinction
and RF power split that gives the best noise figure, then control the two dc bias points
to achieve linearization. This is a significantly more complex control problem than
for the third order only series MZ in that the broadband case requires critical control
of both bias points, and errors in one bias point affect the correct setting of the other.
The link noise figure achievable with broadband linearization is significantly worse,
at least several dB, than for third order only linearization (although the series MZ
has a smaller noise figure penalty than the parallel MZ).

   A further modification of the series MZ [67] is shown in Fig. 4.19. This is no
longer as simple to analyze as the versions above because both outputs of the first
MZ feed both inputs of the second MZ through a directional coupler, and the transfer
function must be calculated from the product of the 2×2 matrices that describe
each of these parts. This structure gives two complementary outputs which can
be linearized; achieving precise balance as well as linearization can lead to quite a
complex control system [68]. This type of modulator is in commercial use in analog
fiber optic links for cable TV.

   The final class of linearized modulator is the ideal type, a single modulator (i.e.,
one requiring no RF power split) that has inherent in its structure the characteristics
of broadband linearization. The only successful modulator of this type so far has
been the directional coupler with extra bias electrodes [69], as shown in Fig. 4.20.
This device is a simple optical structure with a single RF electrode, so no optical
or electrical power splitting needs to be done. It has two bias voltages that must

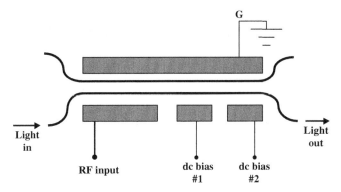

Figure 4.20. Modified directional coupler capable of broadband linearization [69].

be controlled precisely, so it has the same control difficulty as other broadband linearized modulators. This modulator was further analyzed in Ref. 57 and found to be capable of simultaneous second, third and fourth order linearization. Unlike the MZ modulator, the modulation in the directional coupler is distributed along the length of the RF electrode. The consequence of this in this linearized modulator is a tight tolerance on the velocity match between electrical and optical signals [20], tighter than required by frequency response roll off considerations.

4.3.3 Optimization of link performance (noise figure and dynamic range)

New modulator designs, achieving linearization or some other improvement, generate excitement because they offer a solution to one or more of the many issues involved in analog link design. To truly judge whether these represent an improvement or not, though, the dynamic range, noise figure, and any other link performance measures relevant to the intended application must be considered. In this section we will develop criteria for comparing different modulator designs in terms of their effect on the link noise figure and dynamic range. The noise figure and dynamic range are critical in many link applications. There are other measures that are also critical in a smaller number of applications as well, such as gain, maximum output power, phase noise, and sometimes other things. The gain is simply related to a few modulator parameters (see Eq. (4.30)), but the other measures are beyond the scope of this chapter.

To evaluate the noise figure and dynamic range, the noise sources in the link must be added to the analysis presented so far. The output noise of the link is

$$N_{\text{out}} = (f_{\text{in}} + f_{\text{m}})kT_{\text{o}}BG + f_{\text{R}}kT_{\text{o}}B + r(i_{\text{o}}c_0)^2 R_{\text{d}}B + 2ei_{\text{o}}c_0 R_{\text{d}}B$$

$$+ 2e\left[\frac{e}{h\nu}p_{\text{L}}t_{\text{m}}t_{\text{ma}}c_0\right][g_{\text{a}}t_{\text{ad}}\eta_{\text{q}}]^2\left(\frac{g_{\text{a}}-1}{g_{\text{a}}}\right)f_{\text{a}}R_{\text{d}}B. \qquad (4.35)$$

The individual terms and the new variables introduced will be discussed below. Variables introduced earlier were defined in association with Eqs. (4.21), (4.22), (4.24), (4.28), and (4.30). The electrical noise bandwidth B appears in every term, as appropriate for white noise.

The first term is due to the noise present at the link input. This noise power is expressed in terms of the thermal noise from a resistor at the standard temperature $T_o = 290$ K; k is Boltzmann's constant. The input noise power is divided into two sources. External noise, i.e., noise entering the link input from sources outside the link, is represented by f_{in}, where f_{in} is the ratio of the externally generated noise power to kT_o. When calculating the noise figure below, we will use $f_{in} = 1$ as required by the standard definition. There are cases such as the direct connection of an antenna to the modulator where it is simpler to use the actual input noise power directly, so f_{in} may be different than 1, and work with noise temperature (N_{out}/kB) instead of noise figure. The other noise source at the input is thermal noise from the modulator and its matching circuit, represented by f_m. This is discussed in some detail in Ref. 70. One simple example is the traveling wave modulator at high frequency, which has $f_m = 0$ because the thermal noise from the terminating resistor is traveling in the opposite direction to the light and so does not modulate the light. Another example is a lumped element modulator using a resonant circuit that provides a conjugate match, which has $f_m = 1$; some of these modulators can have enough response that this noise source is the dominant contribution to the link output noise.

The second term is the receiver thermal noise. For the *pin* diode detector, the detector type in widest use in analog links, the only significant noise source (other than shot noise) is thermal noise. The primary contributions to this are the thermal noise of any matching resistors connected to the detector, and the equivalent input noise of the amplifier following the detector if this component is included as part of the link. The thermal noise is represented by the parameter f_R, so the output noise of the unity-gain receiver in Fig. 4.12 from this source is $f_R kT_o B$. (Some examples: (1) If the detector is directly connected to the input of an amplifier whose input impedance is 50 Ω and whose input noise resistance is also 50 Ω, then $f_R = F_A - 1$, where F_A is the amplifier's noise figure. An accurate model of this case in real situations must consider both the amplifier's input current noise source and its input voltage noise source separately, as in Ref. 52. (2) If the detector is connected to a 50-Ω resistor at temperature T_o inside its package in parallel with a 50-Ω output connector, then $f_R = 1$ because the internal 50-Ω resistor transfers noise power $kT_o B$ to the matched 50-Ω load.)

The third term is the noise generated by the relative intensity noise (RIN) of the laser. The quantity r is the ratio of noise power in a 1-Hz bandwidth to DC power, when the laser's beam is detected and shot and thermal noise are subtracted from the total detected noise. r is usually quoted as dB/Hz (Eq. (4.35) uses the ratio, not the dB value!). RIN is not actually white noise, and it can have a peak at a few hundred kHz for an Nd:YAG laser or several GHz for a semiconductor laser. If the noise bandwidth is so wide that the RIN level changes within it, the product rB must be replaced by the integral of the RIN over the noise bandwidth. In cases where the RIN is constant over the noise bandwidth, Eq. (4.35) is correct as long as the RIN at the frequency of interest is used. RIN can also arise from multiple optical reflections within the link [71], in which case it is referred to as interferometric intensity noise. The r value to use in this equation is the total optical intensity noise arriving at the detector (with the exception of optical amplifier noise, discussed below).

The fourth term is the shot noise from the random generation of photoelectrons at the detector. The charge on an electron is represented by e.

The final term is the noise associated with an optical amplifier placed between the modulator and the detector. If no amplifier is present, this term is zero. The amplifier's optical gain is g_a and the optical noise figure is f_a. The detector quantum efficiency, the probability of generating an electron from an incident photon, is η_q. This is different than the detector responsivity η defined in association with Eq. (4.21): $\eta = e\eta_q/h\nu$ where h is Planck's constant and ν is the optical frequency. Optical amplifiers used in practical links at the present time (such as the semiconductor optical amplifier or the EDFA) are based on stimulated emission from an inverted population of electronic energy levels. The source of added noise in this type of amplifier is spontaneous emission. The noise term in Eq. (4.35) is the signal–spontaneous beat noise, which dominates over spontaneous–spontaneous beat noise when the input power to the optical amplifier is large enough. For high performance analog links, the optical power input to the amplifier is several hundred μW or more and this assumption is true even without an optical bandpass filter to reduce the spontaneous–spontaneous beat noise. The signal–spontaneous beat noise power can be described by a noise figure f_a, with quantum mechanics limiting the minimum noise figure to $f_a = 2\,(3\,\text{dB})$. The noise figure of a commercial EDFA can approach this quantum limit. Semiconductor optical amplifiers generally have a higher noise figure, typically 6 to 10 dB at the present time. The input noise to the optical amplifier is shot noise due to the random arrival times of photons. The situation is completely analogous to the random generation of electrons at the detector and leads to the linear dependence of the noise on the input power to the optical amplifier, which is $p_L t_m t_{ma} c_0$. For a more detailed discussion of optical amplifier noise, see Ref. 52 or another book discussing optical amplifiers.

The noise figure F is defined as the input signal-to-noise ratio divided by the output signal-to-noise ratio, when the input noise power is kT_0 and both ports are impedance matched. This reduces to the output noise (as given by Eq. (4.35) with

$f_{in} = 1$) divided by the link gain (as given by Eq. (4.30)) times $kT_o B$, which gives

$$F = 1 + f_m + \frac{f_R}{i_o^2 R_d} \frac{1}{R_m c_1^2} + \frac{r}{kT_o} \frac{c_0^2}{R_m c_1^2} + \frac{2e}{i_o kT_o} \frac{c_0}{R_m c_1^2}$$

$$+ \frac{2e}{\left(\dfrac{e}{h\nu} p_L t_m t_{ma}\right) kT_o} f_a \left(\frac{g_a - 1}{g_a}\right) \frac{c_0}{R_m c_1^2}, \qquad (4.36)$$

where all the variables have been defined previously. The optical amplifier noise term was simplified using Eq. (4.22) for i_o and the expression for η given in the preceding paragraph. All of the terms are seen to depend on at least one modulator parameter. The term that dominates will depend not only on the modulator but on other link parameters as well.

The term $1 + f_m$ is due to noise at the link input, and of course nothing within the link can reduce the effect of input noise. For very high performance links, all the other terms in Eq. (4.36) can be made small enough that the term $1 + f_m$ dominates [70]. In this case, it is worthwhile to examine the design of the modulator and matching circuit to make f_m as small as possible.

All of the rest of the terms in Eq. (4.36) can be reduced by increasing the modulator response $R_m c_1^2$. The noise figure improves linearly with increase in $R_m c_1^2$ until the terms depending on it become comparable to the input noise terms. Existing modulators suitable for frequencies above about 1 GHz generally do not have enough response to reach this point, so improving modulator response is currently an active area of research.

There are several factors that affect the relative importance of the last four terms in Eq. (4.36). Each term has a constant that characterizes the magnitude of the noise source: the receiver thermal noise is characterized by f_R, the RIN by r/kT_o, the shot noise by $2e/kT_o$, and the optical amplifier noise by $(2ef_a/kT_o)(g_a - 1)/g_a$. Of these constants, only the one relevant to shot noise is determined solely by fundamental constants of nature. The choice of components used in the link determines the magnitudes of the other noise sources. Whatever the values of these constants, though, the current i_o and the modulator's relative transmission at its bias point c_0 can determine which term dominates the noise.

The receiver noise term is inversely proportional to the square of i_o. The modulator transmission t_m is contained in i_o (see Eq. (4.22)), so it is especially important in cases where this term dominates. The contribution of receiver noise to the noise figure rapidly falls as i_o increases, so this term is often not a major consideration in high performance links where i_o is usually several mA. The receiver noise term is the only noise figure term that depends directly on the link gain (see Eq. (4.30) for gain). Several other terms are also reduced by increasing the quantities $R_m c_1^2$

and i_o which also increase the gain, but the relationship is different. This is an important point to understand because there are cases where noise figure is improved by modulator choices that decrease the link gain.

The RIN term is unaffected by i_o. It sets a limit on the maximum useful i_o in that increases in i_o decrease the other terms until the RIN term dominates; increases in i_o beyond this point do not lower the noise figure. Increasing modulator response reduces this term as mentioned above, but the quantity to minimize is actually $c_0^2/R_m c_1^2$ (this can be loosely interpreted using Eq. (4.32) as maximizing the output modulation depth). If this quantity can be reduced, even at the expense of reducing $R_m c_1^2$, the RIN contribution to the noise figure can be reduced.

The shot noise term is usually the primary contribution to the noise figure in high performance links that do not include an optical amplifier. It is inversely proportional to i_o, so the modulator transmission is important here as well. As with the RIN term, the noise figure can be reduced by a reduction in c_0 as well as by an increase in $R_m c_1^2$, but here there is only a linear dependence on c_0.

In links using an optical amplifier, the final term in Eq. (4.36) is important. The dependence on modulator parameters (namely t_m, c_0, and $R_m c_1^2$) is exactly the same as for the shot noise term. The difference is that instead of an inverse dependence on i_o, there is an inverse dependence on $p_L t_m t_{ma}$, the optical power before the amplifier. When considering optical power levels, it is this power that is the limiting factor on the noise figure; i_o can be made large by making $g_a \gg 1/t_{ad}\eta_q$ so that the receiver thermal noise and detector shot noise terms become small compared to the amplifier noise term. Thus it is important to place the amplifier before as much of the link loss as possible. This leads to consideration of placing the amplifier before the modulator, which is sometimes appropriate but has the disadvantage of requiring much higher amplifier power than placement after the modulator, especially if c_0 is small. (The optical amplifier noise term will have a somewhat different form than given here, too, because this expression is only valid for the amplifier between the modulator and detector.) If the single amplifier analyzed here is replaced by a cascade of several amplifiers with losses between them, the amplifier term becomes more complicated but the basic principles are the same.

The dynamic range is the range of input power over which only the desired signals are above the noise floor and no relevant spurious signals are above the noise floor. The exact definition depends on which signals are desired and which spurious signals are considered relevant, so there are many definitions in use. The definition of dynamic range we will use here is the one in common usage in the optical analog link community: the input signal is two equal-power input frequencies (with the fundamental power being the power of *one* of the tones), and the spurious signals are the two-tone intermodulation signals closest to the fundamentals, i.e., $\omega_1 - \omega_2$ for second order and $2\omega_1 - \omega_2$ for third order dynamic range. In Fig. 4.14, the third

Table 4.2. *Linearity factors* L_{xy}

Distortion order (x)	Standard modulator $(y = x)$	Linearized modulator $(y = x + 2)$
2	$\dfrac{1}{2}\dfrac{c_1^4}{c_2^2}$	$\left\lvert \dfrac{c_1^4}{72c_4} \right\rvert^{\frac{2}{3}}$
3	$\dfrac{2}{3}\left\lvert \dfrac{c_1^3}{c_3} \right\rvert$	$\left\lvert \dfrac{2c_1^5}{25c_5} \right\rvert^{\frac{1}{2}}$

order dynamic range for the standard modulator is diagrammed as an example. The dynamic range combines the nonlinearity and the noise performance in a single measure.

The nonlinearity is generally characterized by the output intercept point, or IP_x, where x is the distortion order (2 or 3, here, but higher orders are sometimes used also). The output intercept point is the output power where the fundamental and the xth-order distortion signals have the same power level, when extrapolated from their small signal values. The IP_3 for the standard modulator is diagrammed in Fig. 4.14. The input intercept point, or IIP_x, is the output intercept point divided by the gain. The intercept point can be found by equating the fundamental output (Eq. (4.29b)) with the desired intermodulation output (Eq. (4.29c) or (4.29e)), then invoking the small signal limit to keep only the term of lowest degree in input power, solving for the input intercept point $IIP_x = P_a(= P_b)$, and finally multiplying the input intercept point by the gain to get the output intercept point

$$IP_{xy} = i_o^2 R_d L_{xy}, \tag{4.37}$$

where the linearity factor L_{xy} is given in Table 4.2 for the various cases under consideration here. We introduce the additional subscript y to indicate the input power dependence of the spurious signal, because it is different for standard and linearized modulators (y is the slope of the spurious signal output power vs. fundamental input power on a log–log plot such as Fig. 4.14). This brings out one of the limitations of this type of characterization for linearized modulators. The concept of using an intercept point assumes the spurious signals have a single value of y (a uniform linear slope on a log–log plot like Fig. 4.14), which is not always the case for linearized modulators, as discussed in the previous section. Furthermore, simple amplifier cascade formulae based on the IP_x [72] make the additional assumption that the xth order distortion signals have a power dependence $y = x$, which also can be violated by linearized modulators (see also Ref. 73 for further discussion of the amplifier + linearized modulator cascade). As long as these limitations are kept in mind, these simple characterizations of the nonlinearity can be useful in comparing different types of modulators.

The output intercept point depends only on the maximum dc power $i_o^2 R_d$ and on the linearity factor. The linearity factor is a dimensionless ratio that characterizes the magnitude of the nonlinear distortion term in the modulator's transfer function relative to the linear term.

The dynamic range can be calculated from the output intercept point and the output noise

$$DR_{xy} = \left[\frac{IP_{xy}}{N_{out}} \right]^{\frac{y-1}{y}}. \tag{4.38}$$

(This comes from solving the triangle on the log–log plot of input and output power, and then converting from log to linear, or dBm to watts). Substituting Eqs. (4.35) and (4.37) into (4.38) gives

$$DR_{xy}$$

$$= \left[\frac{L_{xy}}{(1 + f_m)kT_oR_mc_1^2 + \dfrac{f_RkT_o}{i_o^2R_d} + rc_0^2 + \dfrac{2e}{i_o}c_0 + \dfrac{2h\nu f_a}{p_Lt_mt_{ma}}\dfrac{g_a - 1}{g_a}c_0} \right]^{\frac{y-1}{y}} \cdot \frac{1}{B^{\frac{y-1}{y}}}. \tag{4.39}$$

The dependence on the noise bandwidth has led to the practice of computing the dynamic range in a 1-Hz bandwidth and then quoting it in a normalized unit, for example dB Hz$^{2/3}$ for third order distortion of the standard modulator ($y = 3$) or dB Hz$^{4/5}$ for third order distortion of the linearized modulator ($y = 5$).

The noise figure and the dynamic range are the key performance measures for analog links, and Eqs. (4.36) and (4.39) show the connection between the modulator parameters and these performance measures. When taken all together the relationships are complex and numerical modeling must be used to get the full picture. An example of this type of approach is in Ref. 57 where link parameters (and even some modulator parameters) were held fixed and then various linearized modulators were evaluated. In order to gain more understanding of the reasons behind various trends seen in numerical results and to develop some easy criteria for modulator comparison, the problem must be simplified. This can be done by assuming that one particular noise source dominates, and that the modulator is either a standard modulator or a linearized modulator with no errors (so a single value of the slope y applies at all input RF power levels below compression). With these two assumptions a set of simple relationships can be derived. These are listed in Table 4.3.

Table 4.3. *Modulator figures of merit (bigger is better) for dynamic range and noise figure*

Dominant noise source	Noise figure	Dynamic range
Receiver thermal	$R_m c_1^2 t_m^2$	$\left\{ \left[L_{xy} t_m^2 \right] \left[\dfrac{(p_L t_{ma} g_a t_{ad} \eta)^2 R_d}{f_R k T_o B} \right] \right\}^{\frac{y-1}{y}}$
Shot & optical amplifier	$R_m c_1^2 \dfrac{t_m}{c_0}$	$\left\{ \left[L_{xy} \dfrac{t_m}{c_0} \right] \left[\dfrac{p_L t_{ma} g_a t_{ad} \eta}{2 e B (1 + f_a(g_a - 1) t_{ad} \eta_q)} \right] \right\}^{\frac{y-1}{y}}$
RIN	$R_m c_1^2 \dfrac{1}{c_0^2}$	$\left\{ \left[L_{xy} \dfrac{1}{c_0^2} \right] \left[\dfrac{1}{r B} \right] \right\}^{\frac{y-1}{y}}$
Input noise	$\dfrac{1}{1 + f_m}$	$\left\{ \left[L_{xy} \dfrac{1}{(1 + f_m) R_m c_1^2} \right] \left[\dfrac{1}{k T_o B} \right] \right\}^{\frac{y-1}{y}}$

Table 4.3 is useful for understanding which modulator parameters are most important in various situations. The conditions for dominance of a particular noise source were discussed above, in connection with Eqs. (4.35) and (4.36). The noise figure column consists purely of modulator parameters. The effectiveness of one modulator compared to another is represented accurately by these figures of merit, independent of the values of i_o or other link parameters, provided that one uses the figure of merit appropriate to the dominant noise source. The figures of merit in the dynamic range column each consist of a product of two terms, raised to the $(y - 1)/y$ power. The first term contains the purely modulator-related parameters. If one is interested in comparing modulators with the same value of y, i.e., two linearized modulators or two standard modulators, this term alone is sufficient. If one wants to compare a linearized modulator with a standard modulator, though, the full expression must be used because the values of y will be different for the two modulators. Part of the difference in dynamic range in this case is given by the expressions involving (constant) non-modulator parameters raised to the two different exponents.

The most important concept to understand from Table 4.3 and the equations it arose from is that in almost every case the key performance measures depend on more than one modulator parameter. A linearized modulator may change the third order intermodulation slope from $y = 3$ to $y = 5$, and/or improve L_{xy}, but if it does this with a drastic reduction of transmission t_m then the improvement is probably negligible in the shot-noise-limited case. An improvement in modulator response $R_m c_1^2$ reduces noise figure, but only until the input noise dominates; after that, further increases in $R_m c_1^2$ are undesirable because the dynamic range is decreased. The whole picture must be considered when evaluating a new modulator design.

Another principle evident from Table 4.3 is that the different dominant noise sources can cause different modulator parameters to be important. For example, in the case where receiver thermal noise dominates, performance does not depend on c_0. If RIN is dominant, though, a reduction in c_0 can be very effective in improving performance. In the MZ interferometer, moving the bias point ϕ_0 toward $180°$ reduces $R_m c_1^2$, but it reduces c_0 faster, resulting in a dramatic improvement in noise figure and dynamic range in a RIN-dominated case (Ref. 74; substitute the c_n from Eq. (4.26) to see this).

Space does not permit actually plotting the various trends that come out of the analysis given above, but it is hoped that the reader will apply the methods given here to cases of his own interest to see how modulator design can be varied to improve link performance.

References

1. A. Yariv, *Introduction to Optical Electronics*, Holt, Rinehart and Winston, New York, 1976.
2. T. Tamir, *Guided-Wave Optoelectronics*, Springer-Verlag, New York, 1990.
3. J. Pankove, *Optical Processes in Semiconductors*, Dover Publications, New York, 1971.
4. W. E. Martin, "A new waveguide switch/modulator for integrated optics," *Appl. Phys. Lett.* **26**, 562–3, 1975.
5. R. C. Alferness, "Waveguide electrooptic modulators," *IEEE Trans. Microwave Theory Tech.* **MTT-30**, 1121–37, 1982.
6. R. G. Walker, "High-speed III–V semiconductor intensity modulators," *IEEE J. Quantum Electron.* **27**, 654–67, 1991.
7. A. H. Gnauck, S. K. Korotky, J. J. Veselka, J. Nagel, C. T. Kemmerer, W. J. Minford, and D. T. Moser, "Dispersion penalty reduction using an optical modulator with adjustable chirp," *IEEE Photon. Technol. Lett.* **3**, 916–18, 1991.
8. R. H. Rediker and F. J. Leonberger, "Analysis of integrated-optics near 3 dB coupler and Mach-Zehnder interferometric modulator using four-port scattering matrix," *IEEE J. Quantum Electron.* **QE-18**, 1813–16, 1982.
9. R. A. Soref, D. L. McDaniel, Jr., and B. R. Bennett, "Guided-wave intensity modulators using amplitude-and-phase perturbations," *J. Lightwave Technol.* **6**, 437–44, 1988.
10. K. M. Kissa, P. G. Suchoski, and D. K. Lewis, "Accelerated aging of annealed proton-exchanged waveguides," *J. Lightwave Technol.* **13**, 1521–9, 1995.
11. K. J. Williams and R. D. Esman, "Optically amplified downconverting link with shot-noise-limited performance," *IEEE Photon. Technol. Lett.* **8**, 148–50, 1996.
12. V. Ramaswamy, M. D. Devino, and R. D. Standley, "Balanced bridge modulator switch using Ti-indiffused LiNbO₃ strip waveguides," *Appl. Phys. Lett.* **32**, 644–6, 1978.
13. C. H. Bulmer, W. K. Burns, and A. S. Greenblatt, "Phase tuning by laser ablation of LiNbO₃ interferometric modulators to optimum linearity," *IEEE Photon. Technol. Lett.* **3**, 510–12, 1991.
14. C. H. Bulmer and W. K. Burns, "Linear interferometric modulators in Ti:LiNbO₃," *J. Lightwave Technol.* **LT-2**, 512–21, Aug. 1984.

15. G. E. Betts, L. M. Johnson, and C. H. Cox III, "High-sensitivity lumped-element bandpass modulators in LiNbO$_3$," *J. Lightwave Technol.* **7**, 2078–83, 1989.
16. M. M. Howerton, R. P. Moeller, A. S. Greenblatt, and R. Krahenbuhl, "Fully packaged, broadband LiNbO$_3$ modulator with low drive voltage," *IEEE Photon. Technol. Lett.* **12**, 792–4, 2000.
17. R. A. Becker, "Broad-band guided-wave electrooptic modulators," *IEEE J. Quantum Electron.* **QE-20**, 723–7, 1984.
18. J. P. Donnelly and A. Gopinath, "A comparison of power requirements of traveling-wave LiNbO$_3$ optical couplers and interferometric modulators," *IEEE J. Quantum Electron.* **QE-23**, 30–41, 1987.
19. K. Noguchi, O. Mitomi, and H. Miyazawa, "Millimeter-wave Ti:LiNbO$_3$ optical modulators," *J. Lightwave Technol.* **16**, 615–19, 1998.
20. U. V. Cummings and W. B. Bridges, "Bandwidth of linearized electrooptic modulators," *J. Lightwave Technol.* **16**, 1482–90, 1998.
21. S. Kurazono, K. Iwasaki, and N. Kumagai, "New optical modulator consisting of coupled optical waveguides," *Electron. Commun. Jap.* **55**, 103–9, 1972.
22. H. Kogelnik and R. V. Schmidt, "Switched directional couplers with alternating $\Delta\beta$," *IEEE J. Quantum Electron.* **QE-12**, 396–401, 1976.
23. E. Marom, O. G. Ramer, and S. Ruschin, "Relation between normal-mode and coupled-mode analyses of parallel waveguides," *IEEE J. Quantum Electron.* **QE-20**, 1311–19, 1984.
24. T. R. Halemane and S. K. Korotky, "Distortion characteristics of optical directional coupler modulators," *IEEE Trans. Microwave Theory Tech.* **38**, 669–73, 1990.
25. J. E. Watson, M. A. Milbrodt, K. Bahadori, M. F. Dautartas, C. T. Kemmerer, D. T. Moser, A. W. Schelling, T. O. Murphy, J. J. Veselka, and D. A. Herr, "A low-voltage 8 × 8 TiLiNbO$_3$ switch with a dilated-Benes architecture," *J. Lightwave Technol.* **8**, 794–801, 1990.
26. W. K. Burns, M. M. Howerton, and R. P. Moeller, "Performance and modeling of proton exchanged LiTaO$_3$ branching modulators," *J. Lightwave Technol.* **10**, 1403–8, 1992.
27. Y. Silberberg, P. Perlmutter, and J. E. Baran, "Digital optical switch," *Appl. Phys. Lett.* **51**, 1230–2, 1987.
28. W. K. Burns, "Shaping the digital switch," *IEEE Photon. Technol. Lett.* **4**, 861–3, 1992.
29. W. K. Burns and A. F. Milton, "Mode conversion in planar dielectric separating waveguides," *IEEE J. Quantum Electron.* **QE-11**, 32–9, 1975.
30. S. K. Sheem, "Total internal reflection integrated-optics switch: a theoretical evaluation," *Appl. Opt.* **17**, 3679–87, 1978.
31. A. Neyer, "Operation mechanism of electrooptic multimode X-switches," *IEEE J. Quantum Electron.* **QE-20**, 999–1002, 1984.
32. G. E. Betts and W. S. C. Chang, "Crossing-channel waveguide electrooptic modulators," *IEEE J. Quantum Electron.* **QE-22**, 1027–38, 1986.
33. S. Baba, K. Shimomura, and S. Arai, "A novel integrated twin guide (ITG) optical switch with a built-in TIR region," *IEEE Photon. Technol. Lett.* **4**, 486–8, 1992.
34. A. Neyer and W. Sohler, "High-speed cutoff modulator using a Ti-diffused LiNbO$_3$ channel waveguide," *Appl. Phys. Lett.* **35**, 256–8, 1979.
35. S. A. Hamilton, D. R. Yankelevich, A. Knoesen, R. T. Weverka, R. A. Hill, and G. C. Bjorklund, "Polymer in-line fiber modulators for broadband radio-frequency optical links," *J. Opt. Soc. Am. B* **15**, 740–50, 1998.
36. J. S. Weiner, D. A. B. Miller, and D. S. Chemla, "Quadratic electro-optic effect due to quantum-confined Stark effect in quantum wells," *Appl. Phys. Lett.* **50**, 842–4, 1987.

37. Y. Kim, H. Lee, J. Lee, J. Han, T. W. Oh, and J. Jeong,, "Chirp characteristics of 10Gb/s electroabsorption modulator integrated DFB lasers," *IEEE J. Quantum Electron.* **36**, 900–8, 2000.

38. S. Z. Zhang, Y. J. Chiu, P. Abraham, and J. E. Bowers, "25-GHz polarization-insensitive electroabsorption modulators with traveling-wave electrodes," *IEEE Photon. Technol. Lett.* **11**, 191–3, 1999.

39. *Properties of Lithium Niobate*, INSPEC, The Institution of Electrical Engineers, London, 1989.

40. K. Noguchi, H. Miyazawa, and O. Mitomi, "Frequency-dependent propagation characteristics of coplanar waveguide electrode on 100 GHz Ti:LiNbO₃ optical modulator," *Electron. Lett.* **34**, 661–3, 1998.

41. W. K. Burns, P. H. Klein, E. J. West, and L. E. Plew, "Ti diffusion in Ti:LiNbO₃ planar and channel optical waveguides," *J. Appl. Phys.* **50**, 6175–82, 1979.

42. X. F. Cao, R. V. Ramaswamy, and R. Srivastava, "Characterization of annealed proton exchanged LiNbO₃ waveguides for nonlinear frequency conversion," *J. Lightwave Technol.* **10**, 1302–13, 1992.

43. P. G. Suchoski, T. K. Findakly, and F. J. Leonberger, "Stable low-loss proton-exchanged LiNbO₃ waveguide devices with no electro-optic degradation," *Opt. Lett.* **13**, 1050–52, 1988.

44. E. L. Wooten, K. M. Kissa, A. Yi-Yan, E. J. Murphy, D. A. Lafaw, P. F. Hallemeier, D. Maack, D. V. Attanasio, D. J. Fritz, G. J. McBrien, and D. E. Bossi, "A review of lithium niobate modulators for fiber-optic communications systems," *IEEE J. Sel. Topics Quantum Electron.* **6**, 69–82, 2000.

45. G. E. Betts, F. J. O'Donnell, and K. G. Ray, "Effect of annealing on photorefractive damage in titanium-indiffused LiNbO₃ modulators," *IEEE Photon. Technol. Lett.* **6**, 211–13, 1994.

46. C. H. Bulmer, W. K. Burns, and S. C. Hiser, "Pyroelectric effects in LiNbO₃ channel-waveguide devices," *Appl. Phys. Lett.* **48**, 1036–8, 1986.

47. J. Nayyer and H. Nagata, "Suppression of thermal drifts of high speed Ti:LiNbO₃ optical modulators," *IEEE Photon. Technol. Lett.* **6**, 952–5, 1994.

48. M. Seino, M. Naoyuki, T. Nakazawa, Y. Kubota, and M. Doi, "Optical waveguide device with suppressed dc drift," US Patent 5,214,724, May 25, 1993.

49. G. E. Betts, K. G. Ray, and L. M. Johnson, "Suppression of acoustic effects in lithium niobate integrated-optical modulators," in *Integrated Photonics Research*, 1990 Technical Digest Series Vol. 5, Optical Society of America, Washington, D.C., 1990, pp. 37–8.

50. C. Y. Kuo and E. E. Bergmann, "Second-order distortion and electronic compensation in analog links containing fiber amplifiers," *J. Lightwave Technol.* **10**, 1751–9, 1992.

51. K. J. Williams, R. D. Esman, and M. Dagenais, "Nonlinearities in p-i-n microwave photodetectors," *J. Lightwave Technol.* **14**, 84–96, 1996.

52. S. B. Alexander, *Optical Communication Receiver Design*, SPIE Press, Bellingham, WA, 1997.

53. Uehara, S., "Calibration of optical modulator frequency response with application to signal level control," *Appl. Opt.* **17**, 68–71, 1978.

54. R. B. Childs and V. B. O'Byrne, "Multichannel AM video transmission using a high-power Nd:YAG laser and linearized external modulator," *IEEE J. Sel. Areas Commun.* **8**, 1369–76, 1990.

55. M. Nazarathy, J. Berger, A. J. Ley, I. M. Levi, and Y. Kagan, "Progress in externally modulated AM CATV transmission systems," *J. Lightwave Technol.* **11**, 82–105, 1993.

56. J. C. Twichell and R. J. Helkey, "Linearized optical sampler," US patent 6,028,424, Feb. 22,2000.

57. W. B. Bridges and J. H. Schaffner, "Distortion in linearized electrooptic modulators," *IEEE Trans. Microwave Theory Tech.* **43**, 2184–97, 1995.

58. R. B. Welstand, C. K. Sun, S. A. Pappert, Y. Z. Liu, J. M. Chen, J. T. Zhu, A. L. Kellner, and P. K. L. Yu, "Enhanced linear dynamic range property of Franz–Keldysh effect waveguide modulator," *IEEE Photon. Technol. Lett.* **7**, 751–3, 1995.

59. Z.-Q. Lin and W. S. C. Chang, "Reduction of intermodulation distortion of interferometric optical modulators through incoherent mixing of optical waves," *Electron. Lett.* **26**, 1980–2, 1990.

60. S. K. Korotky and R. M. de Ridder, "Dual parallel modulation schemes for low-distortion analog optical transmission," *IEEE J. Sel. Areas Commun.* **8**, 1377–81, 1990.

61. L. M. Johnson and H. V. Roussell, "Reduction of intermodulation distortion in interferometric optical modulators," *Opt. Lett.* **13**, 928–30, 1988.

62. P.-L. Liu, B. J. Li, and Y. S. Trisno, "In search of a linear electrooptic amplitude modulator," *IEEE Photon. Technol. Lett.* **3**, 144–6, 1991.

63. G. E. Betts, "A linearized modulator for sub-octave-bandpass optical analog links," *IEEE Trans. Microwave Theory Tech.* **42**, 2642–9, 1994.

64. G. E. Betts, F. J. O'Donnell, K. G. Ray, D. K. Lewis, D. E. Bossi, K. Kissa, and G. W. Drake, "Reflective linearized modulator," in *Integrated Photonics Research*, 1996 OSA Technical Digest Series Vol. 6, Optical Society of America, Washington, D.C., 1996, pp. 626–9.

65. W. K. Burns, M. M. Howerton, R. P. Moeller, A. S. Greenblatt, and R. W. McElhanon, "Broad-band reflection traveling-wave LiNbO₃ modulator," *IEEE Photon. Technol. Lett.* **10**, 805–6, 1998.

66. G. J. McBrien and J. D. Farina, "Cascaded optic modulator arrangement," US patent 5,168,534, Dec. 1, 1992.

67. H. Skeie and R. V. Johnson, "Linearization of electro-optic modulators by a cascade coupling of phase modulating electrodes," *Proc. SPIE* **1583**, 153–64, 1991.

68. M. Nazarathy, Y. Kagan, and Y. Simler, "Cascaded optical modulation system with high linearity," US patent 5,278,923, Jan. 11, 1994.

69. M. L. Farwell, Z.-Q. Lin, E. Wooten, and W. S. C. Chang, "An electrooptic intensity modulator with improved linearity," *IEEE Photon. Technol. Lett.* **3**, 792–5, 1991.

70. E. I. Ackerman, C. H. Cox III, G. E. Betts, H. V. Roussell, K. G. Ray, and F. J. O'Donnell, "Input impedance conditions for minimizing the noise figure of an analog optical link," *IEEE Trans. Microwave Theory Tech.* **46**, 2025–31, 1998.

71. J. L. Gimlett and N. K. Cheung, "Effects of phase-to-intensity noise conversion by multiple reflections on gigabit-per-second DFB laser transmission systems," *J. Lightwave Technol.* **7**, 888–95, 1989.

72. S. E. Wilson, "Evaluate the distortion of modular cascades," *Microwaves* 67–70, March 1981.

73. J. H. Schaffner and W. B. Bridges, "Intermodulation distortion in high dynamic range microwave fiber-optic links with linearized modulators," *J. Lightwave Technol.* **11**, 3–6, 1993.

74. M. L. Farwell, W. S. C. Chang, and D. R. Huber, "Increased linear dynamic range by low biasing the Mach–Zehnder modulator," *IEEE Photon. Technol. Lett.* **5**, 779–82, 1993.

5

Broadband traveling wave modulators in LiNbO$_3$

MARTA M. HOWERTON AND WILLIAM K. BURNS

Naval Research Laboratory

5.1 Introduction

Traveling wave electro-optic devices were first proposed in 1963 [1] and demonstrated [2],[3] in LiNbO$_3$ waveguide modulators in the late seventies. The basic requirements for broadband operation were known at the outset: velocity matching between the microwave and optical signals, low electrical loss in the microwave guide, and impedance matching between the microwave waveguide and external electrical connectors. However, these conditions were not immediately realized because the dielectric constants and refractive index of LiNbO$_3$ made velocity mismatch inevitable, within the constraints of existing electrode geometries. Eventually, accurate modeling of the multilayer structure using numerical methods, combined with processing improvements, led to devices which were nearly velocity matched [4],[5]. Electrode thickness increased as the structures evolved, which naturally decreased the microwave loss, further extending device bandwidth.

Impedance matching has been considered a lower priority than velocity matching and low losses and was not achieved in a velocity matched device until etched ridge LiNbO$_3$ devices were introduced [6]–[10]. These devices concurrently satisfy the conditions for broadband operation, and operation beyond 100 GHz has recently been attained.

The development of broadband, low V_π modulators [11]–[13] is a final consideration. While the drive voltage can be decreased by lengthening the electrodes, this also increases electrode loss, resulting in a trade-off between drive voltage and bandwidth. Clearly, requirements for a particular application must be considered in modulator design.

This chapter covers the impressive progress that has been made in broadband LiNbO$_3$ modulators over several decades, and addresses current challenges which face researchers.

5.2 Early work

5.2.1 Basic traveling wave design and velocity mismatch derivation

The basic traveling wave structure is shown in Fig. 5.1 [14]. This design has two parallel stripline electrodes and is known as the coplanar strip (CPS) configuration. It is used because of its low propagation loss and good coupling to external connectors, and has a distinct bandwidth advantage compared to lumped electrodes. Ideally, the electrode appears as an extension of the driving transmission line, so it has the same characteristic impedance. The modulator bandwidth is then limited by the difference in velocity between the microwave and optical waves, or velocity mismatch. Here the hot electrode of length L is placed on top of the optical waveguide, as would normally be the case for z-cut titanium-diffused LiNbO$_3$.

In this section we present the dependence of frequency response on velocity mismatch [14]–[16]. For now assume that the electrode and electrical source are impedance matched and microwave losses are negligible. Consider a single-frequency collinear electrical drive signal of the form

$$V(z, t) = V_0 \sin 2\pi f \left(\frac{z n_\mathrm{m}}{c} - t \right), \tag{5.1}$$

where n_m is the microwave effective index, f is the microwave frequency, and c is the speed of light. The voltage seen at position $z(0 \leq z \leq L)$ along the modulator by photons that enter the active region ($z = 0$) at $t = t_0$ is given by

$$V(z, t_0) = V_0 \sin 2\pi f \left(\frac{z n_\mathrm{m} \delta}{c} - t_0 \right), \tag{5.2}$$

with δ defined as

$$\delta = 1 - n_0/n_\mathrm{m} = \Delta n/n_\mathrm{m}, \tag{5.3}$$

where n_0 is the effective index of the guided optical mode, and $\Delta n = n_\mathrm{m} - n_0$ is the index mismatch. The local change in wave vector of the optical wave induced

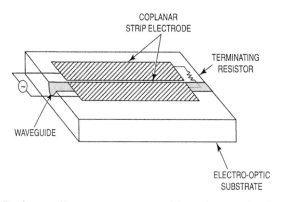

COPLANAR
STRIP ELECTRODE

TERMINATING
RESISTOR

WAVEGUIDE

ELECTRO-OPTIC
SUBSTRATE

Figure 5.1. Basic traveling wave structure with coplanar strip electrodes [14].

by the microwave voltage through the electro-optic effect is given by

$$\Delta\beta(z) = \Delta\beta_0 \sin 2\pi f \left(\frac{z n_m \delta}{c} - t_0 \right), \tag{5.4}$$

where

$$\Delta\beta_0 = -\frac{\pi n_0^3 r \Gamma_{eo} V_0}{\lambda g}. \tag{5.5}$$

Here λ is the free space optical wavelength, r is the relevant electro-optic coefficient, g is the electrode gap, and Γ_{eo} is the electrical–optical overlap integral. Then the total electro-optically induced phase shift $\Delta\phi$ for photons incident at $t = t_0$ is

$$\Delta\phi(t_0) = \int_0^L \Delta\beta(z)\,dz = \Delta\beta_0 L \frac{\sin(\theta/2)}{(\theta/2)} \sin\left(\frac{\theta}{2} - 2\pi f t_0\right), \tag{5.6}$$

where the frequency $\theta(f)$ is given by

$$\theta(f) = 2\pi f n_m \delta L/c. \tag{5.7}$$

Note the $\sin(x)/x$ form of the response. The frequency where $\Delta\phi$ is reduced to 1/2 of its maximum value at $f = 0$ occurs when $\theta/2 = 1.9$, and is given by [15]

$$\Delta f = \frac{1.9c}{\pi n_m L \delta}. \tag{5.8a}$$

The frequency where $\Delta\phi$ is reduced to $1/\sqrt{2}$ of its maximum value occurs when $\theta/2 = 1.4$, and is given by [16]

$$\Delta f = \frac{1.4c}{\pi n_m L \delta}. \tag{5.8b}$$

This occurs when the power in the optical sidebands is reduced by 1/2. Either (5.8a) or (5.8b) may be referred to as the bandwidth, but since definitions differ in the literature [14]–[16] it is important to be specific. For the phase modulator shown here, bandwidth decreases with increasing L and mismatch δ.

For a Mach–Zehnder interferometer the bandwidth derivation proceeds in a similar way. The optical intensity of the interferometer is given by

$$I = I_0 \cos^2 \left(\frac{\Delta\phi + \phi_0}{2} \right), \tag{5.9}$$

where I_0 is the maximum output intensity (independent of frequency), $\Delta\phi$ is the voltage-induced phase difference between interferometer arms, and ϕ_0 is the intrinsic phase imbalance between arms. The frequency dependence of I is completely contained in $\Delta\phi$. $\Delta\beta_0$ is normally larger for the interferometer (by a factor of 1.2 to 2, depending on the electrode configuration) but the value for $\Delta\phi(t_0)$ is still given by (5.6). For a modulator operating in its linear region with $\phi_0 \approx \pi/2$ and small applied voltages, I is reduced to $1/\sqrt{2}$ of its value at $f = 0$ when $\Delta\phi$ is reduced by $1/\sqrt{2}$ [17]. Under these conditions the bandwidth Δf is given by (5.8b) [18]. This criterion is known as the 3 dB electrical bandwidth of the optical output since (optical power) $\propto$ (electrical power)^2. Some authors use the less stringent criteria of 3 dB optical bandwidth, which corresponds to 6 dB electrical bandwidth.

The frequency response of the modulated optical output is known as the optical response, and is discussed in detail in Section 5.4.3. In practice the bandwidth of the optical response is reduced by microwave losses and impedance mismatch, and well as the velocity mismatch which was considered in this derivation.

The importance of velocity matching can be considered in the following way. For $n_0 = n_m$, the optical wave in the waveguide travels at the same speed as the microwave signal in the electrode, so the optical wave sees the same voltage over the entire electrode length. In this case the integrated value of $\Delta\beta(z)$ is proportional to L. Thus, for lossless electrodes, the drive voltage can be reduced by increasing L with no frequency limitation. For $n_0 \neq n_m$, the optical wave and microwave signal are not in step and the useful electrode length is limited by the phase difference or "walkoff." Useful modulation occurs up the distance where the optical wave and microwave signal differ in phase by up to π. Cancellation of modulation occurs when the phase difference is between π and 2π, and the process repeats.

To calculate the velocity mismatch for the modulator in Fig. 5.1, we use the quasi-static approximation for permittivity [19]. The microwave effective index can be expressed as

$$n_m = \epsilon_{re}^{1/2} = \left(\frac{\epsilon_r + 1}{2}\right)^{1/2}. \tag{5.10}$$

ϵ_{re} is the effective dielectric constant, which represents the average of the surrounding materials, $LiNbO_3$ and air. ϵ_r is the relative dielectric constant of $LiNbO_3$, which is given by $(\epsilon_x\epsilon_z)^{1/2}$, where ϵ_x and ϵ_z are the relative dielectric constants for the ordinary and extraordinary axes. Substituting $\epsilon_x = 43$ and $\epsilon_z = 28$ yields $\epsilon_r = 35$ and $n_m = 4.225$. n_0 is 2.15 at 1.3 μm and 2.14 at 1.55 μm, giving $\delta = 0.5$. Thus, the microwave mode is about a factor of 2 slower than the optical mode. This result is valid for thin electrodes without an underlying buffer layer and does not depend on electrode parameters.

While velocity mismatch was considered an inherent limitation in earlier work, researchers eventually overcame the mismatch through improved designs, using artificial velocity matching and "true" velocity matching, to extend frequency response.

5.2.2 Electrode structures

Most TW LiNbO₃ devices use either the coplanar strip (CPS) or coplanar waveguide (CPW) electrode structure, shown in Fig. 5.2. The CPS structure consists of
two parallel striplines having strip width W and separation S. The CPW structure
consists of three parallel electrodes, with center strip width S, separation W, and
semi-infinite outer ground planes. Note that CPS is the complement of CPW, and
that the notation for S and W has a different meaning for each structure. For now,
metal thickness is assumed to be negligible. In real LiNbO₃ devices CPS electrodes
are often asymmetric, where one electrode (the ground plane) is much wider than
the other electrode. In real CPW devices the two outer electrodes are large (typically
≥ 100 µm) compared to the center strip, but their widths are not necessarily equal.
Here we will consider the ideal version of these structures shown in Fig. 5.2.

Analytic expressions for capacitance and characteristic impedance are developed
in Ref. 19 using the conformal mapping technique and assuming an infinite dielectric substrate thickness and negligible metal thickness. The characteristic impedance
Z for each structure is inversely proportional to $\epsilon_{re}^{1/2}$. The dependence of Z on the
ratio S/W is shown in Fig. 5.3 for the CPS and CPW structures, where we used

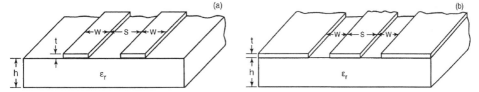

Figure 5.2. (a) Coplanar stripline electrode structure; (b) coplanar waveguide electrode
structure.

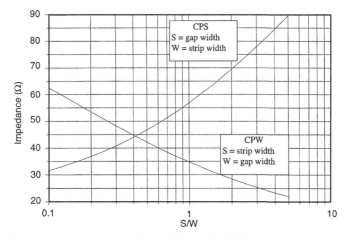

Figure 5.3. Characteristic impedance Z vs. S/W for the CPS and CPW structures in Fig. 5.2,
using $\epsilon_{re}^{1/2} = 35$ for LiNbO₃. Metal thickness is assumed to be negligible.

$\epsilon_{re}^{1/2} = 35$ for LiNbO$_3$. The graph shows that a 50 Ω impedance can be obtained for both structures – with $S/W = 0.65$ for CPS and $S/W = 0.25$ for CPW. However, in a practical LiNbO$_3$ modulator additional constraints make impedance matching difficult. One such constraint is achieving good overlap between the electrical and optical modes (high Γ_{eo}), which is necessary for low drive voltages. For a z-cut device where the electrodes are directly over the waveguides this requires that the width of the hot electrode and Ti waveguide be comparable. The typical Ti wave-guide width is 8 µm at the 1.55 µm wavelength. If the electrode strip width is also 8 µm, then gap widths must be $S = 5$ µm for CPS and $W = 32$ µm for CPW to satisfy the impedance matching condition. For a z-cut interferometer the arm spacing is approximately the same as the gap width, which is clearly too small for the CPS case. For the CPW structure the larger gap width is geometrically feasible, but presents a problem when the additional requirement of velocity matching is imposed.

Analytic expressions which take metal thickness into account are also provided in Ref. 19. For both CPS and CPW electrodes with fixed S/W, Z decreases as metal thickness increases. The model in Ref. 19 is valid when the ratio of metal thickness to electrode gap is less than 0.1 [18], a condition which was normally not met in TW LiNbO$_3$ devices as they evolved. Additionally, early models did not account for a buffer layer. As we shall see in Section 5.3.1, thick electrodes and thick buffer layers were instrumental in improving velocity matching in TW LiNbO$_3$ devices, and numerical techniques were necessary for accurate modeling.

5.2.3 Early broadband traveling wave modulators

Figure 5.4 shows a traveling wave interferometer, which was demonstrated in z-cut Ti:LiNbO$_3$ in 1983 [20]. Velocity mismatch was considered inevitable at the time, so the structure was designed to achieve a 50 Ω impedance. CPS electrodes were deposited on top of a 0.25 µm SiO$_2$ buffer layer. The gold electrodes were 3 µm

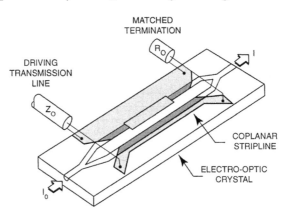

Figure 5.4. 17 GHz traveling wave interferometer designed for impedance matching [20]. Gold electrodes are 3 µm thick.

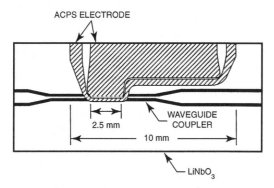

Figure 5.5. Traveling wave directional coupler with 22 GHz bandwidth [21]. The impedance along the active region is 35 Ω, and the tapered electrodes mate with external K connectors.

thick and 6 mm long, and the stripline width and gap were 10 μm and 15 μm, respectively. Careful packaging yielded resonance-free broadband operation to 17 GHz at the 0.83 μm wavelength, beyond the 12 GHz velocity mismatch limit initially calculated by the authors. This led the authors to conclude that the thickness of the electrodes and presence of a buffer layer caused the RF field to extend into the air and SiO$_2$, thereby lowering $n_{\rm m}$.

The directional coupler shown in Fig. 5.5 [21] used tapered electrodes extending to the edge of the LiNbO$_3$ to mate with the external microwave connectors. The gold electrodes were 3–4 μm thick, with a 15 μm strip width and a 5 μm gap width, resulting in a 35 Ω impedance for this CPS design. While the short active region (2.5 mm) contributed to high DC V_π (26 V), an electrical 3 dB bandwidth of 22 GHz was demonstrated at 1.56 μm, as well as response extending to 40 GHz. Device response was mainly limited by velocity mismatch, and to a lesser degree, electrode loss and impedance mismatch.

5.2.4 Artificial velocity matching

5.2.4.1 Periodic phase reversal

Given the difficulty of obtaining true velocity matching ($n_{\rm m} = n_0$), researchers studied ways to achieve artificial velocity matching. One technique which was proposed to overcome this limitation uses a periodic electrode structure to reverse the sign of the electric field, and thus the sign of the electrical phase, when the electrical and optical phase differences become negative [14]. As shown in Fig. 5.6a, this is accomplished by shifting the electrode laterally after an interaction length L_i. In Fig. 5.6b, the local phase shift $\Delta\beta(z)$ is plotted as a function of distance along the modulator for phase reversed and uniform electrodes. Note that in alternating sections the sign of $\Delta\beta(z)$ is flipped for the phase reversed electrode, compared to the uniform electrode. For phase reversed electrodes $\Delta\beta(z)$ remains positive along the modulator length, so each section contributes to the modulator response.

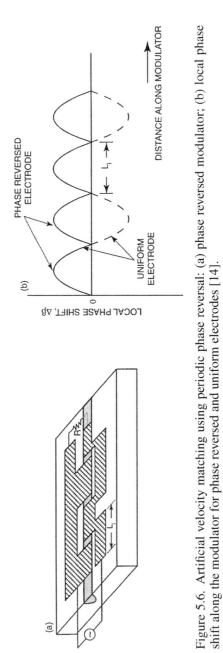

Figure 5.6. Artificial velocity matching using periodic phase reversal: (a) phase reversed modulator; (b) local phase shift along the modulator for phase reversed and uniform electrodes [14].

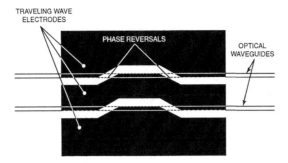

Figure 5.7. Periodic phase reversed modulator with three-section electrodes [23].

In contrast, for uniform electrodes with velocity mismatch the negative $\Delta\beta(z)$ sections decrease the integrated $\Delta\beta$. This degrades modulation response and leads to higher drive voltages. A drawback of periodic structures is that they result in a bandpass response [14].

For a given design frequency f_d, the effective velocity matching condition of the periodic structure is satisfied when

$$L_i = \frac{c}{2n_m f_d \delta} \qquad (5.11)$$

and the resulting bandpass response is centered around f_d.

The first experimental report of periodic phase reversal [23] used the three-section electrode structure shown in Fig. 5.7. These two Ti:LiNbO₃ phase modulators had 3.5 μm thick gold electrodes deposited on a 0.25 μm thick SiO₂ buffer layer, and a total interaction length of 2 cm. Device impedance was 35 Ω. The more efficient device had a 7.5 GHz bandpass response centered at 12.75 GHz. The theoretical fit was made by choosing $n_m = 3.8$ – lower than the commonly accepted value of 4.2 for LiNbO₃. Again, this was an indication that the electrode thickness and buffer layer helped in reducing n_m.

5.2.4.2 Aperiodic phase reversal

Another type of artificial velocity matching, called aperiodic phase reversal, was proposed [24] to overcome the bandpass response of periodically phase reversed structures. With aperiodic phase reversal, the electrodes are laterally offset like the structure in Fig. 5.6a to produce a polarity reversed drive signal; however, the length between transitions is not uniform. Consider first a structure of total electrode length L with only one transition, occurring at length L_i. The frequency response of this device can be tuned by adjusting the ratio L_i/L. For $L_i/L = 0$ the device acts as a conventional low pass modulator; for $L_i/L = 0.5$, it has a bandpass response; and for $L_i/L = 0.25$, it has an "extended" baseband response (Fig. 5.8). By serially combining different L_i values on a common interferometer, an effect similar to

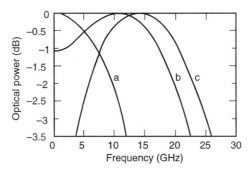

Figure 5.8. Response curves of different types of Mach–Zehnder modulators: (a) low pass modulator; (b) extended low pass modulator; (c) bandpass modulator [24].

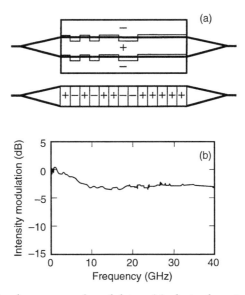

Figure 5.9. Aperiodic phase reversed modulator: (a) electrode pattern and corresponding 13-bit Barker code; (b) experimental frequency response [26].

cascading modulators can be produced. The overall bandwidth can then be tailored based on the response of each unit.

Aperiodic phase reversed structures based on Barker codes were analyzed [25] and implemented [26] in $LiNbO_3$ using a 13-bit sequence. The electrode pattern and frequency response of the Barker code modulator is shown in Fig. 5.9. This packaged device attained better than $-10\,dB$ electrical response at 40 GHz, and a dc V_π of 7.5 V. These results indicated that the technique could simultaneously achieve broadband operation and reasonably low drive voltages (at the time). The drawbacks of this device, like other coded modulators, were nonlinear phase response and a penalty in drive voltage compared to truly velocity matched devices. The effect of electrode losses in periodic and aperiodic phase reversed devices has been treated in Ref. 27.

5.3 True velocity matching

5.3.1 Tailoring the buffer layer and electrode geometry

Early electrode designs were based on analytic models [19], which could not accurately account for metal thickness or buffer layers. Numerical techniques such as the finite element method were needed to accurately model the microwave index of these multilayer structures. As computational methods improved, researchers were able to tailor the geometry to dramatically decrease the microwave effective index toward the optical index, while maintaining nearly 50 Ω impedances and low drive voltages.

To understand the effect of thick buffer layers and thick electrodes, first consider the description of propagation in the CPS and CPW structures shown in Fig. 5.2 [19], where a planar SiO₂ layer may be present on the LiNbO₃. In quasi-static analysis, the mode of propagation is assumed to be pure TEM. Values of characteristic impedance Z and phase constant β depend on capacitance and are given by

$$Z = \frac{1}{c(CC_a)^{1/2}},$$
(5.12a)

$$\beta = \beta_0 \left(\frac{C}{C_a}\right)^{1/2},$$
(5.12b)

where C_a is the capacitance per unit length of the electrode structure with the dielectrics (LiNbO₃ and SiO₂) replaced by air, and C is the capacitance per unit length with the dielectrics present. The velocity of waves in free space is $\beta_0 = \omega/c$. The effective microwave index is given by

$$n_m = \beta/\beta_0 = \left(\frac{C}{C_a}\right)^{1/2},$$
(5.13)

so n_m can be decreased by decreasing C or increasing C_a. Increasing the buffer layer thickness decreases C as the field moves out of the LiNbO₃ and into the buffer layer, which has a lower dielectric constant. Increasing the electrode thickness increases C_a by moving more of the field out of the buffer layer and into the air, effectively increasing the area of the air capacitor formed between the edges of the electrodes. C and C_a also depend strongly on the electrode width to gap width ratio. In addition, to make Z close to 50 Ω, the CC_a product must be considered.

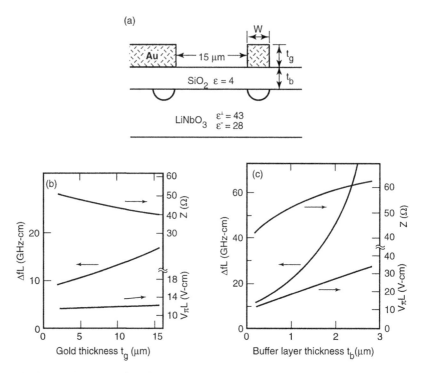

Figure 5.10. (a) Cross-sectional view of coplanar stripline structure used in numerical analysis to generate plots (b) and (c); (b) dependence of bandwidth–length product, voltage–length product, and impedance on electrode thickness t_g; (c) same parameters as in (b) as a function of buffer layer thickness t_b [4].

The first researchers to report near velocity matching in a LiNbO$_3$ modulator used the finite element method to optimize their design [4]. Figure 5.10a depicts a cross-sectional view of the interferometer. The CPS structure has buffer layer thickness t_b and electrode thickness t_g. Figure 5.10b shows how bandwidth–length product ΔfL, voltage–length product $V_\pi L$, and Z depend on t_g, with $t_b = 0.4$ μm. The bandwidth–length product increases by nearly a factor of two as t_g increases from 2 to 16 μm, resulting from improved velocity matching for thicker electrodes. On the other hand, Z moves away from impedance matching as t_g increases beyond ~4 μm, but overall varies within a fairly narrow range (about 40–50 Ω). The voltage–length product is nearly independent of t_g.

In Fig. 5.10c the same parameters are plotted as a function of buffer layer thickness t_b, with $t_g = 9$ μm. The bandwidth–length product depends more strongly on t_b than t_g, and increases sharply as t_b increases from 0.2 to 2.5 μm. However, the penalty for increasing t_b is an increase in the voltage–length product, since the effective voltage seen by the waveguides drops. The variation in Z is roughly 40–60 Ω.

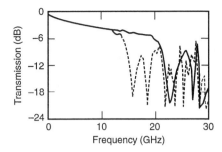

Figure 5.11. Microwave transmission characteristics of the devices reported in Ref. 4 (dashed line, chip width $= 1.18$ μm) and Ref. 5 (solid line, chip width $= 0.57$ μm), which are nearly velocity matched. The chip thickness is 1 mm for both cases. The position of the first loss dip moves to a higher frequency when chip width decreases.

The actual device had a 1 μm thick SiO_2 buffer layer and 10 μm thick gold electrodes. These thicknesses were roughly three times those of conventional traveling wave devices. Electrode strip width was 9 μm, electrode gap was 15 μm, and electrode length was 2 cm. This geometry produced a dramatic reduction in n_m (from 4 to 2.55) compared to existing devices, while simultaneously achieving 50 Ω impedance. The measured electrical 3 dB bandwidth at the 1.55 μm wavelength was 12 GHz, and was not limited by velocity mismatch, but by microwave loss dips. In subsequent work [5] the authors pushed the bandwidth out to 20 GHz, mainly by decreasing chip width by a factor of two, which moved the position of the first dip to a higher frequency. The microwave transmission characteristic for both devices is shown in Fig. 5.11. These sharp dips were attributed to microwave resonances in the chip cross-section at the microwave interface. Additionally, the microwave index of the 20 GHz device was lowered to around 2.35 by using 16 μm thick electrodes. Although the maximum bandwidth based on velocity mismatch calculations was not achieved, these reports represented milestones because they demonstrated the ability to numerically model the complex structures and achieve near velocity match.

Similar success in lowering n_m was obtained by using a shielding plane above the electrodes instead of the thick electrode, also with a thick buffer layer (Fig. 5.12) [28]. Only a thin CPW electrode (~ 4 μm) was used, but the air capacitance was increased by placing a grounded shielding plane 6 μm above the electrodes. Models showed that this geometry, in combination with a 1.2 μm thick buffer layer, could potentially produce values of n_m near 2.2 and Z near 45 Ω. The device bandwidth was 20 GHz, compared to 9 GHz for a comparable device without the shielding plane.

Expressions for C and C_a which explicitly include the buffer layer thickness, electrode thickness, electrode width, and electrode gap are given in Ref. 18, along with plots showing how n_m and Z depend on these parameters. A comprehensive analysis aimed at providing design criteria for optimizing TW LiNbO₃ modulators is also

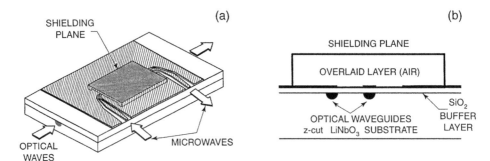

Figure 5.12. Interferometer with CPW electrodes and shielding plane. The shielding plane improves velocity matching [28].

presented. These researchers studied modulators with CPS and CPW electrodes on z-cut LiNbO$_3$ and CPW electrodes on x-cut LiNbO$_3$, starting with waveguide parameters which gave good optical confinement and varying electrode parameters to give a specified bandwidth (10 or 20 GHz). They found that both CPS and CPW structures on z-cut LiNbO$_3$ required significantly less drive power than x-cut CPW structures, mainly because of better velocity and impedance matching, but the difference in performance between z-cut CPS and z-cut CPW structures was small. As we shall see in Section 5.4.1, CPW structures offer an advantage over CPS structures at higher frequencies because of reduced leakage to substrate modes. In practice, most reports on TW LiNbO$_3$ modulators have been for z-cut material. Notable exceptions in the literature are found in Refs. 17, 29, and 30, where interferometers on x-cut LiNbO$_3$ were reported. The z-cut CPW structure is currently the preferred choice for very high frequency operation.

In addition to the many design constraints involved in achieving (or approaching) velocity matching and impedance matching, fabricating narrow, thick electrodes continued to be a challenge. In practice, there was not much room to vary device geometry and still obtain broadband operation. At the 1.3 or 1.55 µm wavelengths Ti strip widths typically range from 7 to 9 µm, and given the importance of good electrical–optical overlap it is not surprising that many researchers reported electrode widths in the same range [4–13],[28],[31–33] for their z-cut LiNbO$_3$ devices. The electrode gaps were all 15 µm. The devices all had CPW electrodes except in Refs. 4 and 5, where CPS electrodes were used.

5.3.2 Effect of electrode wall angle

As electrode thicknesses increased, it became more difficult to accurately measure electrode widths and gaps. This is because the electroplating process used to form the gold layer created non-vertical electrode walls, where the electrodes were wider at the top than at the base. Although the angle itself was not a problem, researchers

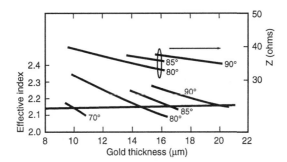

Figure 5.13. Calculated dependence of microwave effective index n_m and impedance on electrode thickness and wall angle [32].

attempting to fine tune the electrode geometry considered the angle in their model. Figure 5.13 shows the calculated dependence of n_m and Z on both electrode thickness and wall angle [32]. The model assumes a CPW structure with electrode strip width $S = 8$ μm and gap width $W = 15$ μm, taken at the electrode base, and a 0.9 μm thick buffer layer. The horizontal line represents n_0, and 90° denotes vertical electrode walls. Clearly both n_m and Z depend strongly on electrode thickness and wall angle. For example, for 16 μm thick electrodes, n_m ranges from 2.25 for a 90° wall to 2.1 for an 80° wall.

Note that if velocity matching is considered a design requirement, it is not possible to simultaneously achieve impedance matching with this structure, even for a wide range of electrode thicknesses and wall angles. Here Z varies from roughly 30 to 40 Ω. On the other hand, it is possible to achieve impedance matching with slightly different structures, as demonstrated in Ref. 4 for a CPS structure, but velocity matching is not ideal (the authors reported $n_m = 2.55$ for 10 μm thick electrodes). As Fig. 5.13 suggests, thicker electrodes can be used to lower n_m, but at the expense of also lowering Z.

In practice it helps to know the expected wall angle in advance so that the proper electroplating thickness can be targeted. In this report the authors assumed an 82° wall angle (based on SEM photographs and self-consistent experimental results). The calculated width at the top of the electrode is then 12.4 μm for an 8 μm base width and 16 μm thickness.

5.4 Microwave loss

5.4.1 Coupling to substrate modes

The microwave loss problem reported in Ref. 5 limited operation to 20 GHz for their device, although the velocity match should have allowed operation to much higher frequencies. Researchers proposed in Ref. 33 that these losses resulted from

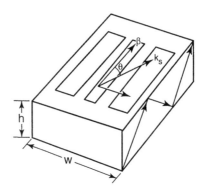

Figure 5.14. Traveling wave structure showing propagation constant of the coplanar waveguide mode (β) and that of the substrate mode (k_s). Leakage occurs when $k_s > \beta$.

coupling between the fundamental coplanar waveguide mode and a substrate mode (surface wave), as predicted earlier in the microwave literature. Not all workers reported similar effects; for example, the devices reported in Refs. 26 and 30 had smooth responses up to at least 40 GHz.

Coplanar transmission lines lose energy to the substrate when the effective index of the substrate mode is larger than that of the coplanar waveguide mode. This happens because the substrate thickness is a significant fraction of the microwave wavelength and can support slab modes which are phase matched to the CPW (or CPS) mode. To understand how this occurs, consider the CPW TW structure shown in Fig. 5.14. The propagation constant along the transmission line is β and the propagation constant of the substrate mode is k_s. Phase match is achieved when $k_s > \beta$, and leakage occurs to both sides of the coplanar guide at an angle given by $\theta = \cos^{-1}(\beta/k_s)$.

Dispersion curves for substrate modes of z-cut LiNbO$_3$ show how the effective substrate index (or propagation constant k_s) depends on h/λ_0, where h is the substrate thickness and λ_0 is the free-space microwave wavelength [33]. As shown in Fig. 5.15 the onset of leakage occurs when $h/\lambda_0 > 0.015$ for the TE$_0$ mode (relevant for a CPS structure with a semi-infinite ground plane on one side) or when $h/\lambda_0 > 0.04$ for the TM$_0$ mode (relevant for a CPW structure with two semi-infinite ground planes). A large area of the CPS surface is unmetallized, and this causes leakage to occur at lower frequencies than for the CPW structure.

In Ref. 33 CPS and CPW structures on LiNbO$_3$ were fabricated having 8 μm strip widths, 15 μm gaps, and 2 to 3 mm wide ground planes. These designs, in combination with thick electrodes and a thick buffer layer, yielded a calculated value of $n_m \approx 2.4$. Figure 5.16 shows the electrical transmission response for the CPS and CPW structures, each having two different substrate thicknesses. (Note the different frequency scales.) It is clear that decreasing the substrate thickness from 0.5 mm to 0.25 mm significantly reduces the microwave loss dips over the

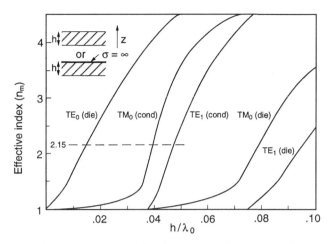

Figure 5.15. Dispersion of substrate modes for z-cut LiNbO$_3$ [33]. h is the substrate thickness and λ_0 is the free-space microwave wavelength.

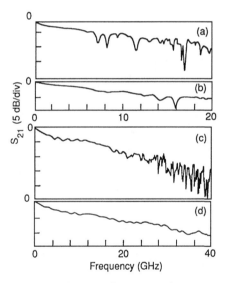

Figure 5.16. Microwave transmission vs. frequency for CPS and CPW structures on LiNbO$_3$: (a) CPS, thickness = 0.5 mm; (b) CPS, thickness = 0.25 mm; (c) CPW, thickness = 0.5 mm; (d) CPW, thickness = 0.25 mm [33].

frequency range shown. Also, the CPW structure has a much better transmission response than the CPS structure. The position of the first loss dip for the 0.50 mm CPW structure in (c) is around 25 GHz. The dip moves beyond 40 GHz for the 0.25 mm CPW structure in (d), and the response is relatively smooth over the entire 0–40 GHz range. This behavior is fairly well-predicted by the dispersion curves. For example, 25 GHz corresponds to $\lambda_0 = 12$ mm; so for the 0.5 mm substrate, $h/\lambda_0 = 0.04$. This is where the onset of leakage is predicted for the CPW structure.

For the 0.25 mm substrate, $h/\lambda_0 = 0.02$, i.e., the substrate is too thin to show leakage at this frequency.

The concept of substrate thinning was extended to a Ti:LiNbO₃ modulator with similar CPW dimensions as noted above, giving the best reported result for LiNbO₃ at the time – −7.5 dB electrical rolloff of the optical response at 40 GHz, with a dc V_π of 5 V [31]. Electrode thickness was 18 μm, leaving only a little room left for improvement in velocity matching (n_m was approximately 2.3). The 35–40 Ω calculated impedance was inevitable under the conditions of near velocity matching, as evidenced by Fig. 5.13, given the buffer layer thickness of 0.9 μm. (A larger t_b could be used to increase Z, but at the expense of also increasing V_π.)

5.4.2 Losses in active and non-active regions

For an unpackaged velocity matched device with no coupling to substrate modes, broadband performance is limited by electrical losses in the coplanar electrode. These consist of conductor losses, which are proportional to $\sqrt{f}$, and dielectric and radiative losses, which are proportional to f [32]. We denote the loss coefficients by α_0 and α_1, and the loss in length L, αL, is approximated by $(\alpha_0\sqrt{f} + \alpha_1 f)L$. Figure 5.17 shows an electrode structure which is designed for packaging. The magnitude of α_0 and α_1 depends on the particular electrode section, i.e., active region, input/output region, taper or bends. The dominant loss mechanism is typically conductor loss in the active region, though at high frequencies (tens of GHz) dielectric and radiative losses in the active region become significant. Losses in the horns are generally small at low frequencies, but depending on the design may also become significant at high frequencies. The taper $\sqrt{f}$ loss and bend f loss are likely to be the largest loss components from the horns.

In Ref. 10 the authors examined losses in a CPW LiNbO₃ modulator which was not designed for packaging. The electrodes had no curved bends, but were tapered along the horns and contacted with probes. In this structure radiative losses

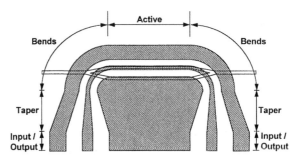

Figure 5.17. Top view of traveling wave interferometer with CPW electrodes [32].

were negligible so α_1 consisted only of dielectric losses. An analytic expression for dielectric loss is given along with experimental data showing total loss up to 110 GHz. For their thick-electrode (29 μm), wide-gap structure conductor loss predominated up to 20 GHz, and both conductor and dielectric loss contributed at higher frequencies.

5.4.3 Dependence of optical response on microwave loss, velocity mismatch, and impedance mismatch

The most comprehensive picture of a TW modulator is provided by transmission data (S_{21}) obtained from a network analyzer. For a purely electrical S_{21} measurement, both the input (port 1) and output (port 2) are at the device input and output. For an optical S_{21} measurement (device optical response) the output is from a photodetector at the optical waveguide output, and the device output is terminated with a 50 Ω load. While the electrical S_{21} measurement includes the effects of microwave loss and impedance mismatch, the optical S_{21} measurement also includes the effect of velocity mismatch.

The normalized optical response (OR) is experimentally obtained by shifting the optical S_{21} vertically to make $OR(f) = 0$ as $f \to 0$. From the OR we define the frequency-dependent drive voltage $V_\pi(f)$ by

$$V_\pi(f) = V_\pi(\text{DC})10^{-OR(f)/20}, \tag{5.14}$$

where $V_\pi(\text{DC})$ is the DC half-wave voltage [32]. In practice, shifting the S_{21} data can give a somewhat ambiguous zero point for the OR because of the network analyzer's low-frequency limit, combined with a sharp initial dip in S_{21}, making it hard to extrapolate back to $f = 0$. For this reason an independent measurement of $V_\pi(\text{DC})$ and $V_\pi(f)$ is helpful.

For non-impedance matched devices, a complete theoretical analysis of the OR [32] must include the effects of both copropagating signals (microwave and optical signals traveling in the same direction) and counterpropagating signals. Thus, reflections due to impedance mismatch at the drive signal/input horn and output horn/termination are included. This greatly complicates the analysis but provides a more accurate model when impedance matching is non-ideal, especially at low frequencies (< 1 GHz) where the counterpropagating signal sharply enhances the OR. At higher frequencies, neglecting the counterpropagating signal has little effect, and a much simpler analysis can be used. In this regime (> 1 GHz) the optical response is approximated by

$$OR(f) = 20 \log[\exp(-\alpha_h L_h)(1 - \Gamma)| + |]. \tag{5.15}$$

$\alpha_h L_h$ represents collective losses in the horns and Γ is the reflection coefficient due

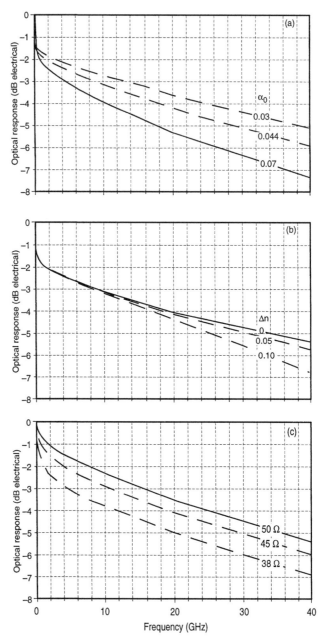

Figure 5.18. Calculated optical response of the modulator in Fig. 5.17 for different active region parameters: (a) vary loss coefficient α_0 $(GHz^{1/2}$ $cm)^{-1}$; (b) vary index mismatch Δn; (c) vary impedance Z. The default case is $\alpha_0 = 0.044$ $(GHz^{1/2}$ $cm)^{-1}$, $\Delta n = 0.05$, and $Z = 38$ Ω. Fixed parameters are $L = 24$ mm and $\alpha_1 = 0.0015 (GHz$ $cm)^{-1}$.

to impedance mismatch at the drive signal/input horn. The parameter $|+|$ is the contribution from the active region and is given by

$$|+| = \left[\frac{1 + e^{-2\alpha L} - 2e^{-\alpha L} \cos\theta}{(\alpha L)^2 + \theta^2} \right]^{1/2} \tag{5.16}$$

where $\theta = \delta L$. For a velocity matched device $\delta = 0$, so (5.16) reduces to

$$|+| = \frac{1 - e^{-\alpha L}}{\alpha L}. \tag{5.17}$$

The effects of electrode loss, impedance, and velocity mismatch on the *OR* are shown in Figs. 5.18 a–c. The calculated response includes the full effects of copropagating and counter-propagating signals for the structure in Fig. 5.17, with active region length $L = 2.4$ cm. For the default case we used $\alpha_0 = 0.044$ (GHz$^{1/2}$ cm)$^{-1}$ and $\alpha_1 = 0.0015$ (GHz cm)$^{-1}$ for $\sqrt{f}$ and f losses in the active region, $Z = 38\ \Omega$, and index mismatch $\Delta n = 0.05$. (For all calculations average loss coefficients for the non-active section were taken as $\alpha_0 = 0.027$ (GHz$^{1/2}$ cm)$^{-1}$ and $\alpha_1 = 0.0047$ (GHz cm)$^{-1}$ over length $L_{\mathrm h} = 0.32$ cm.) In Fig. 5.18a we show the default case and the effect of changing the active region loss coefficient to $\alpha_0 = 0.03$ and 0.07 (GHz$^{1/2}$ cm)$^{-1}$. Note that the higher loss increases the negative slope of the *OR*. By 40 GHz there is about a 2 dB difference in *OR* for this range of α_0.

In Fig. 5.18b we show the effect of varying index mismatch from $\Delta n = 0.05$ for the default case to $\Delta n = 0$ and $\Delta n = 0.10$. Note that the higher Δn increases the negative slope of the *OR*. For this example the penalty incurred at 40 GHz by $\Delta n = 0.10$, compared to $\Delta n = 0$, is about 1.5 dB.

In Fig. 5.18c we show the effect of varying device impedance from 38 Ω for the default case to 45 Ω and 50 Ω. The impedance mismatch with 38 Ω causes a sharp drop in the response at low frequency which is nearly eliminated for the 45 Ω case and completely eliminated at 50 Ω. Improving the mismatch shifts the whole *OR* curve up by a fixed amount while maintaining the same slope, since the electrode loss and velocity mismatch are fixed in this case.

Losses in the non-active region account for approximately 1 dB at 40 GHz in these calculations.

5.4.4 Low frequency acoustic effects

In addition to microwave loss, velocity mismatch, and impedance mismatch, there is an additional effect which can contribute to a drop in the optical response. This drop is associated with the onset of surface acoustic resonances [34],[35] and dispersion in the dielectric constant and electro-optic coefficient [34]. The resonances can occur in the frequency range of tens of MHz to several hundred MHz, and their position and depth depend on substrate orientation and device geometry. In Ref. 35 the authors compared the low frequency behavior of TW LiNbO$_3$ modulators which used the r_{33} electro-optic coefficient and were x-cut, y-propagating or z-cut, y-propagating. Resonances were observed for devices with both crystal orientations around 50 MHz, with a 6 dB variation in *OR* for z-cut devices with the TM polarization and a 2 dB variation in *OR* for the x-cut devices with the TE polarization. Even stronger resonant behavior was reported for the ordinary mode in both x-cut and z-cut LiNbO$_3$. Most reported TW LiNbO$_3$ modulators are z-cut, y-propagating and use the extraordinary mode, and the effect is likely to occur, to some degree, in all devices of this orientation.

Despite the fact that these acoustic resonances were reported in TW LiNbO$_3$ modulators in the late 1980s, this effect has been largely ignored in the literature since then. This is probably because the resonances are easily obscured by low frequency limitations of the optical detector and electronic instrumentation. Unfortunately, these limitations lead to ambiguity in the starting point of the *OR*, which can affect the specified bandwidth, making it appear wider than the "true" (referenced to DC) bandwidth. To overcome this ambiguity it is helpful to directly measure the frequency dependence of V_π in addition to the *OR*. Such results were reported in Ref. 11, where microwave modeling was used to extract the effect of acoustic resonances on bandwidth and drive voltage. For these devices about a 1.5 dB total drop in *OR* was observed in the 0–3 GHz range which could not be accounted for by impedance mismatch, velocity mismatch, or microwave loss.

5.5 Etched ridge modulator

5.5.1 Motivation and design

By the early 1990s two of the basic requirements for broadband operation in TW LiNbO$_3$ modulators had been satisfied. Velocity matching was made possible using thick electrodes, thick buffer layers, and carefully designed structures. Electrode losses were reasonably low, helped in part by the thick electrodes as well as substrate thinning (in some cases) to eliminate substrate mode coupling. The issue remaining to be resolved was impedance mismatch – at the time typical impedances were around 35 to 40 Ω for velocity matched devices. Although increasing the buffer layer thickness to > 2 μm could theoretically produce near 50 Ω impedances in a velocity matched device, it was not normally a realistic option because it severely increased V_π, as noted in Section 5.3.1. This problem was eventually overcome with the introduction of ridges in the LiNbO$_3$ substrate, which provided impedance matching without increasing buffer layer thickness.

Figure 5.19 shows a cross-section of a Mach–Zehnder interferometer with etched ridges and CPW electrodes. The etched ridges are formed by a process which

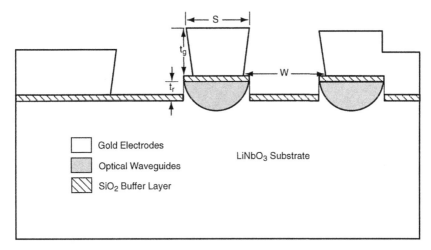

Figure 5.19. Cross-section of a LiNbO₃ etched ridge modulator [13].

removes LiNbO₃ on both sides of the waveguide arms in the active region. Because the LiNbO₃ is replaced by lower dielectric constant materials, air and SiO₂, the capacitance C of the structure decreases, which in turn increases Z (5.12a). Excess loss and distortion of the optical mode are small, since the width of the central unetched portion is about 1–2 μm wider than the undiffused Ti strip width.

In addition to improving impedance matching and microwave loss, the ridge structure is beneficial in reducing the $V_\pi L$ product. This is because the lower dielectric constants of air and SiO₂ cause the electric field to be more concentrated than for the conventional structure. Also, the electric field lines in the waveguide are more nearly parallel to the crystal z-axis, which helps in using the electro-optic coefficient r_{33} more effectively.

The modeled interdependence of electrode gap width W, loss coefficient α_0, electrode thickness t_g, and ridge height t_r is shown in Fig. 5.20a for the modulator in Fig. 5.19, assuming velocity matching and fixed buffer layer thickness. Note that as W increases α_0 decreases. The required t_g increases linearly with W, whereas the required t_r stays fairly constant – between 3 and 4 μm. Increasing the gap width beyond 15 μm produces an increase in $V_\pi L$, as shown in Fig. 5.20b, though this penalty is small compared to the loss benefit.

5.5.2 Performance

The first demonstration of a LiNbO₃ etched ridge modulator [6] used a shielding plane, as in Ref. 28, rather than extremely thick electrodes to obtain velocity matching. The measured optical response implied nearly perfect velocity and impedance matching, along with a loss coefficient of $\alpha_0 = 0.086$ (GHZ$^{1/2}$ cm)$^{-1}$. The 3 dB

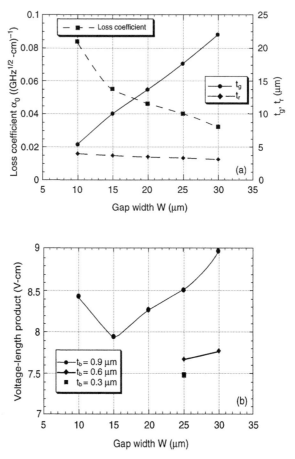

Figure 5.20. (a) Interdependence of loss coefficient α_0, gold thickness t_g, ridge height t_r, and electrode gap width W for the etched ridge modulator in Fig. 5.19. The condition of velocity matching is imposed. (b) Dependence of voltage–length product on gap width [13].

optical bandwidth was 40 GHz ($\sim$10 GHz 3 dB electrical bandwidth) and DC V_π was 3.6 V.

In a subsequent report the same researchers used thicker electrodes (10 μm) without a shielding plane to achieve a 3 dB optical bandwidth of 75 GHz ($\sim$40 GHz 3 dB electrical bandwidth) with a DC V_π of 5.0 V [7]. From experimental data they calculated parameters $n_m = 2.20$, $Z = 47\ \Omega$, and $\alpha_0 = 0.05\ (\mathrm{GHz}^{1/2}\ \mathrm{cm})^{-1}$.

With velocity and impedance matching under control, the design criterion was to further decrease electrode loss while maintaining these matching conditions. This was achieved by widening the electrode gap from 15 to 25 μm while maintaining the 8 μm center electrode width. This in turn required thicker electrodes to maintain velocity matching for a given buffer layer thickness. There was slight penalty in $V_\pi L$ ($\approx$ 10%) with wider electrode gaps, as shown in Fig. 5.20b, which was

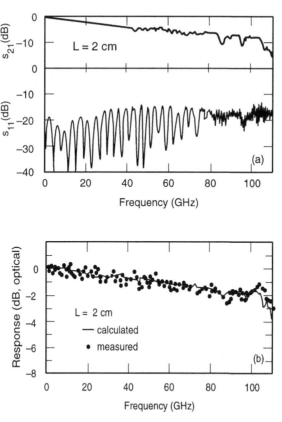

Figure 5.21. Performance of an unpackaged etched ridge modulator with active region length $L = 2$ cm and 25 μm electrode gap width [9]. (a) Transmitted electrical power S_{21} and reflected power S_{11}; (b) optical response.

strongly overshadowed by the loss benefit. Thickness increased by about 10 μm, so fabrication of the narrow, thick electrodes was not trivial.

Response exceeding 100 GHz has been attained in an etched ridge LiNbO₃ modulator using a 25 μm electrode gap and 29 μm thick electrodes [9]. This particular device was not packaged and connectorized – contact was made directly through probes. This provides a bandwidth advantage over comparable but packaged devices. Figure 5.21a shows the S-parameters of this modulator. The electrical transmitted power S_{21} is fairly smooth except for dips around 85 and 90 GHz which are attributed to coupling to substrate modes in the probe region. The reflected power S_{11} is below -15 dB over the entire frequency range, indicating that impedance matching is good. From this data the authors calculated the parameters $n_m = 2.14$, $Z = 47\ \Omega$, and $\alpha_0 = 0.032\ (\text{GHz}^{1/2}\ \text{cm})^{-1}$. The lower loss contributes to a dramatic improvement in the optical response compared to the 15 μm gap devices in Refs. 6 and 7. The optical response is shown in Fig. 5.21b; 3 dB

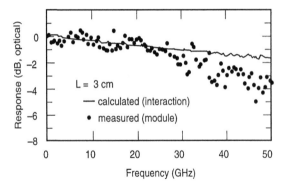

Figure 5.22. Optical response of a packaged etched ridge modulator with $L = 3$ cm and 25 μm electrode gap width [9]. The calculated curve considers the interaction region only, and not the contribution of feeding to the V-connectors, which clearly degrades performance.

optical and electrical bandwidths are 105 GHz and 70 GHz, respectively, and DC V_π is 5.1 V.

In the same study the authors demonstrated another device which was primarily designed for low V_π rather than extremely wideband operation. This device was fabricated using a similar design except the interaction length was increased to 3 cm rather than 2 cm, and it was fully packaged. The longer electrode length helped reduce DC V_π to 3.5 V but also decreased the bandwidth due to higher electrode losses. The optical response, shown in Fig. 5.22, gives measured optical and electrical 3 dB bandwidths of 45 GHz and 30 GHz. The deviation of the measured value from the calculated curve is attributed to losses at the V-connector transition, which were not considered in the calculations. This underscores the impact of packaging on device performance, which is especially important for high bandwidth devices such as these where the basic device design requirements – velocity matching, impedance matching, and low electrical losses – have been attained.

5.6 Trend to low V_π devices

5.6.1 Long single-pass modulator and reflection modulator

Since there is a trade-off between low drive voltage and bandwidth, it is important to consider the desired operating frequencies in modulator design. For applications requiring moderate operating ranges – up to 20 GHz or so – very low drive voltages have been attained using long electrodes. The most straightforward design is a single-pass interferometer similar to the one in Fig. 5.17, with the maximum electrode length allowed for a 3-inch LiNbO$_3$ wafer. An alternative design uses a reflection interferometer, also with maximum length electrodes. The traveling wave reflection interferometer was first reported in Ref. 12, where its performance was compared to a single-pass interferometer having the same chip size.

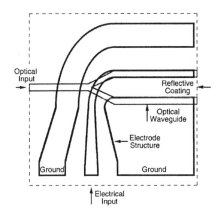

Figure 5.23. Traveling wave reflection interferometer [12].

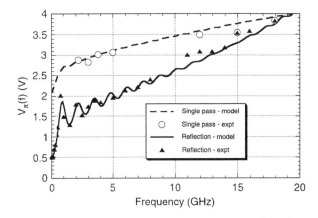

Figure 5.24. Drive voltage as a function of frequency for packaged single-pass and reflection interferometers [12]. The active region length was 4.7 cm for the single-pass device and 5.3 cm (10.6 cm double-pass) for the reflection device.

The reflection interferometer, shown in Fig. 5.23, has a doubled interaction length because it simultaneously reflects both the optical and RF waves before the output branch. The optical wave is reflected by an optical coating and the RF wave is reflected by an open circuit. This occurs at approximately the same point and with the same phase, allowing an optical–electrical interaction in both directions through the device. The optical output is accessed using an optical circulator placed at the front end of the device. This reflection interferometer had a 5.3 cm electrode length and a total interaction length of 10.6 cm, and the single-pass interferometer had a 4.7 cm electrode length. Both devices had a 15/8/15 (gap/strip/gap) CPW configuration with 16 μm thick electrodes.

The measured DC drive voltages at 1.3 μm were 2.1 V for the single-pass interferometer and 0.9 V for the reflection interferometer. In Fig. 5.24 we show $V_\pi(f)$

for frequencies up to 20 GHz for both devices. The data points were direct V_π measurements taken using a two-tone technique and the model was a fit to these points. The model used the following values: $\alpha_0 = 0.053\,(\text{GHz}^{1/2}\,\text{cm})^{-1}$ (from electrical S_{21} measurements); $Z = 38\,\Omega$ (calculated); and an adjustable Δn – giving $\Delta n = 0.06$ for the single-pass device and $\Delta n = 0.05$ for the reflection device. The reflection device is enhanced at frequencies < 0.5 GHz due to constructive interaction between counterpropagating electrical waves, giving a factor of two decrease in V_π relative to the dc value. Overall, the reflection device achieves lower V_π than the single-pass device, with both devices approaching 4.5 V at 20 GHz. The reflection interferometer was the first traveling wave LiNbO$_3$ device reported with drive voltages less than 2 V, and it achieved this up to ~5 GHz.

Because of increased optical insertion loss from the long interaction length and circulator, the reflection device is best suited for applications where plenty of optical power is available. The reflection modulator has been used to demonstrate unamplified link gain [36], taking advantage of its extremely low V_π at low frequencies (~0.5 V) and high optical power from a Nd: YAG laser. 7.5 dB of gain has been achieved at 50 MHz, with positive gain extending to 550 MHz, a region important to cable TV.

5.6.2 Further research on low drive voltage, broadband modulators

Although there is a trade-off between low drive voltage and large bandwidth, low electrode losses (resulting from wider gaps and thicker electrodes) have made it possible to achieve good performance in both these areas. Drive voltages of 7 V at 40 GHz have been reported in a comparison study of packaged devices with and without etched ridges, with the etched ridge device showing about a $^1/_2$ to 1 V advantage over the DC–40 GHz range [11]. Both types of devices had 2.4 cm electrodes (roughly half of "maximum length" for a 3″ water) and 25 μm electrode gaps. The etched ridge device was not perfectly velocity matched ($\Delta n = 0.09$).

Drive voltages of 5 V at 40 GHz have been achieved in longer etched ridge devices (4.1 cm electrodes) with nearly ideal velocity and impedance matching. Figure 5.25 shows the frequency dependent V_π of a long packaged device over the DC–40 GHz range, which includes the experimental V_π inverted from the optical response, direct two-tone V_π measurements, and a fit to the experimental results [37].

5.6.3 Voltage minimization design

In this section we will examine the trade-off between bandwidth and V_π more closely and show how to use this information in designing a modulator. Assume that the device is velocity matched. If we neglect horn losses and consider only the

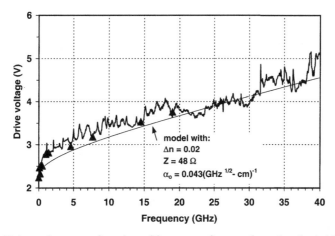

Figure 5.25. Drive voltage as a function of frequency for a packaged etched ridge modulator with active region length $L = 4.1$ cm [37].

$\sqrt{f}$ loss in the active region, (5.15) reduces to

$$OR(f) = 20 \log \left[\frac{(1 - \Gamma)(1 - e^{-\alpha_0 \sqrt{f} L})}{\alpha_0 \sqrt{f} L} \right]. \tag{5.18}$$

The contribution of impedance mismatch can be easily separated from the loss contribution, but for now we will take $\Gamma = 0$. Imposing a 3 dB electrical bandwidth for the optical response gives the requirements

$$\alpha_0 \sqrt{f} L = 0.74 \tag{5.19a}$$

and

$$V_\pi(f) = \sqrt{2} V_\pi(dc) \tag{5.19b}$$

at the 3 dB frequency.

In Fig. 5.26 we plot the required loss coefficient α_0 as a function of interaction length L for various OR bandwidths. The curves clearly show how an increase in length or design bandwidth pushes the loss requirement to lower values.

Consider the following example, where we wish to design an etched ridge modulator with an electrical bandwidth of $f = 20$ GHz, having $L = 4.7$ cm. According to (5.19a) or Fig. 5.26 we require $\alpha_0 = 0.035$ (GHz$^{1/2}$ cm)$^{-1}$. From Fig. 5.20a, this loss coefficient can be achieved with an electrode gap width of $W \approx 28$ μm (and gold thickness $t_g \approx 20$ μm). From Fig. 5.20b, these conditions would yield $V_\pi L = 8.8$ V cm, $V_\pi(dc) = 1.9$ V and $V_\pi(20 \text{ GHz}) = 2.6$ V.

If we incorporate mismatch effects with a 45 Ω impedance, this contributes 0.5 dB in (5.15). The remaining allowance in the active region is 2.5 dB, giving $\alpha_0 \sqrt{f} L = 0.6$ and $\alpha_0 = 0.028$ (GHz$^{1/2}$ cm)$^{-1}$. This requires $W \approx 34$ μm

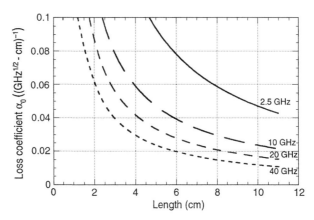

Figure 5.26. Calculated trade-off between loss coefficient α_0 and electrode length L for a given bandwidth requirement.

($t_g \approx 25$ μm), and would produce $V_\pi L = 9.1$ V cm, $V_\pi(\text{dc}) = 1.9$ V and $V_\pi(20 \text{ GHz}) = 2.7$ V. Thus, the slight impedance mismatch does not degrade $V_\pi(f)$ much provided that W and t_g are adjusted to maintain velocity matching. In practice, though, adjusting t_g to achieve this can be challenging.

5.7 Conclusion

Traveling wave $LiNbO_3$ modulators have progressed rapidly, and the basic design problems – achieving velocity and impedance matching and low electrode losses – have been nearly solved. This has resulted in devices that operate beyond 100 GHz. Furthermore, extremely low drive voltages have been obtained in longer devices with more moderate bandwidths – < 4 V at 20 GHz for a reflection modulator and < 5 V at 40 GHz for a single-pass modulator. For the most part, the active region is well-understood. However, the non-active regions have not been studied as extensively, although their effect becomes important as bandwidth increases. The packaging of broadband devices is more complicated than low frequency devices, and this will also require further study. Regardless of the remaining challenges, enormous improvements have been made in broadband, low drive voltage devices. It seems clear that they will have a broad impact in a variety of areas, including microwave links over optical fiber, high frequency electric field sensors, antenna remoting, and instrumentation.

References

1. I. P. Kaminow and J. Liu, "Propagation characteristics of partially loaded two-conductor transmission line for broadband light modulators, *Proc. IEEE*, **51**, 132–6, 1963.

2. M. Izutsu, Y. Yamane, and T. Sueta, "Broadband traveling wave modulator using a LiNbO₃ optical waveguide," *IEEE J. Quantum Electron.*, **QE-13**, 287–90, 1977.

3. M. Izutsu, T. Itoh, and T. Sueta, "10 GHz bandwidth traveling wave LiNbO₃ optical waveguide modulator," *IEEE J. Quantum Electron.*, **QE-14**, 394–5, 1978.

4. M. Seino, N. Mekada, T. Namiki, and H. Nakajima, "33-GHz-cm broadband Ti:LiNbO₃ Mach–Zehnder modulator," *Tech. Dig. ECOC'89*, paper ThB22–5, 1989.

5. M. Seino, N. Mekada, T. Yamane, Y. Kubota, M. Doi, and T. Nakazawa, "20-GHz 3-dB-bandwidth Ti:LiNbO₃ Mach–Zehnder modulator," *Tech. Dig. ECOC '90*, paper ThG1-5, 1990.

6. K. Noguchi, O. Mitomi, K. Kawano, and M. Yanagibashi, "Highly efficient 40-GHz bandwidth Ti:LiNbO₃ optical modulator employing ridge structure," *IEEE Photon. Technol. Lett.*, **5**, 52–4, 1993.

7. K. Noguchi, H. Miyazawa, and O. Mitomi, "75 GHz broadband Ti:LiNbO₃ optical modulator with ridge structure," *Electron. Lett.*, **30**, 949–51, 1994.

8. K. Noguchi, O. Mitomi, H. Miyazawa, and S. Seki, "A broadband Ti:LiNbO₃ optical modulator with a ridge structure," *IEEE J. Quantum Electron.*, **13**, 1164–8, 1995.

9. K. Noguchi, O. Mitomi, and H. Miyazawa, "Millimeter-wave Ti:LiNbO₃ optical modulators," *J. Lightwave Technol.*, **16**, 615–19, 1998.

10. K. Noguchi, H. Miyazawa, and O. Mitomi, "Frequency-dependent propagation characteristics of coplanar waveguide electrode on 100 GHz Ti:LiNbO₃ optical modulator," *Electron. Lett.*, **34**, 662–3, 1998.

11. W. K. Burns, M. M. Howerton, R. P. Moeller, R. Krähenbühl, R. W. McElhanon, and A. S. Greenblatt, "Low drive voltage, broadband LiNbO₃ modulators with and without etched ridges," *IEEE J. Lightwave Technol.*, **17**, 2551–5, 1999.

12. W. K. Burns, M. M. Howerton, R. P. Moeller, A. S. Greenblatt, and R. W. McElhanon, "Broadband reflection traveling wave LiNbO₃ modulator," *IEEE Photon. Technol. Lett.*, **10**, 805–6, 1998.

13. W. K. Burns, "Prospects for low drive voltage LiNbO₃ modulators," *Proc. IEEE Antennas and Propagation Society International Symposium 1997*, pp. 759–61.

14. R. C. Alferness, S. K. Korotky, and E. A. J. Marcatili, "Velocity-matching techniques for integrated optic traveling wave switch/modulators, *IEEE J. Quantum. Electron.*, **QE-20**, 301–9, 1984.

15. R. C. Alferness, "Waveguide electrooptic modulators," *IEEE Trans. Microwave Theory Tech.*," **MTT-30**, 121–37, 1982.

16. S. Martelluci and A. N. Chester, eds., *Integrated Optics*, section by R. V. Schmidt, Plenum Press, New York, 1983.

17. R. A. Becker, "Traveling-wave electro-optic modulator with maximum bandwidth–length product, *Appl. Phys. Lett.*, **45**, 1168–70, 1984.

18. H. Chung, W. S. C. Chang, and E. L. Adler, "Modeling and optimization of traveling-wave LiNbO₃ interferometric modulators," *IEEE J. Quantum. Electron.*, **27**, 608–17, 1991.

19. K. C. Gupta, R. Garg, and I. J. Bahl, *Microstrip Lines and Slotlines*, Artech House, Dedham, MA, Chapters 1 and 7, 1979.

20. C. M. Gee, G. D. Thurmond, and H. W. Yen, "17 GHz bandwidth electro-optic modulator," *Appl. Phys. Lett.*, **43**, 998–1000, 1983.

21. S. K. Korotky, G. Eisenstein, R. S. Tucker, J. J. Veselka, and G. Raybon, "Optical intensity modulation to 40 GHz using a waveguide electro-optic switch," *Appl. Phys. Lett.*, **50**, 1631–3, 1987.

22. D. Erasme, M. G. F. Wilson, "Analysis and optimization of integrated-optic travelling-wave modulators using periodic and non-periodic phase reversals," *Opt. Quantum Electron.*, **18**, 203–11, 1986.

23. D. Erasme, D. A. Humphreys, A. G. Roddie, and M. G. F. Wilson, "Design and performance of phase reversal traveling wave modulators," *J. Lightwave Technol.*, **6**, 933–6, 1988.

24. A. Djupsjöbacka, "Novel type of broadband travelling-wave integrated-optic modulator," *Electron. Lett.*, **21**, 908–9, 1985.

25. M. Nazarathy and D. W. Dolfi, "Velocity-mismatch compensation in traveling-wave modulators using pseudorandom switched-electrode patterns," *J. Opt. Soc. Am. A*, **4**, 1073–9, 1987.

26. D. W. Dolfi, M. Nazarathy, and R. L. Jungerman, "40 GHz electro-optic modulator with 7.5V drive voltage," *Electron. Lett.*, **24**, 528–9, 1988.

27. W. K. Burns, "Analytic output expression for integrated optic phase reversal modulators with microwave loss, *Appl. Opt.* **25**, 3280–3, 1989.

28. K. Kawano, T. Kitoh, H. Jumonji, T. Nozawa, and M. Yanagibashi, "New travelling-wave electrode Mach–Zehnder optical modulator with 20 GHz bandwidth and 4.7 V driving voltage at 1.52 μm wavelength," *Electron. Lett.*, **25**, 1382–3, 1989.

29. D. W. Dolfi, "Travelling-wave 1.3 μm interferometer with high bandwidth, low drive power, and low loss," *Appl. Opt.* **25**, 2479–80, 1986.

30. D. W. Dolfi and T. R. Ranganath, "50 GHz velocity-matched broad wavelength LiNbO$_3$ modulator with multimode active section," *Electron. Lett.*, **28**, 1197–8, 1992.

31. G. K. Gopalakrishnan, C. H. Bulmer, W. K. Burns, R. W. McElhanon, and A. S. Greenblatt, "40 GHz low half-wave voltage Ti:LiNbO$_3$ intensity modulator," *Electron. Lett.*, **28**, 826–7, 1992.

32. G. K. Gopalakrishnan, W. K. Burns, R. W. McElhanon, C. H. Bulmer, and A. S. Greenblatt, "Performance and modeling of broadband LiNbO$_3$ traveling wave modulators," *J. Lightwave Technol.*, **2**, 1807–19, 1994.

33. G. K. Gopalakrishnan, W. K. Burns, and C. H. Bulmer, "Electrical loss mechanisms in traveling wave LiNbO$_3$ optical modulators," *Electron. Lett.*, **28**, 207–8, 1992.

34. J. L. Nightingale, R. A. Becker, R. C. Willis, and J. S. Vrhel, "Characterization of frequency dispersion in Ti-indiffused lithium niobate optical devices," *Appl. Phys. Lett.*, **51**, 716–18, 1987.

35. R. L. Jungerman and C. A. Flory, "Low-frequency acoustic anomalies in lithium niobate Mach–Zehnder interferometers," *Appl. Phys. Lett.*, **53**, 1477–9, 1988.

36. W. K. Burns, M. M. Howerton, and R. P. Moeller, "Broadband, unamplified optical link with RF gain using a LiNbO$_3$ modulator," *IEEE Photon. Technol. Lett.*, **11**, 1656–8, 1999.

37. M. M. Howerton, R. P. Moeller, A. S. Greenblatt, and R. Krähenbühl, "Fully packaged, broadband LiNbO$_3$ modulator with low drive voltage," *IEEE Photon. Technol. Lett.*, **12**, 792–4, 2000.

6

Multiple quantum well electroabsorption modulators for RF photonic links

WILLIAM S. C. CHANG

University of California, San Diego

6.1 Introduction

Performance of multiple quantum well (MQW) electroabsorption (EA) modulators can best be evaluated in terms of a basic RF photonic link. A basic RF photonic link consists of a laser transmitter with its optical intensity modulated by an RF signal, an optical fiber transmission line and a receiver. In the externally modulated link shown in Fig. 6.1, the RF source supplies the signal to the EA modulator. The CW laser radiation is coupled directly or through a pigtailed fiber to the modulator input. The fiber transmission line couples the modulated output to the receiver. The receiver detects the optical radiation and converts the intensity modulation back into RF power. Because of the low transmission loss of fibers, a major advantage of an RF photonic link is its low RF transmission loss, especially for long distances and at high RF frequencies. Many RF channels at widely different frequencies can also share the same optical carrier. More sophisticated links may employ optical or electronic amplification, distribute the modulated optical carrier in a fiber network, down or up convert the RF frequencies using opto-electronic components. However, the fundamental effect of a component such as a modulator can most clearly be understood through the basic link.

In Fig. 6.1, the optical carrier is obtained from CW laser and modulated by an external modulator, based on semiconductors or ferroelectrics such as $LiNbO_3$ or polymers [1,2]. Electroabsorption (EA) semiconductor modulators, which produce RF intensity modulation of an optical carrier, depend either on the Franz–Keldysh effect or the quantum confined Stark effect (QCSE) in multiple quantum well (MQW) heterostructures. We will discuss in this chapter properties of EA modulators in a basic link based on the QCSE in MQW heterostructures. The Franz–Keldysh modulators [3,4,5] have generally lower modulation efficiency and higher saturation optical power. They will not be discussed here.

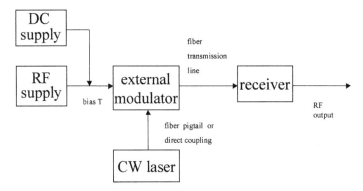

Figure 6.1. Schematic illustration of the externally modulated RF photonic links.

In an MQW EA modulator, if an electric field F is applied to a multiple quantum well (MQW) structure, its absorption of the optical radiation is generally increased [6,7]. As the field is varied due to the applied RF voltage, EA yields a corresponding change in intensity of the laser radiation transmitted through the modulator. This is the principle mechanism of EA modulation.

The modulator is only a part of the RF link [2]. Consequently, the performance of the modulator will be evaluated in terms of (1) the RF link efficiency, (2) the RF bandwidth and (3) the RF spurious free dynamic range (SFDR) of the link (see Chapter 1 for a more detailed discussion of efficiency, bandwidth and SFDR). The bandwidth and the efficiency of MQW EA modulators in RF links have been extensively investigated. How to design EA modulators to maximize the SFDR of photonic links is an issue yet to be fully addressed. Section 6.1.3 will review these figures of merit more in detail. In order to demonstrate clearly the properties of EA modulators, we will consider here only the intrinsic link, meaning a single basic fiber link that does not contain optical and electronic amplifiers.

The primary advantages of MQW EA modulators in RF photonic links are the small size, large RF efficiency (especially at high microwave or millimeter wave frequencies), and suitability for both photonic and opto-electronic integration. EA modulators are typically less than 1 mm long, while $LiNbO_3$ modulators are centimeters long (Chapters 4 and 5). Directly modulated lasers do not function well at frequencies much above 10 GHz (Chapter 3). At higher frequencies, $LiNbO_3$ modulators require much larger RF driving voltage than EA modulators, thereby reducing the RF link efficiency [8]. Thus MQW EA modulators are particularly attractive for high RF frequency applications. Since MQW material structures are made of layers epitaxially grown on InP or GaAs substrates, additional electronic circuits or CW lasers may be fabricated and integrated with the modulator on the same chip.

It is not the intention of this chapter to survey past accomplishments and the many papers published on EA modulators. Instead, this chapter focuses on the understanding and design of MQW EA modulators that will be used to transmit the RF signal. References are used only to illustrate specific points.

In order to design and to understand comprehensively the properties of an EA MQW modulator for RF photonic links, we will analyze its performance in a basic link from many different points of view, such as quantum physics, material science, device physics, circuit and system performance. Readers may prefer to by-pass detailed discussions from certain points of view such as quantum physics when they are not deeply interested in such discussions. Applications such as up- and down-conversion of the RF frequency, phased antenna arrays, etc. will be discussed in other chapters. Note, however, that if EA modulators are used for such applications, their design and performance goals will be different from the ones presented here.

6.1.1 Introduction to the MQW EA modulator

An EA MQW modulator typically consists of a multimode waveguide (WG) that contains the MQW structures, as shown in Fig. 6.2. At its input, the WG is coupled either to a single mode fiber that is pigtailed to a laser or to a photonic integrated laser. At the input port, only the fundamental WG mode is excited. At its output, the waveguide is coupled to another single mode fiber. The radiation will propagate from the input to the output for distances of the order of mm in the fundamental mode without much diffraction loss or mode conversion [9].

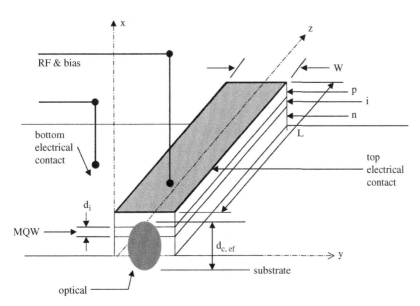

Figure 6.2. Illustration of the basic structure of a MQW modulator.

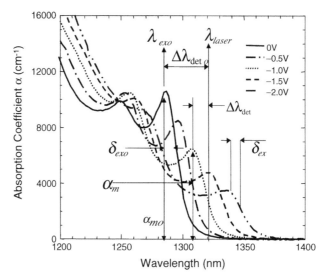

Figure 6.3. Absorption spectra of InAsP/GaInP MQW at different V_{bias} for the TE polarization.

6.1.1.1 Electroabsorption

Electrically, in the lumped circuit representation, the modulator is a reverse biased p-i-n diode [10,11]. In response to a DC voltage V applied to the waveguide, there is an induced electric field in the undoped MQW in the x-direction, $F \cong V/d_i$. In this case, the small built-in potential, V_{bi}, of the p-n junction may be neglected for reasonably large V. As F increases, the exciton absorption peak of the MQW material (λ_{ex}) shifts toward longer wavelengths. Figure 6.3 shows typical low-frequency exciton EA spectra (i.e., the absorption coefficient α as a function of wavelength measured at various voltages) [12]. The α vs. λ curve at $V = 0$ clearly shows a heavy hole exciton absorption peak at λ_{exo}. Here the laser radiation is at λ_{laser}. The laser wavelength λ_{laser} is detuned from λ_{exo} by $\Delta\lambda_{deto}$ toward the long wavelength. The background α at λ larger and smaller than λ_{exo} are caused by (1) the light hole exciton absorption at $\lambda \sim 1250$ nm and (2) the continuing absorption from the bandgap. The data shown in Fig. 6.3 was obtained on an InAsP/GaInP MQW structure grown epitaxially on an InP substrate.

As F increases, three important changes of the absorption curve take place. (a) The exciton peak shifts toward a longer wavelength. This is known as the red shift of the QCSE. The detuning of the laser wavelength from the exciton peak is now $\Delta\lambda_{det}$ (i.e., $\lambda_{laser} - \lambda_{ex}$), $\Delta\lambda_{det}$ is smaller than $\Delta\lambda_{deto}$. (b) The absorption at the exciton peak decreases, i.e., $\alpha_m < \alpha_{mo}$. (c) The linewidth of the exciton absorption broadens, i.e., $\delta_{ex} > \delta_{exo}$. The net result is that the absorption at λ_{laser}

increases when $\Delta\lambda_{\mathrm{det}} > 0$. When $\Delta\lambda_{\mathrm{det}} < 0$, then the absorption may decrease as depicted in Fig. 6.3 for $V > 1.5$ V. The photon energy E is related to its wavelength by $E = h\nu = hc/\lambda$. The detuning wavelength $\Delta\lambda_{\mathrm{det}}$ of the laser is related to the detuning energy, $\Delta\mathrm{det}$, of the laser photon energy from the photon energy at the exciton peak by $\Delta\mathrm{det} = -(hc/\lambda^2)\Delta\lambda_{\mathrm{det}} = hc(1/\lambda_{\mathrm{ex}} - 1/\lambda_{\mathrm{laser}})$.

6.1.1.2 The EA MQW waveguide modulator

Figure 6.2 shows a typical waveguide structure for EA modulators. The WG mode is propagating in the z-direction. The MQW region is the only undoped region. Other semiconductor layers are either p or n doped. $V|_{\mathrm{bias}}$ is the applied DC bias voltage. A DC bias electric field $F_{\mathrm{C}}|_{\mathrm{bias}}$ and a superposed RF electric field $F_{\mathrm{C,RF}}$ (due to the RF voltage $V_{\mathrm{C,RF}}$) are applied to the MQW layers in the x-direction with $F_{\mathrm{C}}|_{\mathrm{bias}} \cong V_{\mathrm{C}}|_{\mathrm{bias}}/d_i = F|_{\mathrm{bias}} \cong V|_{\mathrm{bias}}/d_i$. Here, the subscript C designates the quantity across the i region. V_{RF} is the voltage at the RF source output. V_{RF} and $V_{\mathrm{C,RF}}$ are determined from circuit analysis. In most references, $F_{\mathrm{C,RF}} \cong V_{\mathrm{C,RF}}/d_i$. However, in a more precise analysis, $F_{\mathrm{C,RF}}$ is just proportional to $V_{\mathrm{C,RF}}$, i.e., $V_{\mathrm{C,RF}}/d_{i,\mathrm{eff}}$. The proportionality constant, $d_{i,\mathrm{eff}}$, is obtained from the microwave analysis of the electric field $F_{\mathrm{C,RF}}$ produced by $V_{\mathrm{C,RF}}$ on the electrodes. $d_{i,\mathrm{eff}}$ is a function of x and y. At low frequencies, $d_{i,\mathrm{eff}} \cong d_i = $ constant. Note that, in a more general case not discussed in this chapter, the electrically intrinsic region may be thicker than the MQW layer.

The intensity of a forward propagating WG mode can be described as

$$I(t) = I_0 e^{-\alpha_{\mathrm{eff}} z}, \quad \text{where}$$
$$\alpha_{\mathrm{eff}}(F, \lambda) = \alpha_0 + [\Gamma\alpha|_{\mathrm{bias}}(F|_{\mathrm{bias}}, \lambda)] + \Gamma\Delta\alpha(F_{\mathrm{C,RF}}\cos\omega t, \lambda). \qquad (6.1)$$

Here, α_{eff} is the effective intensity attenuation constant of the WG mode and I_0 is the power of the WG mode at $z = 0$. The attenuation coefficient of the optical WG mode, α_0, at $V_{\mathrm{bias}} = V_{\mathrm{RF}} = 0$ is affected by all loss mechanisms such as residual absorption, scattering and diffraction, and $\omega = 2\pi f$ is the RF angular frequency. From the physical point of view, there is absorption only in the quantum well layers. The barrier layers are transparent because they have a larger bandgap compared to the MQWs. In this chapter and in the literature, "$\alpha|_{\mathrm{bias}}$" represents the increase of the averaged absorption coefficient (averaging the total α of all layers over the thickness of the MQW in the i region) due to the bias electric field $F|_{\mathrm{bias}}$. This bias dependent absorption coefficient, $\alpha|_{\mathrm{bias}}$, is zero when $V|_{\mathrm{bias}}$ is zero. "$\Delta\alpha(t)$" is the averaged instantaneous change of α due to the RF electric field ($F_{\mathrm{C,RF}}\cos\omega t$). It has a flat frequency response from DC to millimeter wave frequencies, and it appears to follow instantaneously the RF electric field [13].

The optical filling factor, Γ, is the fraction of the power of the WG mode in the intrinsic MQW layers,

$$\Gamma = \frac{\displaystyle\iint_{\text{MQW}} |E_{\text{WG}}|^2 dx\, dy}{\displaystyle\iint_{\text{WG}} |E_{\text{WG}}|^2 dx\, dy} = \frac{d_i}{d_{\text{eff}}}, \tag{6.2}$$

E_{WG} is the optical electric field of the WG mode; Γ is independent of z. For the MQW structure shown in Fig. 6.2, d_{eff} is approximately the optical WG mode size in the x-direction.

In summary, when α_{eff} is independent of z, the transmission of the optical radiation after propagating a distance L along z, including the coupling efficiency η_{ff} to fiber and laser, at both $z = 0$ and $z = L$, is

$$I(z = L)/I_0 = T(V_{\text{RF}} \cos \omega t, \lambda) = \eta_{\text{ff}} e^{-\alpha_o L} \eta = \eta_{\text{ins}} \eta,$$

$$\eta = e^{-\Gamma \alpha(F|_{\text{bias}}, \lambda)L} \cdot e^{-\Gamma \Delta \alpha(F_{\text{C,RF}} \cos \omega t)L}. \tag{6.3}$$

When the propagation time of the optical wave is comparable to the period of the microwave signal, α_{eff} will be a function of z. An equivalent expression of Eq. (6.3) is given in Section 6.5 when propagation time of microwave and optical waves must be taken into consideration.

Here, because of the $\alpha|_{\text{bias}}$, we have neglected the reflection effects inside the WG at both ends of the modulator. The optical insertion efficiency η_{ins} of the fiber-coupled modulator is $\eta_{\text{ff}} e^{-\alpha_o L}$. It is the ratio of the output to the input optical power in the fiber at $V = 0$. The insertion loss is $10 \log_{10} \eta_{\text{ins}}$. The insertion efficiency (or the insertion loss) is independent of F_{bias} and $F_{\text{C,RF}}$.

In order to maximize $\Delta \alpha$ for a given $F_{\text{C,RF}}$, the design of MQW structures requires detailed considerations of QW heterojunctions and their fabrication. These are material design and epitaxial growth issues. In addition, in order to maximize dT/dV_{RF} for the RF signal, there must be additional considerations of device parameters (such as the index and the thickness of the waveguide core, L and waveguide width) and biasing condition that will maximize $F_{\text{C,RF}}$ and $d\eta/dF_{\text{C,RF}}$. Of particular significance are the choices made that determine η_{ins}, $\alpha|_{\text{bias}}$, Γ, $d_{\text{i,eff}}$ and L. For example, matching of the WG and the fiber modes is very important for obtaining high η_{ff}. Good design and fabrication of the electrodes, the p-i-n structure and the contacts are important for obtaining small $d_{\text{i,eff}}$, low series resistance and low spurious capacitance. In order to obtain a large $V_{\text{C,RF}}$ within a given bandwidth for a given input RF power, good design of both the driving and the modulator circuit (in addition to device parameters) is required. Section 6.2 will address the device design issues. Section 6.3 will address the material issues.

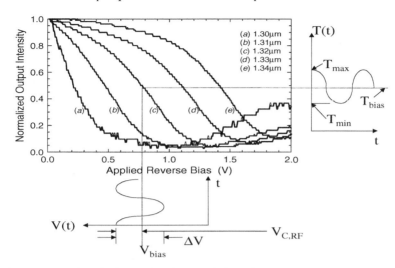

Figure 6.4. Measured optical transfer characteristics of an InAsP/GaInP MQW EA modulator in the TE polarization. The modulator is 3 μm wide and 185 μm long. The insets show the instantaneous (1) input voltage and (2) $I/I_o = T(t)$.

Figure 6.4 shows the normalized measured transmission at DC, i.e., $\eta = T(\frac{V}{d_i}, \lambda)/T(0, \lambda)$, as a function of the applied V at several wavelengths for a typical modulator [14]. Clearly η is not a linear function of V. The total V as a function of time, $V = V_{\mathrm{bias}} + V_{\mathrm{RF}} \cos \omega t$, is shown in the insert below the V axis. T at $\lambda_1 = 1.32$ μm will have a time response shown in the insert on the right side of Fig. 6.4.

In Fig. 6.4, near the bias point and for the $V_{\mathrm{RF}} \cos \omega t$ component,

$$I_o \Delta T = I_o T(0, \lambda) \Delta \eta = I_o \eta_{\mathrm{ff}} e^{-\alpha_o L} \Delta \eta = I_o \eta_{\mathrm{ins}} \cdot \left(\frac{\partial \eta}{\partial V} \right) \bigg|_{\mathrm{bias}} \Delta V \qquad (6.4)$$

in a first order approximation. The slope efficiency is $\Delta T / \Delta V$ (i.e., $\eta_{\mathrm{ins}} \cdot \frac{\Delta \eta}{\Delta V}|_{\mathrm{bias}}$). In terms of Eq. (6.4), $\Delta V = 2V_{\mathrm{RF}}$. On the other hand, the curvature of η with respect to V will also generate an output with frequency components at $n\omega t$, i.e., harmonic distortions. The larger the ΔV, the larger the distortion.

Within a wide range of frequency within which the transit time for the optical wave to propagate through the length of the modulator is much less than the period of the microwave signal, $\Delta \alpha$ in Eq. (6.3) follows instantaneously $V(t)$. In that case, the η and the harmonic distortions are independent of ω. However, $V_{\mathrm{C,RF}}$ is frequency dependent. At frequencies ω so high that the transit time for the optical wave to propagate through the length of the modulator becomes comparable to the period of the microwave signal, the η and the distortions will be frequency dependent, as they are discussed in Section 6.5.

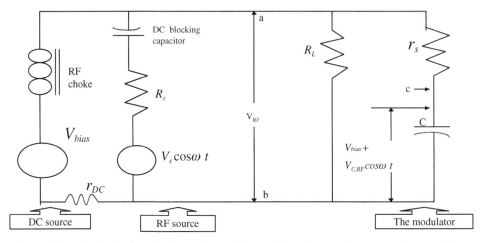

Figure 6.5. The basic circuit of an EA modulator driven by an RF source. The modulator is represented by its capacitance C in series with a resistance r_s. The RF source consists of a voltage source V_s in series with an internal resistance R_s. In order to maximise the RF power transfer to the modulator, R_L is a load resistance placed in parallel with the modulator to match R_s.

6.1.1.3 Electrical considerations of the modulator driving circuit

Electrically, $F_{C,RF}$ is the RF field in the MQW. The total voltage across the i region is $V_C = V|_{bias} + V_{C,RF} \cos \omega t$. Within the frequency range where $\Delta\alpha L$ is independent of ω $(\partial\eta/\partial V)|_{bias}$ is also approximately independent of ω because $d_{i,eff}$ depends only slowly on frequency. Therefore ΔT is proportional to $V_{C,RF}$. The low pass bandwidth of the modulator is the range of RF frequencies from 0 to ω where the $F_{C,RF}$ drops to $1/\sqrt{2}$ of its peak low frequency value. We will show in Section 6.1.3 that such a reduction of $F_{C,RF}$ corresponds to a 3 dB drop in RF link efficiency at ω.

Figure 6.5 shows a commonly used simple input circuit of a p-i-n diode modulator (i.e., r_s in series with C) driven by an RF voltage, V_s [10,11,14]. The equivalent circuit representation of the modulator is accurate when there is only a moderate amount of photocurrent generated by absorption. The internal impedance of the RF source is R_s. The load resistor R_L in parallel with the modulator is used to obtain a better match to the RF source and a broader RC bandwidth. For such a circuit and for $R_L = R_s$,

$$\left|\frac{V_{C,RF}}{V_s}\right| = \frac{1/\omega C}{\sqrt{(R_s + 2r_s)^2 + (2/\omega C)^2}}. \tag{6.5}$$

The 3 dB RF bandwidth is $f_{3dB} = 1/\pi[(R_s + 2r_s)C]$.

Clearly, for large I_0, a modification of the equivalent circuit of the modulator is required in order to include the induced photocurrent [15]. There are many different driving circuits which can be designed to improve the $V_{C,RF}$ within a given bandwidth [2]. For example, integrated impedance transformers have been investigated for improving this ratio at high center frequency over a narrow bandwidth [16,17]. The important point is that $V_{C,RF}$ and the bandwidth are determined from the properties of the entire driving circuit.

6.1.2 Exciton absorption and the quantum confined Stark effect (QCSE)

A double heterostructure for MQW modulators consists of a semiconductor layer with a smaller bandgap, E_Γ, called the well, sandwiched between semiconductor layers with a larger bandgap, E_g, called the barriers. The bandgap discontinuity between the well and the barrier, $E_g - E_\Gamma$, is the sum of the discontinuities of the conduction bands, ΔE_C, and of the valence bands, ΔE_v. The thickness of the well, L_w, is typically 50 to 150 Å thick. The barrier is typically just thick enough (e.g. 50 to 100 Å) to isolate the wells. Figure 6.6a shows a typical one-dimensional potential diagram of the conduction and the valence bands as a function of x at $F = 0$. E_e and E_h are discrete energy levels in the potential energy diagram in the x-direction shown in Fig. 6.6a [6,7,18,19]. E_{el} is the lowest electron energy level, i.e., the ground state, in the conduction band, while E_{hh1} is the highest hole energy level in the valence band. The subscript hh stands for heavy holes, and 1 stands for first order. Other energy levels such as light holes, E_{lh}, in the valence band and excited states in the conduction band are also present. These are not shown in Fig. 6.6.

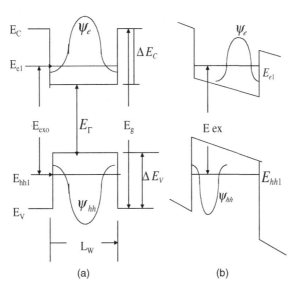

Figure 6.6. A schematic diagram of the envelope functions, the potential diagram and the energy levels of electrons and holes. (a) Zero field. (b) Non-zero field.

The total energy of electrons and holes is the sum of their energy in the x-direction, i.e., $E_e + E_h$, and the energy of an electron–hole pair in the y–z plane, E_{yz}. The quantum mechanical wave functions, ψ_{hh} and ψ_e, for the E_{hh1} and E_{e1} levels shown in Fig. 6.6a, exhibit one-dimensional (x-direction) confinement of electrons and holes.

In order to understand the energy of the electron–hole pair in the y–z plane, let us consider first the energy of electron–hole pair in three dimensions in bulk semiconductors. If electron–hole pairs are created by absorption of light, then they are initially closely spaced. In bulk semiconductors, such electron and hole pairs will experience mutual three-dimensional Coulomb forces similar to those present in a hydrogen atom. The energy of such an electron–hole pair is lower than the energy level of separate electrons and holes; this electron–hole coordination gives rise to a set of energy levels (called exciton levels) located just below the bottom of the conduction band edge. The exciton spectra of bulk materials are observable only at very low temperatures. The situation is different for MQWs.

In the y- and z-directions of MQW, electrons and holes are also subject to the conventional periodic potential of the crystal. The quantum confinement in the x-direction increases the binding energy of the electron–hole pair, i.e., "the exciton". In the limit of a two-dimensional exciton, the binding energy, E_{yz}, is four times the binding energy of the three-dimensional excitons in the bulk material. The binding energy of excitons in a MQW is typically less than 15 meV. It is much smaller than $E_{e1} - E_{hh1}$. Therefore, the induced transition of excitons takes place at a photon energy given approximately by $E_{exo} \cong E_{e1} - E_{hh1}$ shown in Fig. 6.6. Nevertheless, because of the larger binding energy, the induced exciton absorption is observable at room temperature, even under a large F, as shown in Fig. 6.3. Note that there is a linewidth, δ_{ex}, of the exciton absorption line. The basic mechanism of broadening the exciton absorption line, at room temperature, is the LO–phonon interaction. The exciton linewidth is further broadened by lattice imperfections and by impurities that might be present in the epitaxially grown layers. A large exciton linewidth at $F = 0$ (especially a large linewidth found between 4 and 10 K where there is little phonon interaction) is usually an indication of inferior quality of the MQW due to lattice defects and/or a high ionized impurity concentration.

Under the application of an electric field in the x-direction, the potential wells are tilted as shown in Fig. 6.6(b). The quantum mechanical solution for the energy values in the quantum well indicates a reduction in E_{exo}. Therefore the exciton absorption line shifts toward longer wavelengths (i.e., smaller photon energies). This is the quantum confined Stark effect (QCSE) [6,7,18,19]. The amplitude of this exciton absorption depends on the magnitude of the matrix element of the electric dipole, connecting the wavefunctions ψ_e and the ψ_{hh}. Note in Fig. 6.6(b) that the wave functions for E_{e1} and E_{hh1} shift in opposite directions, decreasing the dipole matrix element and the peak α_m. There is also a broadening of δ_{ex} as F is increased. The effects of linewidth broadening and reduction of α_m are evident in Fig. 6.3.

It is well known that there are optical TE and TM modes in the waveguide. TE modes have the dominant electric field polarized in the y-direction. TM modes have the dominant electric field polarized in the x-direction. The polarization direction of the optical electric field affects the magnitude of the dipole matrix of the ψ_e to ψ_{hh} transition. Therefore it is to be expected that the performance of EA MQW modulators will be polarization dependent. The polarization dependence of α is troublesome for RF photonic links whenever the polarization in the input fiber cannot be controlled. Although the laser radiation is linearly polarized and there are polarization maintaining fibers, these fibers are costly. Photonic integrated laser–modulator pairs nicely circumvent this difficulty.

6.1.3 Figures of merit of EA modulators

The performance of any modulator must be evaluated in terms of the RF efficiency and the SFDR that can be achieved in a basic intrinsic link using the modulator.

6.1.3.1 RF efficiency of the link

The effect of slope efficiency of the fundamental signal on the link efficiency is discussed in this section. When distortions are neglected, the slope efficiency of the intensity modulation at a bias voltage V_{bias} is

$$\left.\frac{\Delta T}{\Delta V_{RF}}\right|_{bias} = \frac{V_{C,RF}}{V_{RF}}\left.\frac{\Delta T}{\Delta V_{C,RF}}\right|_{bias} = -\eta_{ins}\frac{V_{C,RF}}{V_{RF}}\left.\frac{\partial\eta}{\partial V_{C,RF}}\right|_{bias} = -\eta_{ins}\left.\frac{\partial\eta}{\partial V_{RF}}\right|_{bias},$$

$$\left.\frac{\partial\eta}{\partial V_{C,RF}}\right|_{bias} = \left\{e^{-\Gamma\alpha(V_{bias}/d_i)L}\left[\Gamma L\left.\frac{\partial(\Delta\alpha)}{\partial V_{C,RF}}\right|_{bias}\right]\right\}. \tag{6.6}$$

The RF efficiency G_{RF} (in dB) for a single fiber link is $10\log_{10}(P_1/P_s)$, where P_s is the RF power input to the modulator and P_1 is the power output from the detector to a load impedance at the frequency ω. According to Eq. (6.6), and assuming passive and lossless impedance matching at the input and at the output [1,2], we obtain

$$\begin{aligned} G_{RF} &= 10\log_{10}P_1 - 10\log_{10}P_s \\ &= 20\log_{10}(\eta_{ins}P_{opt}(\partial\eta/\partial V_{RF})|_{bias}) \\ &\quad + 20\log_{10}(T_{M-D}) + 20\log_{10}(\mathfrak{R}) + 10\log_{10}(R_s R_d) \end{aligned} \tag{6.7}$$

Here, $\mathfrak{R}$ is the responsivity of the detector in A/W. R_s is the source impedance and R_d is the input load impedance of the circuit following the detector. T_{M-D} is the fraction of the optical power emerging from the modulator that will be received by the detector. Equation (6.7) shows clearly that the effect of the modulator on the link efficiency is given by the term,

$$\eta_{\text{ins}} P_{\text{opt}} (\partial \eta / \partial V_{\text{RF}})|_{\text{bias}}.$$

In order to compare EA modulators with LiNbO$_3$ modulators (chapters 1 and 4), note that, for V applied to the electrodes, the $T(V)$ in LiNbO$_3$ Mach–Zehnder modulators has a sinusoidal variation,

$$T_{\text{LiNbO}_3}(V) = \eta_{\text{ins}}^{\text{Li}} \eta^{\text{Li}} = \frac{\eta_{\text{ins}}^{\text{Li}}}{2} \left[1 + \cos\left(\frac{\pi V}{V_\pi}\right) \right].$$

Thus V_π, the voltage swing from maximum T to minimum T, determines $\Delta T / \Delta V$. Normally, in order to achieve the maximum slope efficiency and in order to eliminate even order distortions, the V is biased at $1/2\, V_\pi$. We obtain

$$\left. \frac{\Delta T_{\text{LiNbO}_3}}{\Delta V} \right|_{\text{max}} = -\eta_{\text{ins}}^{\text{Li}} \frac{\partial \eta^{\text{Li}}}{\partial V}, \quad \text{and} \quad V_\pi = -\frac{\pi}{2(\partial \eta^{\text{Li}} / \partial V)}. \tag{6.8}$$

Here, the $\eta_{\text{ins}}^{\text{Li}}$ is the η_{ins} (or T_{FF} in Chapter 1) of LiNbO$_3$ modulators. V_π has been used extensively in the literature to characterize LiNbO$_3$ modulators. The $\partial \eta / \partial V$ of the two types of modulators may be compared more conveniently if an equivalent $V_{\pi,\text{eq}}$ is also used to express the slope efficiency of EA modulators, where

$$V_{\pi,\text{eq}} = \frac{\pi}{2 \left| \dfrac{\partial \eta}{\partial V_{\text{RF}}} \right|_{\text{bias}}}. \tag{6.9}$$

Note that in the traditional comparison of dT/dV, the difference between η_{ins} and $\eta_{\text{ins}}^{\text{Li}}$, has been ignored. Note that the first term in Eq. (6.7) is $20 \log_{10}(\frac{\pi \eta_{\text{ins}} P_{\text{opt}}}{2 V_{\pi,\text{eq}}})$. Thus P_{opt} and η_{ins} are just as significant as V_π. One may always get the same efficiency for modulators that have a larger V_π, provided the ratio of $\eta_{\text{ins}} \cdot P_{\text{opt}}/V_{\pi,\text{eq}}$ is kept constant. Eventually, P_{opt} in an intrinsic link will be limited either by the available laser power or by the saturation power of the modulator or the detector.

6.1.3.2 Nonlinear distortions in EA modulators

Since $F_{\text{C,RF}}$ is equal to $V_{\text{C,RF}}/d_{\text{i,eff}}$, $\Delta\alpha$ is considered a function of $V_{\text{C,RF}} \cos \omega t$, or $\sum_j V_{\text{C,RF}}^{(j)} \cos \omega_j t$ for multiple inputs at different RF frequencies. In short, $V = V|_{\text{bias}} + \sum_j V_{\text{C,RF}}^{(j)} \cos \omega_j t$. $V_{\text{C,RF}}^{(j)}$ is typically much smaller than the bias voltage $V|_{\text{bias}}$. Therefore, the effect of the RF voltages on $\Delta\alpha$ can be expressed as a Taylor series expansion about $T(V|_{\text{bias}})$. Or, considering $\Delta\alpha$ as an implicit function of V, $T(t)$ for a single frequency input is

$$
T(t) \cong \eta_{\text{ins}} e^{-\Gamma \alpha \left(\frac{V|_{\text{bias}}}{d_{i,\text{eff}}}, \lambda \right) L} \left\{ 1 - \left[\frac{\partial (e^{-\Gamma \Delta \alpha L})}{\partial V} \bigg|_{\text{bias}} \right] V_{\text{C,RF}} \cos \omega t \right.
$$

$$
+ \left[\frac{1}{2} \frac{\partial^2 (e^{-\Gamma \Delta \alpha L})}{\partial V^2} \bigg|_{\text{bias}} \right] (V_{\text{C,RF}})^2 \cos^2 \omega t
$$

$$
\left. - \left[\frac{1}{6} \frac{\partial^3 (e^{-\Gamma \Delta \alpha L})}{\partial V^3} \bigg|_{\text{bias}} \right] (V_{\text{C,RF}})^3 \cos^3 \omega t + \cdots \right\}. \tag{6.10}
$$

Note that $\cos^2 \omega t = \frac{1}{2}(\cos 2\omega t + 1)$ and $\cos^3 \omega t = \frac{1}{4}(\cos 3\omega t + 3 \cos \omega t)$. Thus $I(t)$, i.e., $I_o T$, consists of frequencies such as $\omega, 2\omega, 3\omega \ldots$ Neglecting higher order terms with $n > 1$, we obtain Eq. (6.4). If higher order terms with $n > 3$ may be neglected, then the power of $I(t)$ at ω, 2ω and 3ω can be obtained just from the three terms in Eq. (6.10). For example, the sum of the terms containing $\cos \omega t$ times I_o constitutes the optical power modulated at the signal frequency ω. The 3rd harmonic optical power, HM3, is $\{I_o \eta_{\text{ins}} e^{-\Gamma \alpha_{\text{bias}} L} [(1/24) \cdot (\partial^3 (e^{-\Gamma \Delta \alpha L})/\partial V^3)|_{\text{bias}} V_{\text{C,RF}}^3]\}$. If the nth order terms ($n > 3$) can not be neglected, then T will contain $\cos(n\omega t)$. The higher order terms will also contribute more terms that contain $\cos(m\omega t)$, $m \leq n$. The sum of all terms containing $\cos(m\omega t)$ times I_o gives the optical power of the mth order harmonics, designated as HMm. Harmonics at $m > 1$ represent nonlinear distortions.

For many applications, the modulator input consists of RF signals of slightly different ω. In the two equal tone case, $V = V_C|_{\text{bias}} + V_{\text{C,RF}}(\cos \omega_1 t + \cos \omega_2 t)$, the Taylor expansion in Eq. (6.10) is the same, replacing $\cos^n \omega t$ by $(\cos \omega_1 t + \cos \omega_2 t)^n$. There will now be terms with mixed frequency at $\omega_1 - \omega_2$, $\omega_1 + \omega_2$, $2\omega_1, 2\omega_2, 2\omega_1 - \omega_2, 2\omega_2 - \omega_1$, and other higher order mixed frequency terms. The terms at frequencies, $\omega_1 + \omega_2$, $2\omega_1$, $2\omega_2$, $2\omega_1 + \omega_2$, etc., can usually be filtered out electrically in the receiver. However, the third order intermodulation terms at $2\omega_1 - \omega_2$ and $2\omega_2 - \omega_1$, from $(\cos \omega_1 t + \cos \omega_2 t)^3$ and other higher odd order terms, will be too close to ω_1 and ω_2 to be filtered out. Therefore the product of I_o and the third order intermodulation terms, designated as IM3, constitute the optical power of the major intermodulation distortion. When $n > 3$ orders are negligible, IM3/HM3 $= 3$.

6.1.3.3 SFDR of the link

How the SFDR is determined by the slope efficiency, the noise and the nonlinear distortions will be discussed in this section. At the detector output, the RF signal power is P_1. Let the noise power be N that is independent of P_1. Let the RF power of the dominant distortion, HMn (or IMn), be P_n. The SFDR of the link is the range

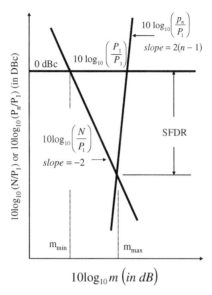

Figure 6.7. The graphical solution for SFDR.

of P_1 that varies from the minimum at $P_1 = N$ to the maximum P_1, where, at that signal level, $P_n = N$. If the detector is also nonlinear, HMn (or IMn) is given by the product of T and the detector response.

Let us consider a graphical solution of SFDR. With an ideal detector, HMn (or IMn) comes only from the modulator. From Eqs. (6.4) and (6.10), P_1 is proportional to $\{I_0[d(e^{-\Gamma \Delta \alpha L})/dV]|_{\text{bias}} V_{\text{C,RF}}\}^2$, and the nth order distortion, P_n, is proportional to $\{I_0[d^n(e^{-\Gamma \Delta \alpha L})/dV^n]|_{\text{bias}}\} V_{\text{C,RF}}^{2n}$. It is customary to define an optical modulation index m, $m = $ (optical power at ω/optical carrier power). In Fig. 6.7, we show a plot of $10 \log_{10}(P_n/P_1)$ and $10 \log_{10}(N/P_1)$ in units of dB$_C$ on the vertical axis as a function of $10 \log_{10} m$ on the horizontal axis. Since $\log a \cdot b = \log a + \log b$ on logarithmic scale, then

$$10 \log_{10} m = \text{constant}_m + 10 \log_{10} V_{\text{C,RF}},$$
$$10 \log_{10}(N/P_1) = \text{constant}_{pl} - 20 \log_{10} V_{\text{C,RF}}$$
$$= \text{constant}_{pl} + 2\text{constant}_m - 2 \cdot (10 \log_{10} m),$$
$$10 \log_{10}(P_n/P_1) = \text{constant}_{pn} + 2(n-1) \cdot (10 \log_{10} m). \tag{6.11}$$

Here, constant_m, constant_{p1} and constant_{pn} include all the terms not dependent on m.

The SPDR in dB$_C$ is given by the change in dB$_C$ from (a) the intersection of the $10 \log_{10}(N/P_1)$ line with the $10 \log_{10}(P_1/P_1)$ line (which is zero dB$_C$) and (b) the intersection of the $10 \log_{10}(N/P_1)$ line with the $10 \log_{10}(P_n/p_1)$ line. The noise floor is determined from $N = NF \cdot g N_{in}$. The noise figure NF of the link is discussed in Chapter 1. Here N_{in} is the noise power at the input, and g is the linear gain equivalent to G_{RF} in Eq. (6.7). The modulator does not add any additional noise. The graphical solution also shows the range of the modulation index for realizing the SFDR. For a specific dominant order of nonlinearity, the effect on SFDR due to any change in the RF link can be calculated. For example, the change of SFDR due to a change of required RF bandwidth Δf (or optical carrier power I_o) can be obtained by shifting the vertical positions of the lines in Fig. 6.7 (i.e., the various constants in Eq. (6.11)), without changing their slopes.

It is interesting to note that the slope of the experimentally measured $10 \log_{10}$ (spurious power/P_1) line will indicate the dominant order of nonlinearity. If the dominant nonlinear distortion comes from more than one order, then the $10 \log_{10}$ (spurious power/P_1) plot will be a curve.

6.2 Analysis and design of p-i-n modulators

It is clear that for a given circuit configuration, the 3 dB low pass bandwidth $f_{3\,dB}$ (from DC to ω) of a p-i-n modulator is affected primarily by its capacitance C. Before a modulator is designed, one picks a value of C required for the application. In order to get a capacitance smaller than or equal to the required C for a specific bandwidth, the condition derived from the parallel plate approximation of the capacitance is

$$L \le (C/\varepsilon W) d_i. \tag{6.12}$$

For a given QW structure and at a given bias voltage, $d\eta/dV_C$ in Eqs. (6.3) and (6.4) can be optimized with respect to both d_i and L, subject to the constraint in Eq. (6.12). When $d_{i,eff} \cong d_i$, we obtain the following optimum $\Delta T/\Delta V$ [14,19].

$$\left.\frac{\Delta T}{\Delta V}\right|_{opt} = \eta_{ins} \frac{V_{C,RF}}{V_{RF}} \left.\frac{d\eta}{dV_{C,RF}}\right|_{opt}, \quad \left.\frac{d\eta}{dV_{C,RF}}\right|_{opt} = \left\{\sqrt{\frac{C}{\varepsilon W d_{eff}}}\right\} \frac{e^{-0.5}}{\sqrt{2}} Q_m,$$

$$Q_m = \left(\frac{\partial\alpha/\partial F}{\sqrt{\alpha}}\right)\bigg|_{bias}. \tag{6.13}$$

Here, $d\eta/dV_{C,RF}|_{opt}$ is evaluated at specific values of d_i and L, designated as d_i^C and L^C,

$$d_i^C = \sqrt{\frac{\varepsilon W d_{eff}}{2C\alpha|_{bias}}} \quad \text{and} \quad L^C = \sqrt{\frac{C d_{eff}}{2\varepsilon W \alpha|_{bias}}}. \tag{6.14}$$

It is interesting to note that $\Delta T / \Delta V$ contains the product of two terms, and they can be optimized separately. The first term, given in the curly bracket, involves the lumped equivalent device parameters, while the second term, Q_m, depends only on the material and the bias. Thus, Q_m is the material figure of merit for the small signal modulation. For efficient modulation, it is very important to get large Q_m. However, it is not unusual for Q_m to vary by a factor of 2 or 3 for the same p-i-n heterostructure grown under different conditions. The precise control of epitaxial growth conditions is very important for obtaining large and reproducible Q_m.

For a given C and Q_m, $\Delta T / \Delta V |_{opt}$ is proportional to $\eta_{ins} / \sqrt{W d_{eff}}$ in Eq. (6.13) η_{ins} is sensitive to mismatch of modes. Large $\Delta T / \Delta V$ is obtained by first designing the d_{eff} and the W of the WG mode to match the fiber mode. For example, passive and doped high index layers surrounding the MQW layer will increase d_{eff} without affecting d_i or C. Finally, the d_i^C and L^C are determined from Eq. (6.14). The optimum d_i and L will depend on $\alpha |_{bias}$ as well as W, d_{eff} and C. Note that the optical filling factor Γ is also affected by d_{eff}.

Any increase of the required 3 dB bandwidth, $f_{3\,dB}$ (i.e., any decrease of C), will demand simultaneously a decrease of L^C and an increase of d_i^C. The net effect is a reduction of $(\Delta T / \Delta V)$, i.e., an increase of $V_{\pi,eq}$, proportional to $\sqrt{C}$. Since the V_π of modulators such as Mach–Zehnder modulators is proportional to L (or C), the MQW EA modulators would have a higher RF efficiency for the same P_{opt} at high RF frequencies than many other types of modulators.

According to Eq. (6.7), an optimized device design requires considerations of the total effect of $d_{i,eff}$, Γ, d_i, and L on $\eta_{ins} \cdot (d\eta/dV)|_{bias} \cdot P_{opt}$. Since the length L is obtained commonly by cleaving the wafer, it cannot be controlled accurately to be L^C. Fortunately, Ref. 9 shows that the actual $d\eta/dV$ is only a slow varying function of L. On the other hand, there may be serious degradation of $\partial\eta/\partial V$ if the optimization requirements are completely ignored. Such an example will be shown in Section 6.5.

Unfortunately, the P_{opt} is often limited by the saturation power of EA modulators. If a large optical power density is applied to the MQW layers, then holes (generated by optical absorption) trapped in the valence band will screen the RF electric field, thereby creating a reduction of $\partial\eta/\partial V$. This is known as the saturation of the slope efficiency. Saturation can be reduced by choosing a semiconductor heterojunction with lower valence band discontinuity (ΔE_v of the well and the barrier) [15,16] and by increasing the V_{bias} to sweep the carriers more effectively out of the MQW layer. However, changes in the material structure and V_{bias} will also affect Q_m. From another point of view, the power density in the quantum well is $P_{opt}\Gamma/W d_i$. The power density in the quantum wells can be reduced by means of a smaller optical filling factor. There is a trade-off between increasing the saturation P_{opt} and decreasing the "$\eta_{ins}(\partial\eta/\partial V)|_{bias}$" as the WG design is changed.

6.3 Growth and characterization of MQW heterostructures

The relationship between Q_m and material characteristics (such as the QCSE shift, the exciton linewidth δ_{ex}, the detuning wavelength $\Delta\lambda_{det}$ and the α_m shown in Fig. 6.3) provides the information for maximizing Q_m.

6.3.1 Selection of material composition

Most MQW modulator structures consist of multiple layers of different III–V compound semiconductors, grown by metal organic chemical vapor deposition (MOCVD) or molecular beam epitaxy (MBE) on GaAs or InP substrates. Figure 6.8 shows the bandgap energy or wavelength versus lattice constant of different binary or ternary alloy compositions [22]. Since the energy of the exciton peak will be close to the bandgap energy, it is important to find the composition of the well material that has a bandgap located either near 0.8 eV (for 1.55 μm applications) or near 0.95 eV (for 1.3 μm applications) In most MQW structures, the lattice constants of all the epitaxial layers (including barriers and buffers) also need to be closely matched to those of the substrates. Otherwise, the lattice mismatch between the layers or between the epitaxial layers and their substrates leads to generation of dislocations. Large numbers of dislocations will significantly degrade the electronic and electro-optical properties of these heterostructures [23,24]. However, below a certain thickness limit, the layers can accommodate the interfacial strain without generating many dislocations. Pseudomorphically strained QW or intermediate buffers between the QW and substrate will allow some compositional deviation from exact lattice matching.

Various compositions of GaAlAs match quite well the GaAs substrate lattice constant. Quantum wells that have the sharpest exciton absorption spectra and large QCSE have been made with GaAs wells and GaAlAs barriers on GaAs substrates. However, the bandgap energy is too high for 1.3 or 1.5 μm lasers. From Fig. 6.8, it is clear that the quaternary alloys, InGaAlAs or GaInAsP, can match the InP substrate lattice constant and have a bandgap near 0.95 eV (1.3 μm). Similar comments can be made for quantum well compositions grown on InP substrates for 1.5 μm applications. In general, the growth of high quality, reproducible, quaternary layers is more difficult than ternary layers [12,19,25]. On InP substrates, strain compensated InAsP/GaInP MQW structures (illustrated in Fig. 6.9) have been grown successfully [12]. For GaAs substrates, ternary compounds such as InGaAs may be chosen so as to have the bandgap around 0.95 eV with a lattice constant slightly mismatched to GaAs [25]. In that case, buffer layers with graded compositions may be grown to accommodate the interfacial lattice mismatch and prevent strain relaxation which can produce both line and point defects [25,26,27,28].

Table 6.1 lists the materials and structures that have been grown in our laboratories and their Q_m (defined in Eq. (6.13)). $Q_m \sim 0.1$ to $0.14\ \mu m^{1/2}V^{-1}$ for most of the

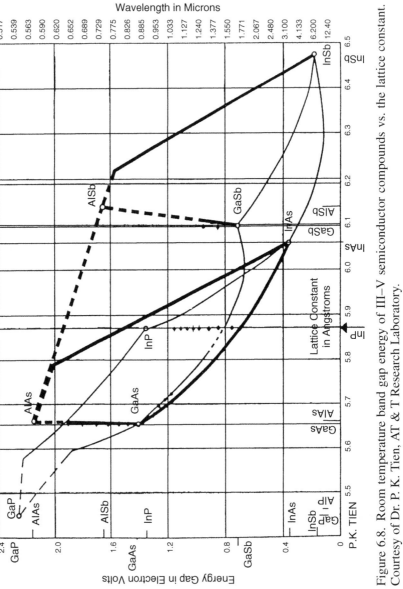

Figure 6.8. Room temperature band gap energy of III–V semiconductor compounds *vs.* the lattice constant. Courtesy of Dr. P. K. Tien, AT & T Research Laboratory.

Table 6.1. *Material figure of merit Q_m for various 1.3 μm MQW electroabsorption modulators*

Material structure	Sample	Number of quantum wells	Well width (Å)	Intrinsic region thickness (μm)	Photon detuning energy (meV)	Material figure of merit ($\mu m^{1/2}V^{-1}$)
Strain-compensated	#1670	11	80	0.20	31	0.14
InAsP/GaInP MQW	#1673	11	95	0.24	26	0.13
on InP substrates	#1459	11	97	0.27	35	0.15
	#1834	6	94	0.13	23	0.12
InGaAlAs/InAlAs	#2366	23	100	0.38	20	0.12
MQW on InP					25	0.11
substrates					30	0.10
					35	0.11
					40	0.10
InGaAs/InAlAs MQW	#3238	10	95	0.20	45	0.14
on GaAs substrates	#3223	20	105	0.42	35	0.13

Modulators have been developed at the University of California, San Diego {14}.

materials shown in the table, to be used at 1.3 μm. Data for Q_m of materials to be used at 1.55 μm is not available.

6.3.2 Materials characterization

The properties of multiple heterostructures depend to a large extent on their alloy composition, on their structural characteristics (line and point defects) and on residual impurities present in their barrier and quantum well layers. All of these parameters are affected by residual strain or strain relaxation introduced during their synthesis or produced during subsequent processing into various device configurations. It is, therefore, of considerable importance to qualify the structural, electronic, optical and electro-optical properties of such materials prior to their fabrication into MQW modulators. A variety of such techniques are available and many of these are listed here:

(1) The key strength of double axis high resolution x-ray diffraction techniques [29,30] is the very high strain sensitivity achievable when appropriate x-ray optics and theoretical simulation are employed. It is rapid, it is non-destructive and samples can be returned to the production process after analysis. High resolution diffraction can provide information concerning the composition and thickness of epilayers, their uniformity and strain relaxation relative to their substrates.

(2) Scanning probe microscopes [31,32] are valuable tools for the characterization of semiconductor surfaces. Scanning tunneling microscopes can provide information on dopant profiling, alloy segregation and on the structural properties of buried interfaces. Atomic force microscopes provide direct surface topographic images including strained

and strain-relaxed layer growth, and scanning capacitance microscopes can be used for determining dopant profiles on a microscopic scale.

(3) Analytical electron microscopy [33,34] is based on the focusing of a fine probe, 1–20 nm in diameter, high energy (100–400 keV) electron beam on a thin foil (20 to 200 nm thick). Inelastic scattering of the beam electrons by core shell electrons creates Auger electronic transitions and characteristic photon emission which can be used for electron energy loss spectroscopic investigations, Auger electron spectrometry and energy dispersive x-ray spectrometry. The advantage of these techniques is that they can detect trace constituents of 10 parts per million with a lateral resolution of 10 nm.

(4) Photoluminescence (PL) [35,36] is a method with widespread applications, particularly for direct bandgap semiconductors. Above bandgap radiation, usually from a laser, is directed on a semiconductor specimen producing electron–hole pairs. After rapidly thermalizing to their lowest energy state these pairs recombine either non-radiatively at defects, surfaces or interfaces or radiatively across the semiconductor bandgap or via intergap energy levels associated with dopants or defects. Scanning PL is routinely used to infer variations in layer composition, quantum well thickness and defect distributions. The nonlinear dependence of the PL intensity of double heterostructures on excitation power can also be used to characterize the junction properties. PL data can also be related quantitatively to p-i-n diode leakage currents. PL line shape analysis is a useful technique for rapid material characterization.

(5) High resolution transmission electron microscopy [37,38] (TEM) can be used for atomic scale characterization of heterostructures, in particular, for investigating their local atomic displacements at and around interfaces. Theoretical simulations can be compared with experimental TEM measurements to infer the presence of line or planar defects extended from the substrate into epilayers during their growth, the presence of microtwins, stacking faults, point defect clusters or stress.

(6) In this test, a p-i-n diode is fabricated on MQW heterostructures grown in the laboratory. It has a ring shape electrode on the top surface to contact the p^+ layer, and a second electrode contacting the n layer outside the ring. The optical radiation can pass through the center of the ring. Such a ring diode can be used first to test its current vs. voltage and capacitance vs. voltage characteristics as a function of frequency. If a tunable optical radiation is transmitted normally through the center window of the ring, then the optical transmission of the same ring diode can be used to obtain the electroabsorption spectra as shown in Fig. 6.3. Such measurements yield the wavelength of the exciton peak, the α_m at the exciton peak, the exciton linewidth δ_{ex}, the residual absorption at $V = 0$ and the QCSE as a function of applied voltage [14,19,25,39]. However, there may be undesirable optical interference patterns due to reflections caused by the dielectric discontinuities between various layers. Photoconductivity, the measurement of the photocurrent of the ring diode as a function of the wavelength of the optical radiation incident on the diode, can provide useful complementary data to those obtained in optical transmission measurements. The effect on the photocurrent due to any optical interference is usually small, and the measurement can also be carried out as a function of temperature down to 4 K.

Because Γ can be calculated reliably and since η_{ins} can be measured separately from the WG modulator at $V = 0$, data obtained from measurements of ring diodes can be used to evaluate $\alpha|_{bias}$, $\partial\alpha/\partial F|_{bias}$, and Q_m.

6.3.3 Q_m and EA characteristics

Small δ_{ex}, large α_m, large QCSE, low impurity concentration in the updoped region and small residual absorption will yield large Q_m. Note that: (1) δ_{exo}, the linewidth at $V = 0$, depends on the crystalline quality of the MQW down to its phonon-broadening limit. (2) The composition and the thickness of the well determine essentially the peak wavelength of exciton absorption, λ_{exo}. For materials listed in Table 6.1, an effective Δdet_o is 25–40 meV from $h\nu_{laser}$. (3) For quantum wells illustrated in Fig. 6.6, QCSE increases with an increase in well width, L_w. QCSE can be calculated with reasonable accuracy from known ΔE_C, ΔE_V and L_W [19,39,40,41]. However, there will also be a field induced broadening of δ_{ex} [42,43] and reduction of α_m at larger fields, limiting the L_W that can be used effectively [19]. Empirically, it is found that the optimum L_W is in the range from 90 to 105 Å, as shown in Table 6.1 [19]. (4) For the same d_i, a decrease of the thickness of the barrier layers, L_b, increases the $\Gamma\alpha$ (i.e., more MQW are packed into the intrinsic region). However, L_b must be sufficiently large to prevent significant overlap of ψ_e and ψ_{hh} from adjacent wells. (5) α_0 is usually controlled by the residual attenuation. Fabrication method such as etching may significantly affect α_0.

In order to predict the performance of a MQW EA modulator, the explicit dependence of α on α_m, δ_{ex}, $\Delta\lambda_{det}$ and F should be known. For $\lambda_1 > \lambda_{exo}$ and yet close to λ_{exo} (with Δdet up to the limit Δdet_u), an empirical approach may be used to represent the exciton absorption spectrum by a Gaussian absorption line with linewidth δ_{ex}, $\alpha(\Delta det) = \alpha_m(F)\exp[-\ln 2(\Delta det/\delta_{ex})^2]$ [19,44,45]. Here Δdet and δ_{ex} are the detuning energy and the linewidth in eV as a function of $F \cdot \alpha_m$ is given by

$$\alpha_m = \sqrt{\frac{\ln 2}{\pi}} \frac{A_{ex0}}{\delta_{ex}(F)} \frac{H(F)}{H(f=0)},$$

$$H(F) = \left|\int \psi_e(F)\psi_{hh}(F)dx\right|^2 \bigg/ \left|\int \psi_e^2(F)dx\right| \cdot \left|\int \psi_{hh}^2(F)dx\right|. \quad (6.15)$$

A_{ex0} is the area of the Gaussian line at $F = 0$. The exciton linewidth is $\delta_{ex}(F) = \delta_{ex0} + \gamma F$. The A_{ex0}, the linewidth δ_{ex0} (at $F = 0$) and γ, are fitted empirically to experimental data [19].

The Gaussian line shape is used only for $\Delta det \leq \Delta det_u$. Following Urbach's model [45,46] of α observed in bulk semiconductors and in GaAs/AlGaAs MQW, an exponential absorption tail may be used to describe the continuing absorption for

Δ det $\geq \Delta$ det$_u$. Here, $\alpha(\lambda) = \alpha_{uo} \exp(-\Delta$ det $/ \Gamma_{uo})$. Γ_{uo} is obtained by fitting the exponential function to the experimental data. The two α functions are matched (i.e., the continuity of $\partial\alpha/\partial E$ and α) at Δ det $= \Delta$ det$_u$, where Δ det$_u = \delta_{ex}^2/(2 \ln 2\Gamma_{uo})$. α_{uo} is the value of α_m in Eq. (6.15) at Δ det$_u$ [19,47]. Experimental verification of this model has been obtained for InGaAlAs/InAlAs quantum wells [19,45]. Representation of the α by analytical functions is necessary in order to predict nonlinear distortions discussed in Section 6.1.3.2.

More specifically, at a given $V_C|_{bias}$, the α_m and the δ_{ex} are slow varying functions of V. In that case, the Gaussian line function is approximated by fixed α_m and δ_{ex} at V_{bias}. The calculation of the nth derivatives in Eq. (6.10) is now an easy chore. For example, it can be shown that, using this approximation, IM3 is zero at a specific V_{bias}. When IM3 $\cong 0$, IM5 becomes the dominant spurious signal. Naturally, the V_{bias} to minimize IM3 may not be the V_{bias} for the best link efficiency.

6.3.4 Growth of the waveguide structure

In addition to MQW layers, WG cladding layers, p and n electric contact layers, and layers with specific doping concentration must also be designed and grown by MBE or MOCVD on the substrate. Most importantly, passive and lossless waveguide layers must be grown in addition to the MQW layer to ensure a high coupling efficiency to the fiber. For photonic integration, InP is the preferred substrate because a 1.3 or 1.55 μm laser structure can be grown on the same substrate. Integrated laser–EA modulator pairs have been demonstrated for telecommunication applications with high coupling efficiency, good polarization control and large efficiency. For opto-electronic integration, GaAs may be the preferred substrate because MIMIC electronic circuits have been developed on GaAs substrates, and InP substrates have lower quality for fabricating electronic circuits. Recently, water-bonding techniques [48] have been developed to avoid hetero-epitaxial growth by removing the MQW and associated layers from the original substrate and then bonding these to a new substrate by a combination of heat and pressure while avoiding, hopefully, most lattice mismatch considerations.

6.4 Fabrication and performance of p-i-n modulators

6.4.1 Fabrication of p-i-n modulators

Details of fabrication procedures will vary considerably depending on the materials used and the processes employed by different laboratories. The fabrication of the InAsP/GaInP MQW EA modulators made at the University of California San Diego are presented here as an example [39,49]. Fabrication of other modulators, using

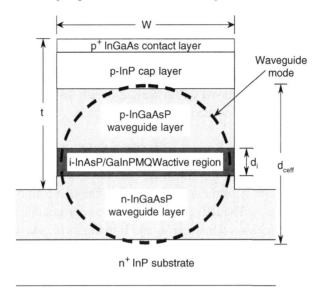

Figure 6.9. Cross-sectional view of the InAsP/GaInP MQW WG modulators. $\lambda_{\text{exciton}} \approx$ 1.27 to 1.28 μm; Δdet $\approx$ 23 to 35 meV at 1.32 μm; QW periods = 6 to 11; $L_{\text{W}} = 95$ Å, $L_{\text{b}} = 120$ Å; $d_{\text{i}} = 0.13$ to 0.25 μm; $\Gamma \approx 6$ to 11%.

InGaAs/InAlAs and InGaAlAs/InAlAs MQW on GaAs substrates, are not presented here [19,25].

Strain compensated structures such as shown in Fig. 6.9 were grown by gas source MBE on (001) oriented S-doped n$^+$ substrates [12,39]. The epitaxial growth temperature was 460 °C, at which the InP growth rate was typically 1 μm/hr. Elemental Ga and In were used as the group III sources, and thermally decomposed AsH$_3$ and PH$_3$ as the group V sources. This growth method allows independent control of the thickness and composition of the InAsP layers. The compressive strain in the InAsP well can be balanced by the tensile strain in the GaInP barriers. Thus a relatively large number of high quality QW periods, e.g., 30, can be grown without exceeding the critical thickness limit for lattice relaxation. The structure for the devices consists of the buffer layer, the n-InGaAsP WG layer, the MQW, the p-InGaAsP WG layer and the p and the p$^+$ contact layers, as shown in Fig. 6.9. The WG layer is designed to match the mode profile of the WG to a lensed or a tapered fiber. The quality of the material structure was first characterized by various methods discussed earlier in Section 6.3, including the measurement of the current–voltage characteristics and the EA spectra of ring diodes.

Fabrication of the modulator started by metalizing an electrode on top of the p$^+$ layer for ohmic contact. Photolithography and CCl$_2$F$_2$ reactive ion beam etching were used to make the waveguide (with the electrode on top) which is approximately 3 to 4 μm wide [49]. The etching depth is approximately 2 μm. Polyimide was spin coated and cured to provide both passivation of the WG as well as low dielectric constant in order to reduce fringing capacitance. For microwave measurements, pads to the top electrode and the bottom n$^+$ layer were made. A coplanar microwave waveguide connects them to the RF source. The substrate was then thinned to $\approx$ 100 μm and cleaved to the desired length. In order to reduce η_{ff}, an antireflection coating was deposited on the cleaved ends.

6.4.2 Measured performance of MQW EA WG p-i-n modulators

Figure 6.10 illustrates a microwave–optical evaluation setup for modulators. The microwave circuit property of the modulator input can be obtained by measuring its S-parameter, S_{11}, at A. The RF efficiency and two-tone nonlinear distortion can be obtained from S_{21} measured between A and B in Fig. 6.10. Many devices of this type have been made and evaluated at UCSD [14]. In all our devices measured at moderate laser power, the equivalent circuit of the modulator consists of a resistance R_s in series with a capacitance as is shown in Fig. 6.5. Recently, it has been shown that another parallel resistance, representing the photocurrent, may have to be added for high P_{opt} [15].

The measured RF efficiency and the *SFDR* of an InAsP/GaInP modulator (with $d_i = 0.14$ μm and $L = 90$ μm) are presented in this section. The measured $d\eta/dV$

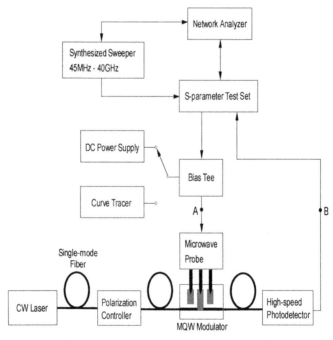

Figure 6.10. Schematic illustration of the high frequency evaluation setup. The S_{21} is measured between reference points A and B.

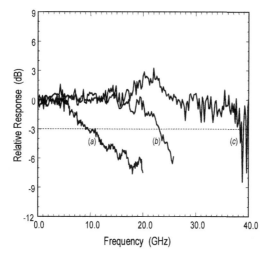

Figure 6.11. Measured small-signal frequency response of InAsP/GaInP MQW EA modulators. The modulators are (a) 370 μm, (b) 180 μm, and (c) 100 μm long with $d_i = 0.27$ μm.

is 1.26 V^{-1}, or $V_{\pi,\text{eq}} = 1.25$ V. Such a low $V_{\pi,\text{eq}}$ is significant, because, according to Eq. (6.7) and assuming $\Re = 0.5$ A/W, we will obtain $G_{\text{RF}} = 0$ dB when $V_{\pi} = 1$ V and $P_{\text{opt}} = 80$ mW. Figure 6.11 shows the relative frequency response of three samples with different L, demonstrating the effect of RC limited bandwidth. Table 6.2 shows the summary of the gain and bandwidth of different p-i-n EA modulators made at UCSD.

For broadband multioctave applications, the dominant spurious signal is the second order distortion. Thus the operating bias is set at $V_b = -0.774$ V to minimize the two-tone IM2 distortion. A RF link gain of -26 dB and noise figure of 27 dB were obtained at 16 mW input optical power. Higher link gain requires larger Q_m, higher η_{ins} or larger optical saturation power. The S_{11} parameter provides the measured value of R_s and C, $R_s = 4\,\Omega$ and $C = 0.23$ pF, consistent with the measured bandwidth of 23 GHz. Figure 6.12 shows the measured $SFDR$ (89 dBHz$^{1/2}$).

For sub-octave applications, the bias is set to just minimize IM3. In that case, RF link gain of -28 dB, $NF = 30$ dB, and $SFDR = 114$ dB Hz$^{4/5}$ were obtained. Table 6.3 shows a comparison of state-of-the-art modulators for 1.3 and 1.5 μm RF photonic link applications. Such $SFDR$ may not be satisfactory for some microwave applications. Larger $SFDR$ can be obtained with larger P_{opt} and with linearization techniques.

6.4.3 Linearization of MQW EA modulators

According to Section 6.1.3.2, a given V_{bias} can be found which will minimize a specific order of nonlinearity with a $\cos n\omega t$ variation. For some applications,

Table 6.2. *Summary of the performance of 1.3 μm lumped element electroabsorption modulators fabricated at University of California, San Diego*

Sample	MQW material	i-region thickness d_i (μm)	Optical insertion loss (dB)	Slope efficiency dT/dV (V^{-1})	Measured RF link gain η_{RF} (dB)	3 dB electrical bandwidth $f_{3\,dB}$ (GHz)	Saturation	Remarks
S3238	InGaAs/InAlAs MQW × 10	0.21	12.5 (uncoated)	0.56	−42	20 (3 μm × 115 μm)	> 17 mW	on GasAs substrate (Lei's structure)
M1425	InAsP/InGaP MQW × 21	0.4923	8.4 (uncoated)	0.28	−38 at 0.58 mA	N/A	observed	on InP substrate (Gerry's structure)
M1459	InAsP/InGaP MQW × 11	0.2699	8.9 (uncoated)	0.55	−32 at 0.86 mA	38 (3 μm × 100 μm)	observed	
M1834	InAsP/InGaP MQW × 6	0.1306	6.3 (AR-coated) Best 4.9	0.87	−26 at 1.0 mA	17 (3 μm × 115 μm)	observed	
M2013	InAsP/InGaP MQW × 8 thinner barrier 74A	0.1378	6.4 (AR-coated) Best 5.9	1.26	−26 at 0.75 mA	23 (3 μm × 90 μm)	N/A	
Rob's	Bulk InGaAsP	0.35	?	?	−30.2	11 ($L = 120$ μm) (without 50 Ω)	> 43 mW	on InP substrate (assume inserting Rob's modulator into KK's link)

Both MQW and Frantz–Keldysh modulators are included, $\eta_{pd} = 0.53$ A/W, $P_{in} = 16$ mW.

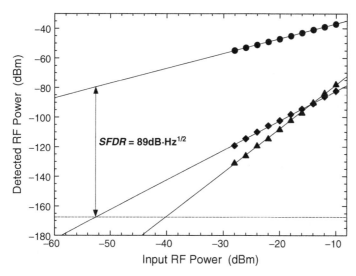

Figure 6.12. Two-tone linearity measurements of the RF photonic link. ● measured funda-mental RF signal; ◆ measured IM2; ▲ measured IM3; — noise floor with 1 Hz bandwidth. The sample is biased to minimize the IM2.

such a method is unsatisfactory. Additional and better methods for linearization are desirable. Figure 6.13 shows an all optical linearization scheme [50] in which the nonlinear distortions of two optical carriers cancel each other. This scheme can be applied to any existing MQW modulators. In order to demonstrate the linearization scheme, a CW Nd:YLF laser is used as the primary source at $\lambda_1 = 1.32$ μm while a tunable semiconductor laser is used as the source at λ_2. A single V_{bias} and input RF signal modulate both optical carriers, at λ_1 and at λ_2. The spectra of the opti-cal light incident on the detector for the CW carrier, the side bands at the signal frequency and the side bands at the third harmonic frequency are illustrated for λ_1 and λ_2. The photodetector recovers two RF signals from the two optically incoher-ent intensity-modulated carriers. The photodetector also functions as a broadband in-phase microwave combiner, coherently adding the detected RF signals. Because of the difference in λ, the inset in Fig. 6.13 shows that the higher order deriva-tives of T will have different magnitudes and signs. Distortion compensation for a particular order (e.g. third) of nonlinearity is achieved by selecting the wavelength and the optical power at λ_2 so that the spurious distortion signal at λ_2 is 180° out of phase with the spurious signal carried by λ_1. Improvement of IM3 *SFDR* by 8 dB has been demonstrated by this method. Similar techniques, such as the split electrode at two bias voltages, will also lead to cancellation of specific orders of distortion.

Table 6.3. *Comparison of state-of-the-art modulators for RF photonic links*

Author	Device	Operating wavelength (μm)	Electrical bandwidth (GHz)	Scaled RF link efficiency $P_{optical} = 10$ mW† (dB)	Scaled suboctave spurious-free dynamic range $P_{optical} = 10$ mW†	Measured suboctave spurious-free dynamic range	Input optical power (mW)
K. K. Loi, W. S. C. Chang *et al.*, UCSD	MQW EA modulator	1.32	23	−27	114 dB Hz$^{4/5}$	114 H dB Hz$^{4/5}$	16
Roger Helkey *et al.*, MIT Lincoln Lab, *IPR'96*, IMH18	Monolithic DFB laser/MQW EA modulator	1.55	Not available	−27	114 dB Hz$^{4/5}$	116 dB Hz$^{4/5}$	14
Shin-Ichi Kancko *et al.*, Mitsubishi Electric Corp., *OFC'98*, ThB4	MQW EA modulator module	1.535	Not available	−36	116 dB Hz$^{4/5}$	123 dB Hz$^{4/5}$	61
Robert Welstand *et al.*, UCSD, *PTL* **8**, 1540 (1996)	Franz–Keldysh InGaAsP EA modulator	1.32	22	−38	117 dB Hz$^{4/5}$	124 dB Hz$^{4/5}$	37
Jan Önnegren *et al.*, *Proc. SPIE* **2560**, 19 (1995)	Franz–Keldysh EA modulator integrated DFB laser	1.55	Not available	−45	97 dB Hz$^{2/3}$	95 dB Hz$^{2/3}$	7
Gary Betts *et al.*, MIT Lincoln Lab, *MTTS'96*	Standard LiNbO$_3$ Mach–Zehnder modulator	1.32	Not available	−41	103 dB Hz$^{2/3}$	108 dB Hz$^{2/3}$	40
Gary Betts *et al.*, MIT Lincoln Lab, *PTL* **8**, 1273 (1996)	Linearized LiNbO$_3$ series Mach–Zehnder modulator	1.32	Not available	−48	106 dB Hz$^{2/3}$	116 dB Hz$^{2/3}$	100

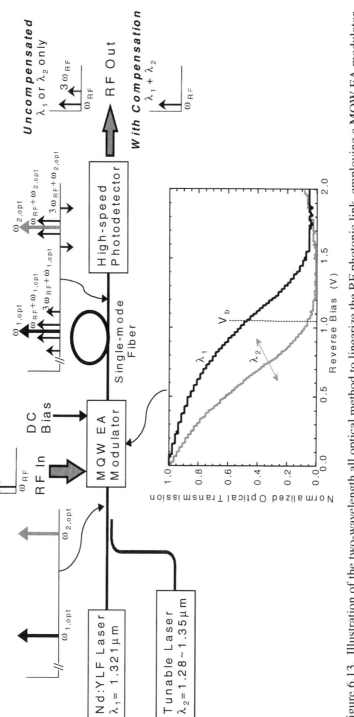

Figure 6.13. Illustration of the two-wavelength all optical method to linearize the RF photonic link, employing a MQW EA modulator.

6.5 Traveling wave EA modulators

The bandwidth of a p-i-n modulator is limited by two factors, (1) the RC time constant in the modulator driving circuit and (2) the transit time of the optical radiation passing through the modulator. In case (1), the $V_{C,RF}/V_s$ at $f_{3\,dB}$ is reduced to $1/\sqrt{2}$ of its peak value at $f = 0$ because of the modulator capacitance C. It is usually the primary bandwidth limitation for p-i-n modulators. Reduction of C will give a larger bandwidth. However, according to Eq. (6.13), there will be a corresponding reduction of $\Delta T/\Delta V|_{opt}$. In order to have a slower decay of $V_{C,RF}$ as a function of frequency, it would be desirable for the modulator to have a constant input impedance such as that obtained in a matched transmission line. In case (2), as the optical wave propagates through the device, $\cos \omega t$ is continuously changing. For sufficiently long propagation distance, $\cos \omega t$ will change sign, canceling the η obtained earlier during the positive cycle. The drop in $T(t)$ due to the time variation of the microwave voltage while the optical carrier propagates through the device limits the modulation frequency. Since the transit time for p-i-n EA modulators ($\sim$100 μm long) is approximately 10^{-12} seconds, the transit time effect is small for $f < 40$ GHz. The transit time effect can be severe at high mm wave frequencies.

   In a traveling wave modulator, the electrodes of the modulator function as a microwave transmission line, in parallel with the optical waveguide. Electrically, this transmission line is terminated at the output port by a matched load. Thus its input impedance will ideally be constant up to very high frequencies. The microwave voltage on the electrode will be the same over very wide bandwidth. The microwave signal in the transmission line and the optical wave in the waveguide propagate approximately in synchronization with each other, thereby reducing the transit time effect. Figure 6.14 illustrates the principles of such a device. In general, since EA modulators are usually less than 1 mm long, the constant input impedance property is much more important to obtaining very wide bandwidth than the transit time reduction property of a traveling wave modulator.

   Let us consider the transit time effect in detail. The microwave electrical field propagating in the $+z$-direction in the i-region can be described as $F^0_{C,RF} e^{-(\alpha_{RF}z/2)}[e^{j(\omega t - \beta_{RF}z)} +$ complex conjugate]. Here α_{RF} is the intensity attenuation coefficient of the microwave, and $\beta_{RF} = \sqrt{\varepsilon_m/\varepsilon_0}\omega/c = n_m\omega/c$. c is the

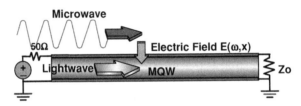

Figure 6.14. Illustration of the operation of a travelling wave modulator.

free space velocity of light. $\varepsilon_m/\varepsilon_o$ is the effective dielectric constant of the microwave mode propagating along the transmission line. At $z = L$, the intensity of the optical wave, propagating with velocity c/n_{eff} from $z = 0$ to $z = L$, is

$$I(z = L)/I_o = \eta_{ins}e^{-\Gamma\alpha_{bias}L}e^{-\Gamma\int_0^L \Delta\alpha(\Delta F)dz} = \eta_{ins}e^{-\Gamma\alpha_{bias}L}e^{-\Gamma\Delta\alpha_{eff}L},$$

$$\text{with} \quad \Delta\alpha_{eff}(\Delta F) = \frac{1}{L}\int_0^L \Delta\alpha(\Delta F)dz, \tag{6.16}$$

$$\text{and} \quad \Delta F = F_{C,RF}^o \cos[\omega t - \delta z]e^{-(\alpha_{RF}/2)z}, \qquad \delta = (n_m - n_{eff})\omega/c. \tag{6.17}$$

From Section 6.1.1.2, we know $F_{C,Rf}^o = V_{C,Rf}^o/d_{i,eff}$. In the first order approximation, the intensity of the optical radiation modulated at $\cos\omega t$ is

$$I_\omega(z = L) = -I_o\left\{\Gamma\eta_{ins}e^{-\Gamma\alpha_{bias}L}\left(\frac{\partial\Delta\alpha_{eff}L}{\partial V_{C,RF}^o}\right)\bigg|_{bias} V_{C,RF}^o \cdot A\cos(\omega t - \phi)\right\}, \tag{6.18}$$

$$A\cos(\omega t - \phi) = \frac{1}{L}\int_0^L \cos(\omega t - \delta z)e^{-(\alpha_{RF}/2)z}dz. \tag{6.19}$$

Equation (6.19) can be evaluated by direct integration. We obtain

$$A^2 = \frac{1/L^2}{[(\alpha_{RF}/2)^2 + \delta^2]}\left\{[1 + e^{-\alpha_{RF}L}] - 2\cos\delta Le^{-\alpha_{RF}L/2}\right\}, \tag{6.20}$$

$$\tan\phi = \frac{\left(-\dfrac{\alpha_{RF}}{2}\sin\delta L - \delta\cos\delta L\right)e^{-\alpha_{RF}L/2} + \delta}{\left(-\dfrac{\alpha_{RF}}{2}\cos\delta L + \delta\sin\delta L\right)e^{-\alpha_{RF}L/2} + \dfrac{\alpha_{RF}}{2}}. \tag{6.21}$$

In summary, the amplitude of I_ω is $-I_o(\Delta T/\Delta V)|_{\omega\approx0}V_{C,RF}\cdot A$. $\Delta T/\Delta V|_{\omega\approx0}$ is given in Eq. (6.10). At low frequencies and at small α_{RF}, $A = 1$ and $\phi = 0$. I_ω/I_o given in Eq. (6.18) agrees with the result expressed in Eq. (6.10). At high frequencies, $A < 1$.

Microwave analysis of the transmission line shows that its electric field pattern is similar to that in a microstrip line, while its magnetic field extends into the buffer layer and the substrate [51,52]. Consequently the microwave signal propagates like a slow wave transmission line with a high microstrip capacitance combined with high inductance. Its microwave phase velocity and impedance are frequency dependent.

Figure 6.15a shows the cross-sectional view of an InAsP/GaInP MQW traveling wave modulator fabricated at UCSD and the calculated RF field pattern of the microwave mode. Within the P-I-N region, there are a 0.3 μm thick undoped MQW

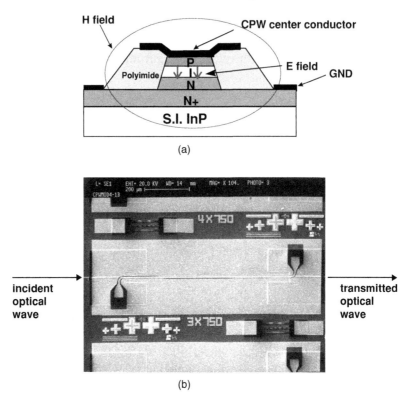

(a)

(b)

Figure 6.15. Illustration and photograph of a travelling wave modulator. (a) Field distribution and the cross-sectional view of the modulator. (b) The top view of the travelling wave modulator with a 3 μm ridge waveguide and a 750 μm long electrode. The contact pads for the microwave electrode are shown on both sides of the waveguide.

layer and two doped InGaAsP passive waveguide layers, 0.6 μm thick. Note that the center conductor is wider than the optical waveguide, so that the RF attenuation of the electrodes can be reduced. However, a wide center conductor will also affect the inductance. Figure 6.15b shows the top view of a traveling wave EA modulator fabricated on this structure. Figure 6.16 shows the measured and simulated performance of such a traveling wave modulator with a 500 μm long electrode. In this figure the vertical axis is the RF link efficiency in dBm. Note the agreement between the simulated and the measured efficiency at frequencies > 5 GHz. The highest frequency is limited by the frequency response of the measuring equipment. Because of the lack of control of the cleavage position, the waveguide length is actually longer than 500 μm in Fig. 6.15b. Because of the high doping of the contact layers, the device behaved like a modulator with an electrode as long as the waveguide at low frequencies, it is longer than 500 μm. This accounts for the difference of the measured and modeled efficiency at frequencies < 5 GHz [52].

Equation (6.18) implies that, at low frequencies, a traveling wave modulator should be as efficient as the p-i-n modulators discussed in Section 6.2 that have

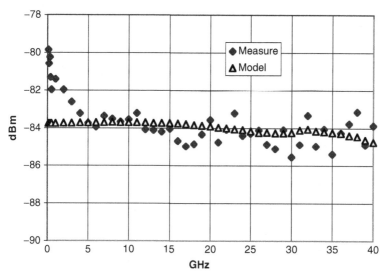

Figure 6.16. Measured and modeled frequency response of a 500 μm long traveling wave modulator.

the same α, $\delta\alpha/\delta F$, Γ, L, d_i, and $d_{i,\text{eff}}$. Since $A \leq 1$, the modulation efficiency of traveling wave modulators also decreases at higher frequencies. However, the efficiency reduction at high frequencies is much slower for traveling wave than for p-i-n modulators.

We learned that $\Delta T/\Delta V|_{\text{opt}}$ is larger for larger L and smaller d_i. It may appear that, in order to take full advantage of the traveling wave structure, one should use as large an L and as small a d_i as possible. However, the characteristic impedance of the microwave transmission line will be very low for very small d_i. It is difficult to match a low impedance to the source. There is a design trade-off between achieving low microwave attenuation and the microwave characteristic impedance that can be matched effectively to the RF source and the need for obtaining the highest modulation efficiency. In considering such design trade-off, it may be necessary to use $d_{i,\text{eff}}$ much larger than the thickness of the MQW layer. A challenge is to design the material structure and the electrode, i.e., the microwave transmission line, so that there will be a high characteristic impedance, low attenuation and high low-frequency RF efficiency. For short EA modulators, the matching of the phase velocity to the optical wave and the low attenuation of the microwave transmission line are not very important.

When the efficiency shown in Fig. 6.16 is compared to the efficiency of devices shown in Table 6.2, the large drop in efficiency ($\sim$40 dB) is caused by: (1) A low η_{ins} of this device ($\sim$20 dB). (2) A thick d_i to achieve higher microwave impedance ($\sim$10 dB). This d_i is much larger than the d_i^C for $L^C = 500$ μm. (3) Impedance mismatch to the RF source without a matching circuit ($\sim$10 dB). Therefore, Fig. 6.16

can only be used as a demonstration of the frequency variation of the traveling wave modulator, not its RF efficiency or SFDR. With a much improved design of the microwave transmission line and η_{ins}, a RF efficiency of -35 dB and 50 GHz bandwidth were achieved in a traveling wave EA modulator using the Franz–Keldysh effect of the bulk semiconductor material [53]. If the same design is used on a MQW EA traveling wave modulator, an RF efficiency higher than -35 dB and a similar bandwidth may be expected.

In conclusion, for lower frequency applications e.g., $f_{3\,dB} < 20$ GHz, it is probably better to use the p-i-n design. For very high frequency and bandwidth, e.g., $f_{3\,dB} > 40$ GHz, it may be better to use the traveling wave design. The effect of transit time, i.e., the mismatch of the microwave and optical phase velocities, is not important for $L \leq 1$ mm and $f < 40$ GHz.

6.6 EA modulation in a resonator

Although research results on resonant EA modulation have not yet been reported EA modulation can clearly enhanced when the MQW material is enclosed within a resonator. $V_{\pi,eq}$ much less than 1 V has been predicted.

In the simplest formulation, consider that the waveguide modulator has two mirrors with intensity reflectivity R at $z = 0$ and $z = L$. This constitues a Fabry–Perot (FP) cavity in the z-direction, containing the MQW WG. Assuming $\eta_{ins} = 1$ and assuming L is tuned to achieve resonance at λ (i.e., maximum T) at $V = 0$, the optical power reflection, $R(V_{C,RF})$, and transmission, $T(V_{C,RF})$, of the FP WG cavity is [47]:

$$\text{For} \quad \theta' = \frac{4\pi \, \Delta n_{eff} L}{\lambda},$$

$$R(V_{C,RF}) = \frac{R + Re^{-2\Gamma\alpha_{bias}L} \, e^{-2\Gamma\Delta\alpha L} - 2Re^{-\Gamma\alpha_{bias}L} \, e^{-\Gamma\Delta\alpha L} \cos\theta'}{1 + R^2 e^{-2\Gamma\alpha_{bias}L} e^{-2\Gamma\Delta\alpha L} - 2Re^{-\Gamma\alpha_{bias}L} \, e^{-\Gamma\Delta\alpha L} \cos\theta'},$$

$$T(V_{C,RF}) = \frac{(1 - R)^2 e^{-\Gamma\alpha_{bias}L} \, e^{-\Gamma\Delta\alpha L}}{1 + R^2 e^{-2\Gamma\alpha_{bias}L} \, e^{-2\Gamma\Delta\alpha L} - 2Re^{-\Gamma\alpha_{bias}L} \, e^{-\Gamma\Delta\alpha L} \cos\theta'}. \qquad (6.22)$$

The power absorbed by the modulator is "$1 - T(V) - R(V)$",

$$1 - T(V) - R(V) = \frac{(1 - R)(1 - e^{-\Gamma\alpha_{bias}L} \, e^{-\Gamma\Delta\alpha L})(1 + Re^{-\Gamma\alpha_{bias}L} \, e^{-\Gamma\Delta\alpha L})}{1 + R^2 e^{-2\Gamma\alpha_{bias}L} \, e^{-2\Gamma\Delta\alpha_L} - 2Re^{-\Gamma\alpha_{bias}L} \, e^{-\Gamma\Delta\alpha L} \cos\theta'}.$$

As the applied voltage is increased, there will also be a change in the n_{eff} (i.e., Δn_{eff}) of the WG mode, in addition to $\Delta\alpha$ of EA. The change of the real part, $n(\lambda)$, of the complex index as a function of applied electric field is called the electro-refraction (ER) effect. Intensity modulation of the optical radiation can be achieved either through the T or the R. The $n(\lambda)$ of MQW is related to its imaginary counterpart $k(\lambda)$, i.e., $\alpha(\lambda)(\lambda/4\pi)$, by the Kramers–Kronig relation. Note that the magnitude and the ratio of Δn and Δk depend on the $\Delta\lambda_{det}$, the exciton line shape, and the α_m.

A number of the properties of resonant modulators can be derived immediately from Eq. (6.22). Several important issues that need to be considered are listed below.

(1) From Eq. (6.13), we know that $V_{\pi,eq}$ is inversely proportional to Q_m. At bias voltages where Q_m is large, there will be a significant amount of α_{bias}. The loaded q-factor of the resonator (at $\Delta\alpha = 0$ and at $\Delta n_{eff} = 0$) will be reduced by α_{bias} no matter how high the unloaded q-factor of the passive resonator is, in the absence of the EA material. For this reason, a very high q-factor of the unloaded resonator does not provide a significant advantage. It is more important to have a resonator (one that can be fabricated easily) with just reasonably high q-factor and low η_{ins}.

(2) The larger the q-factor of the loaded resonator, the more sensitive is the reflection (or transmission) of the resonator with respect to $\Delta\alpha$ and Δn_{eff}. In order to take more advantage of the high q, $\Gamma\alpha_{bias}L$ must be very small. At bias voltage where α_{bias} is significant, one can meet this requirement by using small ΓL. Γ is limited by the thickness of the MQW layer. The thinnest quantum well is a single well, $L_w \sim 100$ Å thick. How short L can be is limited by fabrication technologies.

(3) α_{bias} and $\Delta\alpha$ can also be reduced by a larger $\Delta\lambda_{det}$ or by the material design.

(4) At very high loaded q-factors, the resonance is also very sensitive to the laser wavelength and thermal fluctuations. There is a limit of the usefulness of the high q-factor, determined by the laser wavelength stability, the temperature variation, the fabrication tolerance of the resonance wavelength, etc.

(5) For devices with small d_i, the electrical capacitance may be large, thereby limiting the RF bandwidth of the modulator. The decay time of the optical energy in the loaded resonator is another factor that may limit the bandwidth of the device.

Two limiting cases of the FP WG MQW modulator are: (1) $\Delta n_{eff} \approx 0$, for EA only,

$$R(V) = \frac{R(1 - e^{-\Gamma\alpha_{bias}L}e^{-\Gamma\Delta\alpha L})^2}{(1 - Re^{-\Gamma\alpha_{bias}L}e^{-\Gamma\Delta\alpha L})^2},$$

$$T(V) = \frac{(1 - R)^2 e^{-\Gamma\alpha_{bias}L}e^{-\Gamma\Delta\alpha L}}{(1 - Re^{-\Gamma\alpha_{bias}L}e^{-\Gamma\Delta\alpha L})^2}. \tag{6.23}$$

(2) $\Delta\alpha \approx 0$, for ER only,

$$R(V) = \frac{R(1 - e^{-\Gamma\alpha_{bias}L})^2 + 2Re^{-\Gamma_{bias}L}(1 - \cos\theta')}{(1 - Re^{-\Gamma\alpha_{bias}L})^2 + 2Re^{-\Gamma\alpha_{bias}L}(1 - \cos\theta')},$$

$$T(V) = \frac{(1 - R)^2 e^{-\Gamma\alpha_{bias}L}}{(1 - Re^{-\Gamma\alpha_{bias}L})^2 + 2Re^{-\Gamma\alpha_{bias}L}(1 - \cos\theta')}. \tag{6.24}$$

References

1. D. Ackerman, C. Cox III and N. Rizza, Editors, *Selected Papers on Analog Fiber-Optic Links*, SPIE Milestone Series, V. MS-149, 1998.
2. C. H. Cox III, *Analog Optical Links: Theory and Practice*, Cambridge University Press, to be published.
3. D. A. B. Miller, D. S. Chemla, T. C. Damen, A. C. Gossard, W. Wiegmann, T. H. Wood, and C. A. Burrus, "Band-edge electroabsorption in quantum well structures: the quantum-confined Stark effect," *Phys. Rev. Lett.*, **53**, 2173, 1984.
4. C. Rolland, G. Mak, K. L. Prosyk, C. M. Maritan, and N. Puertz, "High speed and low loss, bulk electroabsorption waveguide modulators at 1.3 μm," *IEEE Photon. Technol. Lett.*, **3**, 894, 1991.
5. R. B. Welstand, "High linearity modulation and detection of semiconductor electroabsorption waveguides," Ph. D. Thesis, University of California San Diego, 1997.
6. D. A. B. Miller, D. S. Chemla, T. C. Damen, A. C. Gossard, W. Wiegmann, T. H. Wood, and C. A. Burns, "Electric field dependence of optical absorption near the bandgap of quantum well structures," *Phys. Rev. B*, **32**, 1043, 1985.
7. D. A. B. Miller, J. S. Weiner, and D. S. Chemla, "Electric-field dependence of linear optical properties in quantum well structures: Waveguide electroabsorption and sum rules," *IEEE, J. Quantum Electron.*, **QE-22**, 816, 1986.
8. W. K. Burns, M. M. Howerton, R. P. Moeller, A. S. Greenblatt, and R. W. McElhanon, "Broad-band reflection traveling wave LiNbO$_3$ modulator," *IEEE Photon. Technol. Lett.*, **10**, 805, 1998.
9. K. K. Loi, J. H. Hodiak, X. B. Mei, C. W. Tu, W. S. C. Chang, D. T. Nicols, L. J. Lembo, and J. C. Brock," Low loss 1.3 μm MQW electroadbsorption modulators for high-linearity analog optical links," *IEEE Photon. Technol. Lett.*, **10**, 1998.
10. G. Metzler and T. Schwander, "RF small-signal equivalent circuit of MQW InGaAs/InAlAs electroabsorption modulator," *Electron. Lett.*, **33**, 1822, 1997.
11. K. K. Loi, X. B. Mei, J. H. Hodiak, C. W. Tu, and W. S. C. Chang, "38 GHz bandwidth 1.3 μm MQW elecroabsorption modulators for RF photonic links," *Electron. Lett.*, **10**, 1018, 1998.
12. X. B. Mei, K. K. Loi, H. H. Wieder, W. S. C. Chang, and C. W. Tu, "Strain-compensated InAsP/GaInP multiple quantum wells for 1.3 μm waveguide modulators," *Appl. Phys. Lett.*, **68**, 90, 1996.
13. T. Ido, S. Tanaka, M. Suzuki, and H Inoue, "MQW electroabsorption optical modulator for 40 Gbt/s modulation," *Electron. Lett.*, **31**, 2124, 1995.
14. K. K. Loi, "Multiple-quantum-well waveguide modulators at 1.3 μm wavelength for analog fiber-optic links," Ph. D. thesis, University of California San Diego, 1998.
15. G. L. Li, P. K. L. Yu, S. A. Pappert, and C. K. Sun, "The effects of photocurrent on microwave properties of electroabsorption modulators," *IEEE MTT-S International Microwave Symposium Digest*, Paper WE2B-2, Anaheim, CA, June 1999.
16. H. H. Liao, X. B. Mei, K. K. Loi, C. W. Tu, P. K. L. Yu, P. M. Asbeck, and W. S. C. Chang, "Design of millimeter wave optical modulators with monolithically integrated narrow band impedance matching circuits," *Proc. SPIE*, **3006**, 318, 1997.
17. H. H. Liao, "Novel microwave structures for ultra high frequency operation of MQW electroabsorption waveguide modulators," Ph. D. Thesis, University of California San Diego, 1997.
18. D. S. Chemla, D. A. B. Miller, P. W. Smith, A. G. Gossard, and W. Wiegmann, "Room temperature excitonic non linear absorption and refraction in Ga|As/AlGaAs multiple quantum well structures," *IEEE J. Quantum Electron.*, **QE-20**, 265, 1984.

19. An-Nien Cheng, "Quaternary InGaAlAs/InAlAs quantum wells for 1.3 μm electroabsorption modulators," Ph. D. Thesis, University of California San Diego, 1994.

20. T. H. Wood, J. Z. Pastalan, C. A. Burrus, B. C. Johnson, B. I. Miller, J. L. Demiguel, U. Koren, and M. G. Young, "Electric field screening by photogenerated holes in MQWs: A new mechanism for absorption saturation," *Appl. Phys. Lett.*, **57**, 1081, 1990.

21. A. M. Fox, D. A. B. Miller, G. Livescu, J. E. Cunningham, and W. Y. Jan, "Quantum well carrier sweep out: Relation to electroabsorption and exciton saturation," *IEEE J. Quantum Electron.*, **27**, 2281, 1991.

22. Courtesy of Dr. P. K. Tien, AT&T Research Laboratory, unpublished.

23. J. W. Mathews and A. E. Blakeslee, "Defects in epitaxial multilayers," *J. Crys. Growth*, **27**, 118, 1974.

24. R. People and J. C. Bean, "Calculations of critical thickness versus lattice mismatch for GeSi strained-layer heterostructures," *Appl. Phys. Lett.* **47**, 322, 1985.

25. L. Shen, "InGaAs/InAlAs quantum wells for 1.3 μm electroabsorption modulators on GaAs substrates," Ph. D. Thesis, University of California San Diego, 1997.

26. L. Shen, H. H. Wieder, and W. S. C. Chang, "Electroabsorption at 1.3 μm on GaAs substrates using a step-graded low temperature grown InAlAs buffer," *IEEE Photon. Technol. Lett.*, **8**, 352, 1996.

27. S. M. Lord, "Growth of high indium content InGaAs on GaAs substrates for optical applications," Ph. D. Thesis, Stanford University, 1993.

28. H. C. Chui and J. S. Harris Jr., "Growth studies on the $In_{0.5}Ga_{0.5}As/Al$ GaAs quantum wells grown on GaAs with a linearly graded InGaAs buffer," *J. Vac. Soc. Technol. B*, **12**, 1019, 1994.

29. B. K. Tanner and D. K. Brown, "Advanced x-ray scattering techniques for the characterization of semiconductor material," *J. Cryst. Growth*, **126**, 1, 1993.

30. P. F. Fewster, "X-ray diffraction from low dimensional structures," *Semicond, Sci. Technol.* **8**, 1915, 1993.

31. R. M. Feenstra, "Cross-sectional scanning tunneling microscopy of III–V semiconductor structures," *Semicond. Sci. Technol.* **9**, 2157, 1994.

32. J. A. Stroscio and W. J. Kaiser, *Scanning Tunneling Microscopy*, Academic Press, Boston, Chapters 5 and 6, 1993.

33. D. C. Joy, A. D. Romig, and J. I. Goldstein, *Principles of Analytical Electron Microscopy*, Plenum Press, New York, 1986.

34. R. D. Leapman and D. E. Newbury, "Trace element analysis at nanometer spatial resolution by parallel-detection electron energy loss spectroscopy," *Anal. Chem.*, **65**, 2409, 1993.

35. H. Hovel, "Scanning photoluminescence of semiconductors," *Semicond. Sci. Technol.*, **7**, A1, 1992.

36. C. J. Miner, *Rapid Non-destructive Scanning of Compound Semiconductor Wafers and Epitaxial Layers, Semiconductor Characterization, Present Status and Future Needs*, ed. W. M. Bullis, D. C. Seiler, and A. C. Diebold, AIP Press, 1996, p. 605.

37. J. H. van der Merwe, "Misfit dislocation generation in epitaxial layers,"*Crit. Rev. Solid State Mater. Sci.*, **17**, 187, 1991, CRC Press.

38. A. Nandedkar and J. Narayan, "Atomic structure of dislocations in silicon, germanium and diamond," *Philos. Mag.*, **A62**, 873, 1990.

39. Xiaobing Mei, "InAsP/GaInP strain-compensated multiple quantum wells and their optical modulator applications," Ph. D. Thesis, University of California, San Diego, 1997.

40. G. Bastard, "Theoretical investigation of superlattice band structure in the envelope function approximation," *Phys. Rev.*, **B25**, 7584, 1982.

41. D. A. B. Miller, D. S. Chemla, T. C. Damen, A. C. Gossard, W. Wiegmann, T. H. Wood, and C. A. Burrus, *Phys. Rev.*, **B32**, 1043, 1985.

42. F. Y. Juang, J. Singh, P. K. Bhatacharya, K. Bajema, and R. Merlin, "Field dependent linewidths and photoluminescence energies in GaAs-AlGaAs multi-quantum well modulator," *Appl. Phys. Lett.*, **48**. 1246, 1986.

43. S. Hong and J. Singh, "Excitonic energies and inhomogeneous line broadening effects in InAsAs-InGaAs modulator structures," *J. Appl. Phys.*, **42**, 1994, 1987.

44. D. A. B. Miller, D. S. Chemla, D. J. Eilenberger, P. W. Smith, A. C. Gossard, W. Wiegmann, T. H. Wood, and C. A. Burrus, "Large room-temperature optical nonlinearity in GaAs/Ga$_{1-x}$As$_x$ multiple quantum well structures," *Appl. Phys. Lett.*, **41**, 679, 1982.

45. A.-N. Cheng, H. H. Wieder, and W. S. C. Chang, "Electroabsorption in lattice-matched InGaAlAs-InAlAs quantum wells at 1.3 μm," *IEEE Photon. Technol. Lett.*, **7**, 1159, 1995.

46. T. He, P. Ehrhart, and P. Mauffels, "Optical band gap and Urbach tail in Y-doped BaCeO$_3$," *J. Appl. Phys.*, **79**, 3129, 1996.

47. J. H. Hodiak, "Design of fiber-coupled surface-normal Fabry Perot electroabsorption modulators for analog applications," Ph. D. Thesis, University of California San Diego, 1999.

48. Y. Okuno, K. Uomi, M. Aoki, and T. Tsuchia, "Direct wafer bonding of III–V compound semiconductors for free-material and free-orientation integration," *IEEE J. Quantum Electron.* **33**, 959, 1997.

49. K. K. Loi, I. Sakamoto, X. B. Mei, C. W. Tu, and W. S. C. Chang, "High efficiency 1.3 μm InAsP/GaInP MQW electroabsorption waveguide modulators for microwave fiber optic links," *IEEE Photon. Technol. Lett.*, **8**, 626, 1996.

50. K. K. Loi, J. H. Hodiak, C. W. Tu and W. S. C. Chang, "Linearization of 1.3 μm MQW electroabsorption modulators using an all-optical frequency-insensitive technique," *IEEE Photon. Technol. Lett.*, **10**, 964, 1998.

51. H. H. Liao, X. B. Mei, P. M. Asbeck, C. W. Tu, and W. S. C. Chang, "Microwave structures for traveling wave MQW electroabsorption modulators for wide band 1.3 μm photonic links," *Proc. SPIE*, **3006**, 291, 1997.

52. H. H. Liao, "Novel microwave structures for ultra high frequency operation of MQW electroabsorption waveguide modulators," Ph. D. Thesis, University of California San Diego, 1997.

53. G. L. Li, S. A. Pappert, C. K. Sun, W. S. C. Chang, and P. K. L. Yu, "50 GHz traveling-wave InGaAsP/InP electroabsorption modulator: measurement and analysis," *IEEE Trans. Microwave Theory Tech.*, special issue, *Microwave and Millimeter Wave Photonics*, 2001.

7

Polymer modulators for RF photonics

TIMOTHY VAN ECK

Lockheed Martin Space Systems Company

7.1 Benefits of polymer modulators

Polymer electro-optic modulators offer several important advantages over more mature technologies such as lithium niobate interferometric modulators or semiconductor electroabsorption modulators. Velocity matching between RF and optical waves is much simpler because there is a close match between the dielectric constants at optical and radio frequencies, enabling flat response to higher frequencies. The relatively low dielectric constant enables 50-Ω drive electrodes to be easily achieved with simultaneous velocity matching. Sensitivity as much as an order of magnitude greater than in lithium niobate may be achievable, primarily through material engineering to achieve very large electro-optic coefficients, but also through the drive field concentration enabled by parallel-plate or microstrip drive electrodes. Greater sensitivity can result in greater RF gain and smaller noise figure of an RF link incorporating an external modulator. While the electro-optic effect employed is resonant, the wavelength of operation is far from resonance, so that there is little wavelength sensitivity or thermal sensitivity, in contrast to electroabsorption modulators which are operated close to resonance. The refractive index of polymer materials is nearly matched to that of glass optical fibers, enabling small Fresnel losses at interfaces. The material cost of polymer modulators may be low because they can be made of thin films of high-value polymer material on top of inexpensive substrates. Because polymer waveguide layers can be fabricated on a variety of substrates, including flexible substrates, by spin-coating, integration of polymer waveguides with other components is possible, and conformable devices could be made. Although polymer modulator technology is not as mature as lithium niobate or electroabsorption technology, its many potential advantages are compelling.

Figure 7.1. Two implementations of Mach–Zehnder modulators, with the input channel on the left and output channel(s) on the right. (a) Standard Mach–Zehnder. (b) Balanced Mach–Zehnder.

7.2 Benefits for RF links

One of the advantages of polymer modulators is the potential for much higher sensitivity (lower half-wave voltage) than other technologies, which can enable RF photonic links with gain and small noise figure. Figure 7.1 illustrates two Mach–Zehnder modulator designs, the first with a single output channel, and the second with two complementary output channels, referred to as a balanced Mach–Zehnder. Two important figures of merit in RF links are the gain and noise figure, while two of the figures of merit for a photonic modulator are the half-wave voltage, V_π and the insertion loss. The RF gain for the single-channel case can be expressed* as follows [1]:

$$G = \frac{\pi^2 i_B^2 R_m R_d}{V_\pi^2},\tag{7.1}$$

where V_π is the half-wave voltage of the modulator, i_B is the photodetector DC current, R_m is the modulator input impedance, and R_d is the photodetector load resistance. The noise figure can be expressed as

$$F = 1 + f_m + \frac{V_\pi^2}{R_m \pi^2}\left(\frac{1}{i_B^2 R_d} + \frac{2e}{i_B k_B T_0} + \frac{r}{k_B T_0}\right),\tag{7.2}$$

where k_B is Boltzmann's constant, T_0 is the standard temperature, 290 K, e is the electron charge, r is the relative intensity noise expressed as a power ratio per Hz, and f_m is the fraction of the modulator input thermal noise that contributes to available noise power at the output. The first two terms of Eq. (7.2) represent noise sources at the input end, the external input noise and modulator input noise, respectively. The terms in brackets represent noise sources at the output end; from left to right, thermal noise, shot noise, and relative intensity noise. Most of the parameters involved in the gain and noise figure would not depend on the choice of modulator material except for the half-wave voltage, V_π, which could be considerably less for a polymer

---

* Expressions for link gain and noise figure depend on the details of the link design, particularly the circuit impedances, but different link designs often differ in the gain and individual noise terms only by multiplicative factors. Equations (7.1) and (7.2) follow the link design in Ref. 1 for a Mach–Zehnder modulator biased for 50% DC transmission.

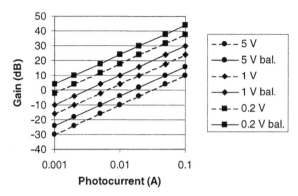

Figure 7.2. The computed RF gain of a photonic link, as a function of detected photocurrent, for several values of the modulator half-wave voltage. Results are shown for the standard single-output Mach–Zehnder (dashed lines), and also for the two-output balanced Mach–Zehnder (solid lines).

modulator than for lithium niobate modulators of the same length. Equation (7.1) shows that the gain is proportional to the inverse square of the half-wave voltage, and Eq. (7.2) shows that all the contributions to noise figure originating at the output end are proportional to the square of the half-wave voltage.

For the balanced Mach–Zehnder feeding a balanced photodetector receiver [2,3], the signal current is doubled, while the DC current is largely cancelled, so the gain and input noise are increased by 6 dB, the output thermal noise is unchanged, the shot noise increases by only 3 dB, and most of the relative intensity noise is cancelled:

$$G = \frac{4\pi^2 i_B^2 R_m R_d}{V_\pi^2},$$ (7.3)

$$F = 1 + f_m + \frac{V_\pi^2}{4R_m\pi^2}\left(\frac{1}{i_B^2 R_d} + \frac{4e}{i_B k_B T_0} + \frac{2Cr}{k_B T_0}\right),$$ (7.4)

where C is a number between zero and one expressing the effectiveness of relative intensity noise cancellation. The gain and noise figure are graphed in Fig. 7.2 and Fig. 7.3, respectively, as a function of optical output power, for several selected values of the half-wave voltage.* Desirable gain and noise figure are available by two methods, reducing half-wave voltage, or increasing the optical power to the photodetector, either by increasing the laser power or by minimizing optical losses. A key advantage of electro-optic polymer materials for electro-optic modulators is the potential for low half-wave voltage. For lithium niobate, the electro-optic coefficient is fixed, and the half-wave voltage can be reduced only by increasing the modulator length, which reduces its bandwidth because of RF loss in the electrode.

---

* Assumptions: $R_m = R_d = 50\ \Omega$, $f_m = 0$, $r = -165$ dB/Hz, and $C = 10^{-2}$.

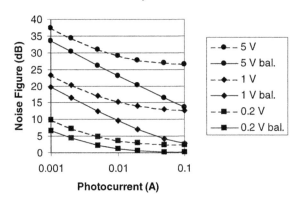

Figure 7.3. The computed noise figure of a photonic link, as a function of detected photocurrent, for several values of the modulator half-wave voltage. Results are shown for the standard single-output Mach–Zehnder (dashed lines), and also for the two-output balanced Mach–Zehnder (solid lines).

For electro-optic polymer materials, there is still great potential for increasing the electro-optic effect through material engineering, and the development of high-bandwidth modulators with V_π as low as 1 V is likely [4]. Approaches to reducing V_π will be discussed in Section 7.6. The other approach to increasing gain and reducing noise figure is to increase optical power. The term i_B^2/V_π^2 in the gain shows that lowering V_π and increasing i_B are equally effective in increasing the gain. However, the terms in the noise figure expressions show that lowering V_π is more effective in reducing noise, since it is equally effective in reducing thermal, shot, and intensity noise in the photodetector, while increasing i_B is less effective in reducing shot noise, and does not reduce intensity noise at all. In many RF systems, power and weight of power supplies must be minimized, so it is much more attractive to reduce the modulator half-wave voltage, which costs nothing to the system designer, than to increase optical power.

7.3 Electro-optic polymer materials

Electro-optic polymer materials are amorphous organic materials composed of an electro-optically active component, called a chromophore, and a polymer matrix. The material is made electro-optic by giving the ensemble of chromophore molecules an average alignment, usually by poling, which gives the material the characteristics of the ∞mm point group [5]. Of most interest is the largest electro-optic coefficient, r_{33}, but the poled material will also have $r_{13}, r_{23}, r_{42}, r_{51}$ electro-optic coefficients [6]. The chromophore material may be mixed with a polymer host (guest–host), or may be chemically bonded to the polymer chain (attached, or cross-linked). Poling must be done at elevated temperature, near or higher than the glass transition temperature, and the material must then be cooled to freeze in the alignment. After poling the material must not be returned to temperatures near the glass transition temperature, or the chromophore alignment will be lost.

7.3.1 Chromophores

A chromophore, or dye, is an organic molecule, such as [7] the examples in Fig. 7.4, with an electron donor group on one end and an electron accepting group on the other, and a bridge with π-electron structure connecting the two. Such a molecule will be an electrical dipole because in the ground state it will typically have a net positive charge on the donor end and a net negative charge on the acceptor end. It also has color because the transition from the ground state to the first excited state has an energy corresponding to a visible optical wavelength. This transition also makes the molecule polarizable by an incident optical field. Of interest for electro-optic purposes is the first hyperpolarizability, denoted β. With two excitation fields, one optical and one RF, the hyperpolarizability gives rise to a microscopic optical polarization, which is manifested as a macroscopic linear electro-optic, or Pockels, effect, only if there is an average alignment of the molecules in a particular direction.

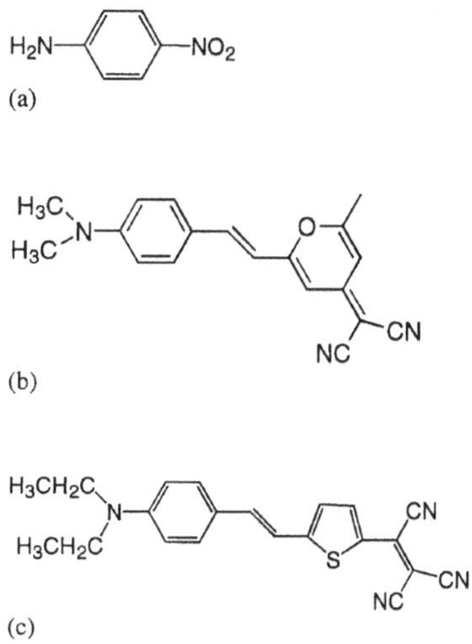

Figure 7.4. Three chromophore molecules with molecular hyperpolarizabilities, each with the electron donor group on the left side and the electron acceptor group on the right side. (a) p-nitroaniline, with simple donor and acceptor groups and a simple benzene ring bridge. (b) DCM, with stronger donor and acceptor groups and a longer bridge. (c) RT9800, a high-activity chromophore with strong donor, strong acceptor, and very long bridge, reprinted with permission from Y. M. Cai and A. K.-Y. Jen, "Thermally stable poled polyquinoline thin film with very large electro-optic response", *Appl. Phys. Lett.*, **67**, 299 (1995), copyright 1995, American Institute of Physics.

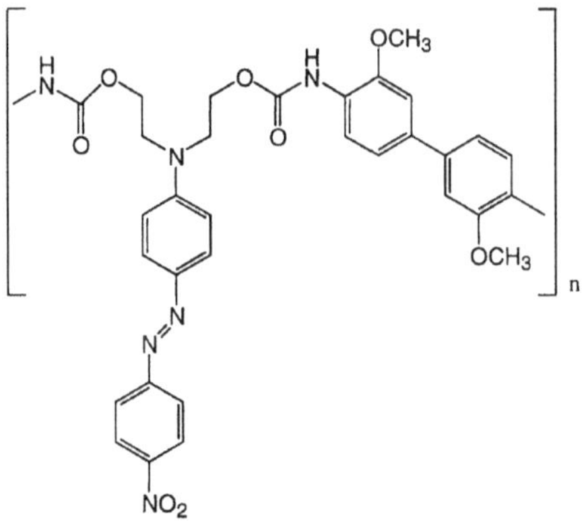

Figure 7.5. A typical attached polymer, PUDR19. The horizontal structure represents the polymer backbone, a polyurethane, with brackets indicating a repeated unit of the polymer chain. The vertical structure represents a chromophore, DR19. In this example the chromophore is part of the polymer backbone. Reprinted with permission from Ref. 10. Copyright 1998 American Chemical Society.

7.3.2 Guest–host and attached polymers

There are two major methods of incorporating chromophore materials into polymers, guest–host and attachment. Guest–host materials are simply mixtures of chromophore material and polymer. The host polymer material dominates most of the properties of the material during processing and after a thin film is fabricated. The guest chromophore material is mixed into the host in a moderate proportion, often limited to 15% to 35% by weight. In attached polymers, the chromophore material is chemically attached to the polymer chain, as demonstrated in Fig. 7.5. This can provide for a higher concentration of chromophore in the composite material, up to about 65% by weight [8], as well as better material stability.

7.3.3 Thermoplastic, thermoset, and crosslinked polymers

The microscopic rigidity of some polymers is described by the thermoplastic or thermoset models, while others are described by the crosslinking model. Thermoplastic polymers are characterized by a glass transition temperature, T_g. At temperatures below T_g, the material is in a glassy state, meaning that it will not flow significantly, and at temperatures above T_g, the material is in a rubbery state, meaning that it will flow, but much more slowly than a liquid. The glassy state is also characterized by a relatively fixed free volume, while the rubbery state is characterized by a free volume that increases with increasing temperature. Thermoplastic polymers are characterized by a T_g that is independent of thermal history. A related type of polymer is the thermoset polymers in which heating causes permanent change, in many cases manifested as a T_g that increases when the material is exposed to higher temperatures [9]. Alignment of the chromophores to create the electro-optic coefficient must be done at a temperature near or above T_g, but T_g must be well above any temperature at which the modulator will be operated or stored, to avoid relaxation of the chromophore alignment.

Another approach to preserving the chromophore alignment is crosslinking after or during poling. Crosslinking involves creating a high density of chemical bonds between different segments of the polymer chain, giving it much more long-term stability and rigidity. Crosslinking may be accomplished by incorporating a crosslinking site in the polymer, in the chromophore, or by adding a crosslinking component. Crosslinking may be initiated either by elevated temperature or by exposure to radiation, such as ultraviolet radiation. Additional alignment stability can be achieved by attaching the chromophore to the polymer at one end and providing a second attachment point at the opposite end of the chromophore, to be activated after or during poling [10]. The challenge of thermal crosslinking is the competing requirements of poling at elevated temperature, for which molecular mobility is needed, and crosslinking, which permanently restricts molecular mobility [9,11].

7.3.4 Cladding materials

Any planar integrated optic waveguide requires cladding layers above and below the core layer. For polymer waveguides, the cladding material may be any material that is compatible with the core, and is most often another polymer material. The cladding material need not be electro-optic. The selection of a cladding material is determined by compatibility of the material and the processing needed to form the core and cladding layers. The cladding must have a refractive index slightly less than that of the core so that light will be confined to the core and single mode waveguides can be made with practical dimensions, and the cladding should have low optical absorption and scattering at the wavelength of operation. The cladding layer should be mechanically and thermally compatible with the core, and deposition of any layer must not dissolve, crack, or otherwise adversely impact any underlying layer. The cladding should not have electrical conductivity less than the core, and it is preferable for the cladding conductivity at the poling temperature to be significantly higher than the core conductivity.

7.4 Methods of fabrication

7.4.1 Device design

A cross-section of a typical polymer channel waveguide is shown in Fig. 7.6. The substrate is typically silicon, chosen for its high surface quality at low cost and

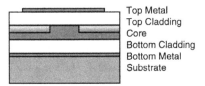

Figure 7.6. Cross-section of a polymer waveguide, with metal ground plane on top of the substrate, and top metal electrode on top of the polymer layers. The core is laterally patterned, with the shaded area representing a graded index.

compatibility with semiconductor processing equipment, but other substrates have been used as well, such as GaAs, SF/duroid®* [12], and Mylar®† [13]. After deposition and patterning of a bottom metal layer, the bottom cladding, core, and top cladding are deposited one at a time, followed by deposition and patterning of the top metal pattern. At some point in the process one of the polymer layers is patterned to cause the lateral confinement needed to produce channel waveguides. The channel waveguide patterning is done photolithographically so that many waveguide devices can be defined simultaneously. In the usual configuration, the largest electro-optic coefficient, r_{33}, is in the direction normal to the surface. Consequently, the optical mode should be polarized normal to the polymer surface, and the electrical drive field should be applied normal to the polymer surface. This drive field is usually achieved by applying a voltage between parallel metal planes immediately above and below the polymer layers, or for high-frequency operation, a microstrip structure consisting of a ground plane immediately below the polymer layers, and a thin metal strip immediately above.

The same modulator configurations that are used in lithium niobate modulators can also be used in polymer modulators, including the single-output Mach–Zehnder modulator and dual-output differential Mach–Zehnder modulator shown in Fig. 7.1, as well as directional couplers and 2×2 switches. The optical phase in the two arms can be modulated independently; simple modulation can be achieved by applying a voltage to one arm and grounding the other, or a difference signal can be generated by applying two different signals to the two arms. The dual-output modulator is more difficult to produce, but produces two complementary output signals, which when detected with a differential detector conserve optical power, increase sensitivity, and can cancel laser relative intensity noise.

7.4.2 Polymer deposition

The bottom cladding, core, and top cladding layers are deposited in sequence, each being fully cured before the next layer is applied. Each layer is typically deposited from a liquid solution by spin coating on a commercial photoresist spinner, then

---

* SF/duroid® electrical insulation is a licensed trademark of Rogers Corporation.
† Mylar® polyester film is a DuPont trademark.

heated to remove residual solvent. For some polymer materials, thermal curing also occurs during the heating step. Before depositing another polymer layer on top, each layer must be hard enough to resist dissolution or cracking during deposition of the subsequent layer.

7.4.3 Waveguide patterning and electrode fabrication

There are several methods of creating lateral optical confinement, including selective photobleaching and dry etching of the polymer core layer. Selective photobleaching has the advantage of being a planar process; the refractive index of the core layer is patterned, and all layers retain their planar surfaces. After all three polymer layers are deposited, a photomask is patterned on top of the top cladding layer, as illustrated in Fig. 7.7, with the opaque areas located where channel waveguides are desired. Then the entire wafer is exposed to uniform ultraviolet radiation, normally by flood exposure under the mercury lamp in a mask aligner. The ultraviolet passes through the top cladding, which is mostly transparent, and is absorbed strongly by the chromophore molecules near the top of the core layer. The chromophore molecule is photochemically altered, resulting in a local reduction of the refractive index of the core material. Because of the strong UV absorption near the top interface, the resulting refractive index distribution is thought to be graded, as illustrated in Fig. 7.7, with the lowest index at the top surface. This results in a waveguiding structure similar to a rib waveguide. The photobleaching could be done to the core layer before deposition of the top cladding, but that would expose the core–cladding interface, where the optical intensity will be high during device operation, to possibly harmful photolithographic processing (photoresist developer and metal etchant). By making use of the transparency of the top cladding layer, only the top surface of the top cladding, where the optical power will be negligible during device operation, will be exposed to photolithographic processing.

Lateral confinement can also be realized by mechanical patterning of the core or bottom cladding layer, primarily by dry etch methods such as plasma etching,

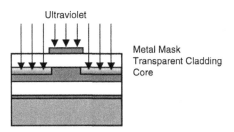

Figure 7.7. During channel waveguide patterning by selective photobleaching, the entire wafer is exposed to ultraviolet radiation, the metal mask blocks the ultraviolet where the channels are desired, and in the unmasked areas the ultraviolet passes through the transparent cladding and reduces the refractive index near the top of the core layer.

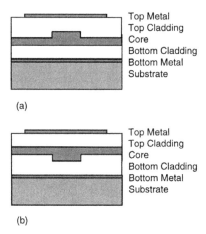

Figure 7.8. (a) Etched rib waveguide, in which a shallow ridge is etched in the core layer, then covered with the top cladding. (b) Etched trench waveguide, in which a shallow trench is etched in the bottom cladding layer, then covered by the core layer.

reactive ion etching, or ion milling. The core layer may be patterned so that it is totally surrounded by lower-index cladding material, or it may be patterned as a rib waveguide, i.e., only a shallow etch is performed, just deep enough for lateral confinement. For either configuration, the core may be patterned by etching ribs into the core material, or trenches may be etched into the bottom cladding and filled with core material, as illustrated in Fig. 7.8. In either case, any polymer layers deposited on top of the etched layer may have surface distortions due to the flow conditions of the polymer over a non-flat surface. Such surface distortions can be minimized by making the etch depth shallow rather than deep, or by using a planarizing material. In addition, surface distortions in the top surface of the top cladding may be less important than distortions in the interface between the core and top cladding; this fact tends to favor the rib etch over the trench etch. However, the trench etch has the advantage that it does not involve photolithographic processing directly on top of the core layer.

Patterning of the bottom metal, if necessary, may be done with standard wet or dry etching processes. For the top metal electrodes, thin metal is adequate for operation up to about 100–1000 MHz, and thick metal is required for operation at higher frequencies. Thin metal may be deposited by evaporation and patterned by wet etching. Thick metal may be deposited more efficiently by electroplating. Since dimensional control of fine metal structures during wet etching is difficult for thick metal layers, thick metal may be plated through a photolithographically defined mask layer.

7.4.4 Poling

The polymer core layer is made electro-optic by incorporating chromophore molecules with microscopic permanent dipole moments and optical hyperpolarizabilities, and aligning the dipole moments. As deposited by spin-coating, the molecules are randomly oriented, so that the net sum of all of the microscopic hyperpolarizabilities is approximately zero. The dipole moments, and thus the components of optical hyperpolarizability that are parallel to the dipole moments, can be aligned, or poled, with an applied electric field, if the chromphore molecules are temporarily made rotationally mobile by heating the polymer to a temperature near its glass transition temperature, T_g. To maintain the alignment, the temperature is reduced while the poling field is maintained until the temperature is at least 50–100 degrees below the glass transition temperature. The alignment of the chromphores is not in thermodynamic equilibrium, but at sufficiently low temperatures will take many years to relax to the equilibrium, unaligned condition. Thus the electro-optic effect in the modulator can be maintained for many years.

In the contact poling technique, poling electrodes are required immediately above and below the polymer layers. Usually this can be done with the same electrodes that will be used as drive electrodes. The wafer or die containing the modulator is heated to the poling temperature, and while the bottom metal is grounded, the poling voltage is applied to the top electrode, and after a period of time, the temperature is reduced to near room temperature, then the poling voltage is removed.

In the corona poling technique, no top electrode is used during poling. The voltage is applied to the top polymer surface by means of a corona field. The corona field deposits a high density of ions onto the top surface, and the ions have relatively low mobility so they create a surface charge layer, resulting in a strong electric field inside the polymer layers. The corona field is created by applying a high voltage of several thousand volts to the emitter, which could be a single sharp point, a single thin wire, or a grid of thin wires, made out of tungsten or some other suitable metal material. If the emitter is a single point, then the corona charge will be non-uniformly distributed on the polymer surface, but a relatively uniform field can be created by using an array of points or wires.

One advantage of contact poling is that the voltage applied across all the polymer layers is known, and anomalies in the field can be detected by monitoring the current. Another advantage is that the poling field is uniform across the entire area covered by the electrode. One advantage of corona poling is that the localized electrical breakdowns that occasionally occur due to defects are limited locally, because of the relative immobility of the surface charge creating the poling field. With contact poling, the metal poling electrode provides good conductivity across the surface, which allows local breakdowns to affect the poling field over the entire poled area.

7.4.5 Endface preparation

For efficient coupling of optical power into and out of a modulator, optical-quality flatness at the endface is desired. Endfaces have been prepared by cleaving, sawing, and polishing. Cleaving depends on having a substrate with cleaving planes perpendicular to the surface, such as $\langle 100 \rangle$ silicon, and propagating a cleave from the substrate through all of the layers of the modulator. Excellent cleaved endfaces can be achieved, but repeatability and accurate placement of the cleave can be difficult. Optical-quality endfaces can also be prepared by sawing with a silicon dicing saw, and these saw cuts can be placed very accurately and repeatably. Optical-quality polishing can be challenging because the polymer is much softer than most substrates, but can be accomplished by sandwiching the polymer layers between two layers of the substrate material.

7.4.6 Packaging

The major elements of packaging are the creation of robust, low-loss, broadband connections to optical fiber and electrical cable or waveguide; protection from the environment, including heat, water vapor, vibration, and radiation; and design of a package that does not itself have electrical or mechanical resonances that would interfere with the device operation. The optical mode size in both fiber and waveguide are measured in micrometers, and misalignments as small as one micrometer can cause significant loss, so the positional alignment tolerances between fiber and polymer waveguide are usually less than one micrometer. Because the polymer modulator uses a specific optical polarization, polarization-maintaining fiber must be used, at least on the input, and the fiber axis must also be aligned accurately with the optical polarization required by the waveguide. Because the optical fiber has a small cross-section, the optical connection should be made robust by permanently holding the optical fiber in a much larger mechanical structure prior to attaching it to the polymer waveguide and its substrate. For the electrical contact, an RF field must be launched with low loss into the microstrip drive electrode, usually through a transition region as discussed in the next section. It is desirable to make this connection with the sort of robust electrical connection that is commonly used in electronic systems, such as a wire bond, and for the electrical waveguides on either side of the connection to have similar size, shape, and electromagnetic field patterns, to minimize RF loss. One attractive feature of the polymer modulator is the use of the microstrip drive electrode, which has good confinement of the RF field under the top electrode, in comparison with coplanar electrodes.

7.5 Frequency response

For a traveling wave modulator, the fundamental frequency response is limited by velocity mismatch and by frequency-dependent RF loss. One great advantage of polymer electro-optic modulators is the relatively small natural velocity mismatch, which can be tuned to near zero with 50-Ω transmission line structures that are easily designed and fabricated. For that reason, the frequency response of polymer modulators is generally limited primarily by the RF loss of the drive electrode.

The RF loss, $A(f)$, is expressed as a power loss measured in dB/cm, and is related to the amplitude attenuation constant $\alpha(f)$, in cm^{-1}, by $\alpha(f) = A(f) \times \ln(10)/20$, where f is the frequency. $A(f)$ is usually dominated by skin effect loss which scales as the square root of the frequency. The difference in propagation constant between optical and RF waves is

$$K = \frac{(\sqrt{\varepsilon_{re}} - n_{eff})\omega}{c},$$ (7.5)

where ε_{re} is the effective dielectric constant of the RF drive electrode structure, n_{eff} is the effective refractive index of the optical waveguide, $\omega = 2\pi f$ and c is the speed of light. For a device electrode of length L, the integrated phase modulation along the length of the electrode is [14]

$$\Delta\phi(f) = \Delta\phi_0 R(f),$$ (7.6)

where

$$\Delta\phi_0 = \frac{\pi n^3 r_{33} \Gamma V L}{\lambda h},$$ (7.7)

$$R(f) = \frac{1}{L} \int_0^L \exp(-\alpha(f)z)\sin(Kz - \omega t)\,dz$$

$$= \frac{1}{L}\left(\frac{\exp(-2\alpha(f)L) - 2\exp(-\alpha(f)L)\cos(KL) + 1}{\alpha(f)^2 + K^2}\right)^{1/2}\sin(\Phi - \omega t),$$ (7.8)

V is the applied voltage, Γ is the overlap integral between the optical mode and the electro-optic core layer, λ is the optical wavelength, h is the vertical separation between electrodes (the polymer thickness), and Φ is a phase factor described by Teng [14] which will be ignored here.

In the limit of zero velocity mismatch, Eq. (7.8) reduces to [15]

$$R(f) = \frac{1 - \exp(-\alpha(f)L)}{\alpha(f)L},$$ (7.9)

which has its maximum value 1 at $\alpha = 0$ and decays monotonically for increasing α.

In the limit of zero RF attenuation, Eq. (7.8) reduces to

$$R(f) = \frac{\sin(KL/2)}{KL/2},$$ (7.10)

which has its maximum value 1 for $K = 0$ and has a series of nulls thereafter.

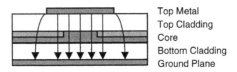

Figure 7.9. The microstrip electrode concentrates RF electric field in the vicinity of the electro-optic channel waveguide.

For a traveling wave modulator the optical waveguide must be embedded in a microwave transmission line. The usual choice of transmission line for polymer modulators is the microstrip line. With the ground plane immediately under the waveguide cladding and the microstrip immediately on top, the microstrip line provides excellent concentration of the drive electric field in the electro-optic waveguide, and thus very efficient conversion of RF drive power to the electric field that causes modulation. Microstrip lines, illustrated in Fig. 7.9, are TEM-like transmission lines characterized by a dielectric height h, dielectric relative permittivity ε_r, conductor width W, and conductor thickness t. Empirical design equations are available for computing the characteristic impedance Z_{0m}, effective dielectric constant ε_{re}, and propagation loss $A_c(f)$ [16]. For the case $W/h > 1$, the following equations are used (see Ref. 16 for the case $W/h < 1$).

$$Z_{0m} = \frac{\eta}{\sqrt{\varepsilon_{re}}} \left[\frac{W_e}{h} + 1.393 + 0.667 \ln\left(\frac{W_e}{h} + 1.444 \right) \right]^{-1}, \quad (7.11)$$

$$\varepsilon_{re} = \frac{\varepsilon_r + 1}{2} + \frac{\varepsilon_r - 1}{2} \left[\left(1 + 12\frac{h}{W} \right)^{-\frac{1}{2}} \right] - \frac{\varepsilon_r - 1}{4.6} \frac{t/h}{\sqrt{W/h}}, \quad (7.12)$$

$$A_c(f) = 20 \, \log(e) \frac{R_s(f)Z_{0m}\varepsilon_{re}(f)}{\eta^2 h} \left(1 + \frac{h}{W_e} \left(1 + \frac{1.25}{\pi} \ln\left(\frac{2h}{t} \right) \right) \right)$$
$$\times \left(\frac{W_e}{h} + \frac{0.667W_e/h}{W_e/h + 1.444} \right), \quad (7.13)$$

where

$$\frac{W_e}{h} = \frac{W}{h} + \frac{1.25}{\pi} \frac{t}{h} \left[1 + \ln\left(\frac{2h}{t} \right) \right], \quad (7.14)$$

η is the free-space characteristic impedance, $R_s(f)$ is the skin depth, and $\varepsilon_{re}(f)$ is the frequency-dependent effective dielectric constant given in Ref. 16. For example, taking the microstrip width W to be the most freely varied parameter, for a polymer with $\varepsilon_r = 3.4$, total polymer thickness $h = 8$ μm, and a conductor thickness $t = 4$ μm, the dependence of characteristic impedance Z_{0m} and effective dielectric constant ε_{re} on microstrip width are plotted in Fig. 7.10. For a characteristic impedance of 50 Ω, the conductor width should be chosen to be 16.1 μm, and

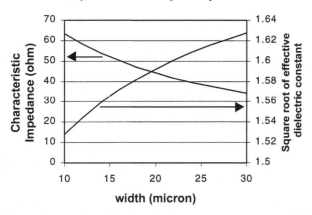

Figure 7.10. With a polymer material, the microstrip structure can easily be designed for both 50-Ω characteristic impedance and small velocity mismatch.

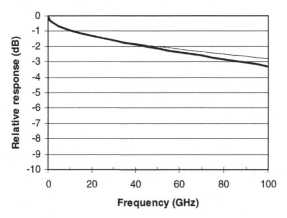

Figure 7.11. Computed frequency response for 1-cm electrode length. The thin line shows the calculated response for zero velocity mismatch, and the thick line includes the effect of velocity mismatch for the numerical example in the text.

the square root of the RF effective dielectric constant will be 1.57. For an optical waveguide with effective index of 1.63 and an electrode length of 1 cm, the 3-dB electrical bandwidth (1.5-dB optical) would be over 260 GHz, in the absence of any RF loss. Figure 7.10 shows that the microstrip line can be designed for 50-Ω matching with minimal impact on the velocity mismatch. In the absence of any velocity mismatch, the optical response would drop 1.5 dB, corresponding to 3-dB electrical bandwidth, at the frequency at which $\alpha L = 6.4$ dB. This does not take account of any frequency dependence of the microstrip line equivalent circuit. The computed frequency response for a 1-cm electrode length, including only the skin effect RF loss and velocity mismatch, is shown for this example in Fig. 7.11. The computation suggests a 3-dB bandwidth of over 80 GHz, with the velocity mismatch contributing only a small fraction of the total degradation of response at high frequency.

The most widely used design equations were developed to match experimental results for conventional microstrip lines [16], which are characterized by substrates with thickness from 100 to 500 μm and dielectric constants from 10 to 13. There are also design equations developed to describe thin film microstrip lines for multilayer interconnects, [17,18] for which the "substrate" is a deposited insulator of thickness 1 to 25 μm and the dielectric constant ranges from 3 to 5, but the conductor thickness is only 1 to 2 μm. The microstrip lines used to drive polymer modulators use a polymer thin film of 5–15 μm as the "substrate", with dielectric constants of 2.5 to 4, and metal thickness of 3–8 μm, and may be better described by the latter set of design equations.

While the microstrip structure described above works very well for driving a modulator, the strip width is much narrower than the post diameter or tab width on commonly available RF connectors. It is useful to create a transition structure from the microstrip to the outside contact. One design is shown in Fig. 7.12. At the connector end is a coplanar transmission line with feature widths comparable to the RF connector widths. In the transition region the center conductor gradually becomes narrower until it matches the width of the microstrip conductor. The gaps between center conductor and ground become narrower as needed to maintain 50-Ω impedance along the length of the coplanar line, until the ground half-planes come together underneath the polymer to become the microstrip ground plane. This transition structure can be designed so that 50-Ω impedance is maintained at all points to minimize back-reflection. For higher frequencies, antipodal finline transition structures have been developed to adapt from millimeter-wave waveguide to microstrip, which will cover approximately the same frequency range as the RF waveguide it is designed for [13,19].

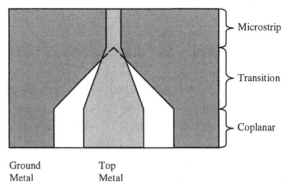

Figure 7.12. The adiabatic transition from the microstrip electrode to the coplanar structure. The wide center electrode of the coplanar structure is tapered to the width of the microstrip, while the gap between the center and ground electrodes is also tapered to maintain 50-Ω impedance. The two halves of the ground electrode are joined underneath the microstrip.

There have been several demonstrations of broadband operation of polymer modulators that show little frequency rolloff at high frequencies. A common element of many of these demonstrations is that the fundamental frequency rolloff discussed above is masked somewhat by resonances due to input and output connections to the driving microstrip line. Since the polymer modulator microstrip electrode does not follow a conventional microstrip geometry, there are no commercial connectors optimized for launching into it; commercial connectors generally have much larger dimensions than the 50-Ω microstrip on the optical waveguide structure. Teng [14] demonstrated optical intensity modulation with a Mach–Zehnder modulator from 10 MHz to 40 GHz with less than 3 dB electrical rolloff, with a drive electrode 1.2 cm long; see Fig. 7.13. Shi [20] and Wang [21,22] showed the electro-optic phase response from 2 to 60 GHz, with zero rolloff from 2 to 26 GHz, a drop of about 6 dB optical from 26 to 34 GHz, followed by no consistent rolloff from 34 to 60 GHz; see Fig. 7.14. In addition, operation from 74 to 113 GHz has been demonstrated with better than 3-dB electro-optic response

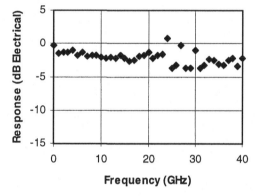

Figure 7.13. Teng's modulator shows less than 3-dB rolloff from 0 to 40 GHz. Reprinted with permission from C. C. Teng, "Traveling-wave polymeric optical intensity modulator with more than 40 GHz of 3-dB electrical bandwidth", *Appl. Phys. Lett.*, **60**, 1538–40, 1992, copyright 1992, American Institute of Physics.

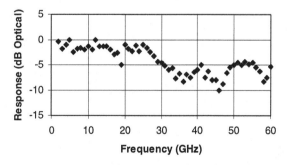

Figure 7.14. Wang's modulators show no rolloff from 2 to 26 GHz, and no consistent rolloff from 34 to 60 GHz. From Ref. 20, with permission.

across the measurement band, using commercial W-band coplanar probes to couple RF power into the driving microstrip line [12].

7.6 Approaches to low half-wave voltage

7.6.1 Dependence of V_π on material and device parameters

For modulators in RF systems it is desirable to minimize the half-wave voltage, V_π, for the purpose of maximizing RF gain and sensitivity, and minimizing noise figure and power consumption. The value of V_π is determined as follows:

$$V_\pi(f) = \frac{\lambda h}{\Gamma n^3 r_{33} R(f) L},\qquad(7.15)$$

where
  λ = optical wavelength,
  h = effective polymer thickness,
  n = refractive index,
  r_{33} = electro-optic coefficient,
  $R(f)$ = effect of electrode attenuation (see previous section),
  L = length of electro-optic interaction,

and Γ is a fill factor that accounts for both non-uniform distribution of electric field in the microstrip stucture and (usually) confinement of the electro-optic effect to the core layer.

The half-wave voltage can be minimized by improvements in both geometrical factors and the material factors, n and r_{33}. The major improvements to be made in V_π will result from increasing the material parameter r_{33}. Significant improvements may also be obtained by optimizing geometrical factors.

7.6.2 Geometrical factors

The polymer thickness h and interaction length L are the two factors that explicitly influence V_π, but $R(f)$ depends on L, while h and Γ may be related.

For small V_π, small h is desired, but h is limited by the evanescent tail of the optical power into the cladding, particularly by resistive loss in the metal ground plane or drive electrode at the metal–cladding interface. The penetration of this tail depends on the refractive index difference between core and cladding, and on the core layer thickness, as well as the core width to a lesser extent. The fraction of the optical power that extends into the claddings, $1 - \Gamma$, is reduced by making the light better confined in the core, which is accomplished by increasing the core–cladding index difference and increasing the core thickness and width, without increasing them so much that the waveguide supports higher-order modes. In addition, the exponential attenuation constant in the cladding increases as the difference between effective index and cladding index increases. For a given choice of relative refractive indices and core thickness, the optical power at the metal surface is reduced by increasing the cladding thickness, which in turn would increase V_π. Thus, for small V_π, the waveguide parameters would be selected so that the second waveguide mode is just cut off, and the cladding is just thicker than that thickness at which resistive loss would become significant.

The half-wave voltage can also be decreased by increasing the electro-optic interaction length, L, until the decrease of $R(f)$ cancels the effect of increasing L in the desired frequency range.

As an example, if an r_{33} coefficient of 130 pm/V can be achieved in a polymer modulator, then with the following assumptions,

$\lambda = 1.3 \ \mu m$,
$h = 8 \ \mu m$,
$\Gamma = 0.9$,
$n = 1.7$,
$R(f) = 0.9$,
$L = 2 \ cm$,

the figure of merit $V_\pi = 1.0$ V could be achieved with the standard microstrip design.

Another device design that can reduce V_π is the push-pull configuration. The concept is to cause equal but opposite phase modulation in the two Mach–Zehnder arms, thus doubling the sensitivity, and halving V_π. In one configuration [23] the two arms of the Mach–Zehnder are poled with opposite polarity by two different poling electrodes, then driven with the same polarity by the same microstrip electrode, as illustrated in Fig. 7.15. This creates conflicting requirements on the lateral separation between the two arms; a large separation is desired to avoid arcing between the poling electrodes which are held at high voltage with opposite polarities during poling, and also to avoid optical coupling between the two arms; but a small separation is desired to minimize the microstrip width, and thus the polymer thickness, to maximize electric field in the microstrip structure while still maintaining 50-Ω impedance. These conflicting requirements can be partially ameliorated by the use of a modified microstrip drive electrode [4,24] with a split ground, which also serves as the separately controlled poling electrodes, as illustrated in Fig. 7.16. Not only are the opposite-polarity poling fields isolated from each other by polymer material, rather than air, so that the breakdown field can be much higher, but the top electrode can be wider than in the microstrip design, since the characteristic impedance is dominated, to first order, by the overlap between top and

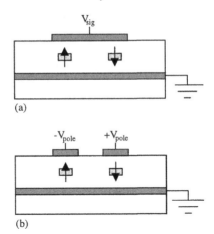

Figure 7.15. (a) Push-pull modulator design with standard microstrip drive electrode. (b) To achieve opposite polarity of the electro-optic effect in the two arms, poling was performed with two different poling electrodes.

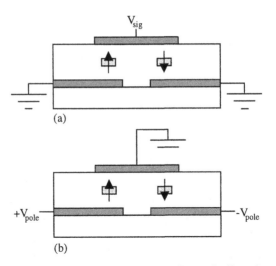

Figure 7.16. (a) Push-pull modulator design with microstrip line-slot ground drive electrode. (b) To achieve opposite polarity of the electro-optic effect in the two arms, during poling the top electrode is grounded, and opposite polarities of the poling voltage are applied to each side of the split ground electrode.

bottom electrodes. This modified microstrip electrode was used to demonstrate an RF modulator [4,24] with half-wave voltage of only 0.8 V, a 3-cm electrode length, a polymer thickness of 10 μm, and characteristic impedance of 38 Ω. Achieving a characteristic impedance of 50 Ω is challenging for waveguide structures [25] with polymer layers thinner than 10 μm. This experimental 0.8-V modulator was made with PMMA as the host material. PMMA does not provide adequate thermal stability, but the split ground microstrip and comparable values of the r_{33} coefficient could be reproduced with more stable materials. Another electrode design allowing a large separation between push-pull poling electrodes involves splitting the 50-Ω microstrip feed into two parallel 25-Ω microstrip lines [26,27].

7.6.3 Material factors

The half-wave voltage depends largely on the polymer material electro-optic coefficient, r_{33}, which in turn depends mainly on the chromophore hyperpolarizability, the number density of chromophores in the material, and how efficiently they are aligned. A simple model is often used that relates the macroscopic r_{33} coefficient to the microscopic hyperpolarizability β_{eo} of the chromophore molecule. This model is based on the assumptions that only the chromophore ground state and first excited state contribute to the electro-optic effect, and that the molecules do not interact with each other [5,28]:

$$r_{33} = \frac{2 f_{dc} f_{eo}^2}{n^4} N \beta_{eo} L_3 \left(\frac{\mu f_{dc} E_{pol}}{k_B T_{pol}} \right), \tag{7.16}$$

$$f_{dc} = \frac{\varepsilon(n^2 + 2)}{n^2 + 2\varepsilon}, \qquad f_{eo} = \frac{n^2 + 2}{3}, \tag{7.17}$$

$$\beta_{eo} = \beta_0 \frac{3 - \lambda_0^2/\lambda^2}{3\left(1 - \lambda_0^2/\lambda^2\right)^2}, \tag{7.18}$$

$$\beta_0 = \beta_{shg}\left(1 - \lambda_0^2/\lambda_{shg}^2\right)\left(1 - 4\lambda_0^2/\lambda_{shg}^2\right), \tag{7.19}$$

where
f_{dc} = low-frequency field factor,
f_{eo} = optical-frequency field factor,
n = refractive index,
N = chromophore number density,
L_3 = third order Langevin function,
μ = molecular permanent dipole moment,
E_{pol} = poling electric field,
k_B = Boltzmann constant,
T_{pol} = poling temperature,
ε = low-frequency dielectric constant,
λ = optical wavelength,
λ_0 = resonance wavelength,
λ_{shg} = wavelength of second harmonic generation measurement,
β_0 = non-resonant hyperpolarizability,
β_{shg} = hyperpolarizability measured by second harmonic generation.

In practice, it is easier to characterize second harmonic generation of the chromophore in solution than the electro-optic coefficient in a poled film, so the non-resonant hyperpolarizability β_0 is often derived from the measured hyperpolarizability β_{shg} responsible for second harmonic generation. In the literature the product $\mu\beta_{shg}$ is often reported rather than β_{eo} or β_0 because it is directly determined in the second harmonic generation measurement.

As seen in Eq. (7.15) above, the material parameter that determines V_π is not r_{33}, but $n^3 r_{33}$:

$$n^3 r_{33} = \frac{2 f_{dc} f_{eo}^2}{n} N \beta_{eo} L_3 \left(\frac{\mu f_{dc} E_{pol}}{k_B T_{pol}} \right). \tag{7.20}$$

Equation (7.15) shows explicitly that V_π scales with the optical wavelength, λ. However, the electro-optic coefficient and refractive index are both wavelength dependent. The electro-optic coefficient, as shown in Eqs. (7.16)–(7.18), increases as the resonance wavelength approaches the wavelength of operation. The refractive index also increases as the resonance wavelength approaches the wavelength of operation, but since most of the refractive index is non-resonant, this is a less significant effect. Thus, for a given core material, as the wavelength is increased, such as from 1.3 μm to 1.55 μm, the increase of V_π seen explicitly in Eq. (7.15) is compounded by decreases in both r_{33} and n^3.

The value of the r_{33} electro-optic coefficient depends significantly on three major factors; the molecular hyperpolarizability β_{eo}, the number density N, and the alignment order parameter expressed by the third order Langevin function L_3.

The first major factor in Eq. (7.16) is the molecular hyperpolarizability, β_{eo}. To have a large value of β_{eo}, the molecule generally must have an electron donor group on one end, an electron acceptor group on the other end, and a π-bonded bridge between the two that electronically couples them. The value of molecular hyperpolarizability depends on several factors, such as the length and nature of the bridge, the electron donor strength, and electron acceptor strength. In recent years tremendous progress has been made in increasing the magnitude of β_0 by molecular engineering. A standard chromophore used in an early generation of electro-optic polymers, DANS, has a β_0 value of 55×10^{-30} esu [29], and recent developments have produced new chromophores such as FTC with β_0 values as large as 635×10^{-30} esu [10].

Equation (7.18) shows that β_{eo} has a wavelength dispersion determined by the resonance wavelength, λ_0, as illustrated in Fig. (7.17). The hyperpolarizability β_0 is related to more fundamental quantities [29] as follows:

$$\beta_0 \propto \frac{(\mu_{ee} - \mu_{gg})\mu_{ge}^2}{E_{ge}^2}, \tag{7.21}$$

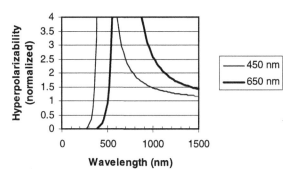

Figure 7.17. Wavelength dispersion of the hyperpolarizability, β_{eo}, normalized by β_0. The lighter line indicates a typical resonance wavelength of 450 nm, the darker line, a red-shifted resonance wavelength of 650 nm. The curves show that β_{eo} is enhanced at the operating wavelength for the chromophore with the longer-wavelength resonance.

where μ_{gg} and μ_{ee} are the dipole moments of the ground and excited states, respectively, and μ_{ge} and E_{ge} are the transition dipole moment and transition energy from the ground to the excited state, respectively. The transition energy term in the denominator indicates the trend of chromophores with larger values of β_0 also having longer resonance wavelengths, further increasing the value of β_{eo}. Figure 7.17 shows that for a fixed wavelength, like 1300 nm, the resonant contribution to β_{eo} is larger when the resonance wavelength is longer. Because there is a broad absorption line centered at the resonance wavelength with a long absorption tail, the molecules with higher β_0 tend to have significant absorption at shorter wavelengths, and thus may be restricted to use at longer wavelengths.

The second major factor in Eq. (7.16) is the number density of chromophores, N. This is related to the weight fraction, F_w, of the chromophore material in the polymer composite, the composite mass density ρ, and the chromophore molecular weight, M; $N = N_A F_w \rho / M$, where N_A is Avogadro's number. The chromophores with higher β tend to have larger values of molecular weight, so the quantity $\mu \beta / M$ is often cited as a figure of merit. Fortunately, the molecular weight does not tend to scale with the value of β, so that the higher-β chromophores also tend to have higher values of this figure of merit. Guest–host polymer composites, in which the guest chromophore is dissolved in the host polymer, have dye loadings limited on the order of 15–35% by the limitations of solubility. The chromophore molecule can also be chemically attached to a polymer backbone in several different configurations, and the chemical attachment can help improve the overall solubility and increase the concentration of the chromophore. Attached polymers have been reported in which the chromophore makes up 65% of the mass in the composite material [8], for a number density of approximately 16×10^{20} molecules per cm^3.

The third major factor in Eq. (7.16) is the order parameter describing the alignment of the chromophore molecules by the poling field, which can also be expressed as an expectation value related to the angle θ between the poling field and the permanent dipole moment of each chromophore molecule:

$$L_3(x) = \langle \cos^3 \theta \rangle. \tag{7.22}$$

This parameter has a value between -1 and 1. The L_3 function is linear for small poling fields, and saturates very gradually for larger poling fields. In practice, the poling field may be limited by the onset of electrical breakdown during poling, at 200–400 V/μm, depending on the cleanliness of the fabrication conditions and quality of the polymer thin films, or by conductivity mismatches between core and cladding materials [30], which tend to be more disadvantageous at higher temperatures and higher poling fields. Values of this parameter as larger as 0.3 to 0.5 have been achieved in poled polymer films.

A simple calculation of r_{33}, based on the FTC chromophore [10], and on the following assumptions, shows what might be anticipated for the electro-optic coefficient of polymer modulators:

$\beta_{eo} = 780 \times 10^{-30}$ esu,
$N = 5.4 \times 10^{20}$ cm^{-3},
$\langle \cos^3 \theta \rangle = 0.5$,
$n = 1.7$,
$f_{dc} = 1.72$,
$f_{eo} = 1.63$,
$r_{33} = 129$ pm/V.

In practice, such high values of r_{33} have been difficult to achieve, because the parameters listed are not all independent, and the model breaks down for large concentrations of chromophores, where their response to the poling field is not independent, and for large poling fields. Dalton *et al.*, have developed models to describe chromophore aggregation due to the electric field interactions between chromophores [31]. This aggregation would commonly result in chromophores pairing with their permanent dipole moments oppositely aligned, effectively cancelling their contributions to the bulk electro-optic coefficient. Dalton *et al.* have also shown the effectiveness of a solution to the aggregation effect by adding long chemical spacer groups to the molecular bridge, tending to prevent molecules from coming so close together that they would permanently pair up [10,31].

Electro-optic r_{33} coefficients as large as 90 pm/V at 1300 nm wavelength have been reported with a β_{eo} value of approximately 1100×10^{-30} esu, number density of 3×10^{20} cm^{-3}, and alignment order parameter of approximately 0.5 [4,32].* †

---

* The product of mu-beta measured by second harmonic generation in Ref. 32 was in the range of 14 000 to 19 000 for three related chromophores. The value of the hyperpolarizability can be quite sensitive to the local molecular environment. Thus, the value of the hyperpolarizability measured in liquid solution may underestimate, or overestimate, the value in a solid film in a modulator.

† In Ref. 4, the electro-optic coefficient at 1.06 μm was reported as approximately 125 pm/V, corresponding to 90 pm/V at 1.3 μm.

At this writing, several efforts continue to achieve dramatically larger values of r_{33}, and much smaller values of V_π in RF modulators. These efforts include development of chromophores with larger values of β and spacer groups to limit aggregation, chemical attachment of chromophores to polymer chains, and efforts to increase the poling field inside the core material.

7.7 Summary

Polymer electro-optic modulators for RF have many potential advantages over more mature electro-optic technologies. The polymer material is used as a thin film on a variety of substrates, potentially enabling conformal devices, integration with semiconductor electronics or opto-electronics, and low cost. The low dielectric constant of polymer materials enables nearly flat frequency response to nearly 100 GHz (with a 1-cm electrode length) and simple simultaneous realization of 50-Ω drive impedance and velocity matching. The efficiency of the microstrip drive electrode structure and the potential for material engineering enable very low values of half-wave voltage, and consequently high RF gain and low noise figure. Modulators with half-wave voltage of 0.8 V, with voltage–length product of 2.4 V cm, have already been achieved, and further reduction of the half-wave voltage is expected.

References

1. G. E. Betts, L. M. Johnson, and C. H. Cox, III, "Optimization of externally modulated analog optical links", *Proc. SPIE* **1562**, 281–302, 1991.
2. E. Ackerman, S. Wanuga, J. MacDonald, and J. Prince, "Balanced receiver external modulation fiber-optic link architecture with reduced noise figure", in *IEEE Microwave Theory Tech. 1993 Symp. Dig.*, pp. 723–6.
3. M. S. Islam, T. Chau, S. Mathai, T. Itoh, M. C. Wu, D. L. Sivco, and A. Y. Cho, "Distributed balanced photodetectors for broadband noise suppression", *IEEE Trans. Microwave Theory Tech.* **47**, 1282–8, 1999.
4. Y. Shi, C. Zhang, H. Zhang, J. H. Bechtel, L. R. Dalton, B. Robinson, and W. H. Steier, "Low (sub-1-volt) halfwave voltage polymeric electro-optic modulators achieved by controlling chromophore shape", *Science*, **288**, 119–22, 2000.
5. K. D. Singer, M. G. Kuzyk, J. E. Sohn, "Second-order nonlinear-optical processes in orientationally ordered materials: relationship between molecular and macroscopic properties" *J. Opt. Soc. Am. B*, **4**, 968–76, 1987.
6. H. S. Nalwa, T. Watanabe, and S. Miyata, "Organic materials for second-order nonlinear optics", in *Nonlinear Optics of Organic Molecules and Polymers*, eds. H. S. Nalwa and S. Miyata, CRC Press, Boca Raton, 1997, p. 249.
7. Y. M. Cai and A. K.-Y. Jen, "Thermally stable poled polyquinoline thin film with very large electro-optic response", *Appl. Phys. Lett.* **67**, 299, 1995.
8. C. C. Teng, "High-speed electro-optic modulators from nonlinear optical polymers", in *Nonlinear Optics of Organic Molecules and Polymers*, eds. H. S. Nalwa and S. Miyata, CRC Press, Boca Raton, 1997, pp. 445–6.

9. R. J. Roe, "Glass transition", in *Encyclopedia of Polymer Science and Engineering*, 2nd Edn., Vol. 7, ed. J. I. Kroschwitz, John Wiley & Sons, New York, 1987, pp. 531–44.

10. L. Dalton, A. Harper, A. Ren, F. Wang, G. Todorova, J. Chen, C. Zhang, and M. Lee, "Polymeric electro-optic modulators: from chromophore design to integration with semiconductor very large scale integration electronics and silica fiber optics", *Ind. Eng. Chem. Res.* **38**, 8–33, 1999.

11. S. S. H. Mao, Y. Ra, L. Guo, C. Zhang, and L. R. Dalton, "Progress toward device-quality second-order nonlinear optical materials 1. Influence of composition and processing conditions on nonlinearity, temporal stability, and optical loss", *Chem. Mater.* **10**, 147, 1998.

12. D. Chen, H. R. Fetterman, A. Chen, W. H. Steier, L. R. Dalton, W. Wang, and Y. Shi, "Demonstration of 110 GHz electro-optic polymer modulators", *Appl. Phys. Lett.* **70**, 3335–7, 1997.

13. D. Chen, D. Bhattacharya, A. Udupa, B. Tsap, H. R. Fetterman, A. Chen, S.-S. Lee, J. Chen, W. H. Steier, and L. R. Dalton, "High-frequency polymer modulators with integrated finline transitions and low V_π", *IEEE Photon. Technol. Lett.* **11**, 54–6, 1999.

14. C. C. Teng, "Traveling-wave polymeric optical intensity modulator with more than 40 GHz of 3-dB electrical bandwidth", *Appl. Phys. Lett.* **60**, 1538–40, 1992.

15. R. C. Alferness, "Waveguide electrooptic modulators", *IEEE Trans. Microwave Theory Tech.* **30**, 1121–37, 1982.

16. K. C. Gupta, R. Garg, I. Bahl, and P. Bhartia, *Microstrip Lines and Slotlines*, 2nd Edn., Artech House, Boston, pp. 102–11, 1996.

17. H. Y. Lee and T. Itoh, "Phenomenological loss equivalence method for planar quasi-TEM transmission lines with a thin normal conductor or superconductor", *IEEE Trans. Microwave Theory Tech.*, **37**, 1904–9, 1989.

18. G. E. Ponchak and A. N. Downey, "Characterization of thin film microstrip lines on polyimide", *IEEE Trans. Components, Packag. Manuf. Technol. B*, **21**, 171–6, 1998.

19. K. C. Gupta, R. Garg, I. Bahl, and P. Bhartia, *Microstrip Lines and Slotlines*, 2nd Edn., Artech House, Boston, pp. 364–8, 1996.

20. Y. Shi, W. Wang, J. H. Bechtel, A. Chen, S. Garner, S. Kalluri, W. H. Steier, D. Chen, H. R. Fetterman, L. R. Dalton, and L. Yu, "Fabrication and characterization of high-speed polyurethane-disperse red 19 integrated electrooptic modulators for analog system applications", *IEEE J. Sel. Topics Quantum Electron.*, **2**, 289–99, 1996.

21. W. Wang, D. Chen, H. R. Fetterman, Y. Shi, W. H. Steier, and L. R. Dalton, "40-GHz polymer electrooptic phase modulators", *IEEE Photon. Technol. Lett.* **7**, 638–40, 1995.

22. W. Wang, D. Chen, H. R. Fetterman, Y. Shi, W. H. Steier, and L. R. Dalton, "Optical heterodyne detection of 60 GHz electro-optic modulation from polymer waveguide modulators", *Appl. Phys. Lett.* **67**, 1806–8, 1995.

23. W. Wang, Y. Shi, D. J. Olson, W. Lin, and J. H. Bechtel, "Push-pull poled polymer Mach–Zehnder modulators with a single microstrip line electrode", *IEEE Photon. Technol. Lett.* **11**, 51–3, 1999.

24. Y. Shi, W. Lin, D. J. Olson, J. H. Bechtel, and W. Wang, "Microstrip line-slot ground electrode for high-speed optical push-pull polymer modulators", paper FB3-1, *Organic Thin Films for Photonics Applications 1999*, Santa Clara, California, September 1999.

25. D. G. Girton, S. L. Kwiatkowski, G. F. Lipscomb, and R. S. Lytel, *Appl. Phys. Lett.* **58**, 1730, 1991.

26. K. H. Hahn, D. W. Dolfi, R. S. Moshrefzadeh, P. A. Pedersen, and C. V. Francis, "Novel two-arm microwave transmission line for high-speed electro-optic polymer modulators", *Electron. Lett.*, **30**, 1220–2, 1994.
27. S. Ermer, D. G. Girton, L. S. Dries, R. E. Taylor, W. Eades, T. E. Van Eck, A. S. Moss, and W. W. Anderson, "Low-voltage electro-optic modulation using amorphous polycarbonate host material", *Proc. SPIE* **3949**, 148–55, 2000.
28. K. D. Singer, S. J. Lalama, J. E. Sohn, and R. D. Small, "Electro-optic organic materials", in *Nonlinear Optical Properties of Organic Molecules and Crystals*, Vol. 1, eds. D. S. Chemla and J. Zyss, Academic Press, Orlando, 1987, p. 462.
29. S. R. Marder, L.-T. Cheng, B. G. Tiemann, and D. N. Beratan, "Structure/property relationships for molecular second-order nonlinear optics", in *Nonlinear Optical Properties of Organic Materials IV*, ed. Kenneth D. Singer, *Proc. SPIE*, **1560**, 86–97.
30. D. G. Girton, W. W. Anderson, J. A. Marley, T. E. Van Eck, and S. Ermer, "Current flow in doped and undoped electro-optic polymer films during poling", in *Organic Thin Films for Photonics Applications*, Vol. 21, 1995 OSA Technical Digest Series, Optical Society of America, Washington DC, 1995, pp. 470–473.
31. A. Harper, S. Sun, L. R. Dalton, S. M. Garner, A. Chen, S. Kalluri, W. H. Steier, and B. H. Robinson, "Translating microscopic optical nonlinearity into macroscopic optical nonlinearity: the role of chromophore–chromophore electrostatic interactions", *J. Opt. Soc. Am. B*, **15**, 329–37 (1998).
32. Cheng Zhang, Ph. D. thesis, University of Southern California, May 1999, p. 161.

8

Photodiodes for high performance analog links

P. K. L. YU

University of California, San Diego

MING C. WU

University of California, Los Angeles

8.1 Introduction

The optical detector plays an important role in an analog fiber link as its performance determines the baseline characteristics of the link. In the past, the development of the detector has evolved around on-off operation, high speed and high responsivity aspects due to its extensive uses in digital fiber links [1–3]. However, for analog applications, as discussed in Chapter 1, the detector has to incorporate additional designs to ensure high power and high linear dynamic range operation, both of which are essential for meeting the high link gain and low noise figure requirements.

The organization of this chapter goes as follows. In the remainder of this section an overall view of optical detection is presented including key terminology and figures of merit. Section 8.2 compares the baseline properties of different photodiodes. Section 8.3 briefly discusses the noise sources for photodiodes, with special attention to shot noise and noise attributed to laser relative intensity noise. These noise sources are dependent on the optical power. Section 8.4 highlights the research status of photodiode nonlinearity. Section 8.5 presents some recent advances in photodiodes related to analog fiber link applications.

8.1.1 Definitions

Semiconductor materials are commonly employed for optical detectors mainly because the resultant detectors are usually small, low noise, operate with a low voltage, and can be easily assembled with the rest of the receiver circuits. Photons interact with electrons in semiconductors and are absorbed in processes whereby electrons make a transition from a lower to a higher energy state. These electrons are detected in an external circuit in the form of either current or voltage signal. The absorption is critically dependent on the photon energy and a characteristic absorption spectrum is associated with each material. The absorption spectrum can be sensitive to the electric field and temperature variations.

The absorptivity α_o, defined as the incremental decrease in optical intensity per unit length per unit incident intensity, is commonly referred to as the absorption coefficient of the material. The photon energy $h\nu$ (where h is the Planck's constant and ν is the frequency) and the wavelength λ are simply related by $\lambda = \frac{1.24}{h\nu}$ where λ is measured in units of micrometers and energy $h\nu$ is in electron volts. The absorptivity describes the power decay of an optical beam as a function of its propagation distance inside the material. In general, a large α_o is desirable for obtaining high optical-to-electrical power conversion efficiency. A small α_o means a thick absorption region is needed for significant absorption to take place. However, too large an α_o results in a small penetration depth, and many of the electron and hole pairs generated can be lost via near-surface recombination and thus their existence is not detected in the external circuit.

The bandgap energy of crystalline silicon corresponds to wavelengths around 1.1 μm, so it is mainly used for photodetection in the 0.8–1.0 μm spectral range. For the spectral range of 1.3–1.55 μm, it is more practical to use germanium and compound semiconductors such as InGaAs(P) and InAlGaAs that have bandgap energies corresponding to wavelengths ∼1.6 μm. The Germanium photodetector is not popular at present because of its high dark current, i.e., the current that leaks through the device.

An important performance parameter of an optical detector at a given wavelength is the external quantum efficiency, η. It is defined as the ratio of the number of electrons generated to the number of incident photons before any photogain occurs. For a given structure, η takes into account the surface optical reflection loss and other losses in the detector. In general, η depends on the absorptivity of the materials and dimensions of the absorption region. Photogain M can result from carrier injection in semiconductor materials, as in the case of photoconductive devices, or from impact ionization, as in the case of avalanche photodiodes [4].

The responsivity R of a detector is the ratio of the output electrical response to the input optical power and can be expressed as

$$R = M\eta \frac{q}{h\nu} \text{ A/W}, \tag{8.1}$$

where q is the electronic charge. For applications where a relatively low level of optical power is incident on the photodetector, the responsivity remains unchanged as the optical power is varied slightly. The relation between R (or η) and wavelength is usually referred to as the spectral response characteristic. For applications where a high level of optical power is incident on the detector, R can drop due to the carrier screening effect, as discussed in Section 8.4 [5]. Nonlinear response can also arise as a consequence of detection saturation.

The response speed of a photodetector can be very important for many analog and digital applications. It is defined as the frequency where the detected electrical power is half that near DC. The characteristic 3-dB cut-off frequency f_{3dB} is inversely proportional to the response time of the detector. However, the f_{3dB} can vary with optical power in some structures, and can be greatly reduced at high power, as discussed in Section 8.4.

The response time τ depends on the transit time of carriers, the carrier diffusion, the carrier multiplication in the semiconductor, as well as the circuit time constant. On one hand, materials with a long minority carrier diffusion length assure the collection of carriers generated outside the high field region and thus the photodetector can achieve high quantum efficiency. On the other hand, the detector response time can be degraded by the slow diffusion process. For instance, the average diffusion velocity v_{diff} for minority carriers in a p-type material is equal to $L_n/\tau_n = D_n/L_n$ for a thick p-type region. In Section 8.5.4 we mention a novel ultra-high speed detector that makes use of optimized diffusion limited transport [6].

The response time τ can be estimated from the f_{3dB} obtained from frequency modulation response measurement using a modulated optical carrier generated by direct modulation or by external modulation of laser light, or simply by optical mixing of two laser sources whose emission wavelengths are temperature tuned. Alternatively, τ can be determined from the time-domain response of the detector illuminated with short optical pulses, such as those generated by mode-locked lasers, or those from a gain switched laser driven by a comb generator. Provided that the rise time and the fall time of the optical pulse are much shorter than those of the detector, τ can be estimated from both the rise time and the fall time of the detected current pulse. For a very high speed detector (>40 GHz), a high speed sampling oscilloscope is usually used to measure the total time-domain response, including the detector response time, the effect of the finite optical pulse width and the finite response time of the sampling circuit. The detector response can be de-embedded from the total time-domain response [7,8].

For detectors with photogain, an additional figure of merit – the so-called gain–bandwidth (MB) product is defined. It is simply the product of the 3-dB bandwidth and the DC gain. In many cases, the MB product is closely related to device geometry and device parameters.

The quality of the detected signals can be degraded due to the presence of noise. The noise performance of a photodetector is a major consideration in determining the overall system dynamic range for analog applications, and in determining the bit-error-rate for digital applications. Noise can arise at the transmitter, the electrical and optical amplifiers, and the receiver. For fiber optic links, the common noise sources are the laser intensity noise, the Poisson (shot) noise due to the quantized nature of photons and the finite response time of the detector, dark current noise due to generation and recombination of carriers in the detection region, and thermal (Johnson) noise due to random motion of carriers in the resistive load of the detector. It should be noted that, for free space analog links, the background radiation constitutes an additional noise source. Three commonly used measurements of noise performance are the signal-to-noise (S/N) ratio, noise equivalent power (NEP), and specific detectivity (D^*) [1]. The S/N ratio is simply the ratio of detected electrical signal power to the noise power at the output of the detector (or receiver). The maximum S/N places a limit on the dynamic range of the link. The NEP of a detector is the optical power that needs to be incident on the detector so that the S/N equals unity for a given wavelength, detector temperature, and resolution bandwidth. The specific detectivity is the inverse of the NEP normalized to the square root of the detector area A and the bandwidth B (it should be noted that while the NEP is normally used to characterize the sensitivity of a photodiode, D^* is used when the incident flux is larger than the detector as is commonly the case for free space optical links).

For high performance photodetectors, materials with good crystal quality and with high purity are usually employed so as to minimize trapping and recombination of photogenerated carriers at defect sites. In photodetectors where heterostructure is used, close matching of crystal lattice constants along the material interface is usually a safeguard against premature failure of detectors, as interfacial states or traps (which lead to an internal potential barrier) can be the source of gradual degradation of the dark current and other detector characteristics. In addition, two material-related issues are commonly encountered in detector designs: namely the doping distribution in the detection region of the device and the nature of the metal–semiconductor contact. The doping profile affects the effective width of the depletion region and can be tailored to fine tune the carrier transport. For instance, a low doping level of the depletion region is often desirable for photodiodes for fast sweep out of carriers at low bias voltage as well as for small device capacitance. A low-doped material can be useful for making high linearity and high power photodiodes (see Section 8.4). The properties of the metal–semiconductor contact can affect carrier injection (and thus gain), the response speed, and the saturation optical power levels (and thus linearity) of the detector, as the contact itself can be viewed as a potential barrier.

For a valence-band-to-conduction-band (or simply band-to-band) transition of electrons that involves the absorption of a photon – which is called the intrinsic absorption process – the photon energy needs to be larger than that of the bandgap for large α_0. However, there are several ways to detect photons with energy less than the bandgap energy. For instance, for far-infrared detection, such as that in extrinsically doped silicon impurity band photoconductor, the photoabsorption occurs as the electron makes a transition from an impurity level within the band gap to the nearby band. Free-carrier absorption can take place when the electron makes a transition within the same band while a photon is being absorbed. The Franz–Keldysh effect in bulk semiconductors [9] and the quantum confined Stark effect in quantum wells exploit the electric-field dependent shift of the absorption edge for below bandgap absorption [10]. For semiconductors, the band-to-band absorption process so far results in the largest absorptivity simply because of the large number of valance band and conduction band states available for optical transitions.

8.1.2 Receiver figures of merit for analog links

For photodiodes used in analog fiber links, the photodiode slope efficiency, s_D, equivalent to the responsivity defined earlier, is the most relevant receiver parameter in the link power budget and linearity calculation. With reference to Eq. (8.1), s_D is simply the product of the external quantum efficiency and the ratio of the electron charge to the photon energy.

The maximum value of s_D occurs when $\eta = 1$, i.e., when every incident photon is converted into an electron that flows in the external circuit. It should be noted that, in addition to optical power dependence, s_D has a wavelength dependence, mostly due to that of η and ν. Ideally, it is desirable to maintain the same s_D at all power levels and at all frequencies, in order to simplify the receiver design. Also, saturation of s_D in photodiodes is usually accompanied by the increase of harmonic distortions, and this limits the spurious free dynamic range of the analog links. The 1-dB compression point P_{1dB} of photodetection is referred to as the input optical power at which the output is 1 dB below that at a constant s_D.

For narrow band (single-octave) operation, the existence of the two-tone third order intermodulation distortions (IMD3) of the photodetector is more relevant than those of the second order intermodulation distortions. The figure of merit is the output referenced third order intercept point IP3 (dBm) that is obtained from $1.5 \times$ fundamental signals (dBm) $- 0.5 \times$ IMD3 (dBm). The characterization of the nonlinear distortion is discussed in Section 8.4.

8.2 Overview of photodetector structures for analog links

Photodiodes typically have an absorption region where a strong electric field is applied so that the photogenerated carriers can be swept out to produce a current or voltage signal in an external circuit. Common types of photodiodes are p-n junction photodiodes, p-intrinsic-n (PIN) photodiodes, Schottky junction photodiodes, and avalanche photodiodes (APDs). So far the PIN and Schottky diodes have received the most attention for analog link applications and are focused on here.

8.2.1 P-N and PIN photodiodes

A p-n photodiode usually consists of a simple p^+-n (or n^+-p) junction. For surface normal detection, the optical signal is incident at the top (or the bottom) surface. For some devices, such as optical switches and waveguide photodiodes, the optical beam can be incident parallel to the junction [11]. In normal operation, the diode is reverse-biased to separate the photogenerated electron and hole pairs collected at the depletion region. The external circuit load detects a signal current as the photogenerated carriers traverse the depletion region. The induced current stops when the carriers reach the bulk material outside the depletion region. This phenomenon gives rise to the transit time limit of the response speed. For a one-dimensional velocity distribution $v(x)$ of the carriers and the thickness of the depletion region W, the carrier transit time T_{tr} can be estimated from:

$$T_{tr} = \int_0^W \frac{dx}{v(x)}. \tag{8.2}$$

For III–V semiconductors, each micrometer of the depletion region is equivalent to about 10 ps of transit time when the carriers are drifted at peak velocity ($\sim 10^7$ cm/s).

The external quantum efficiency of a top-illuminated photodiode with a thin p region is given by

$$\eta = (1 - R_o)\left[e^{-\alpha_o W_1}(1 - e^{-\alpha_o W}) + \left(e^{-W_1/L_n} - e^{-\alpha_o W_1}\right)\left(\frac{\alpha_o}{\alpha_o + 1/L_n}\right)\right.$$
$$\left. + e^{-\alpha_o(W_1+W)}\left(\frac{\alpha_o}{\alpha_o + 1/L_p}\right)\right], \tag{8.3}$$

where R_o denotes the reflectivity at the air–semiconductor interface, W_1 is the p region thickness, and L_n and L_p are diffusion lengths of electrons in the p region and holes in the n region, respectively. The first term inside the square bracket of Eq. (8.3) accounts for the photogeneration in the depletion region; the second and third terms account for those in the p and n regions, respectively. For homostructure p-n junction photodiodes, it is difficult to optimize the quantum efficiency and the transit time at the same time due to the fact that the depletion width usually depends on the applied electric field. This can be further complicated by the nonlinear relationship between the carrier velocities and the electric field.

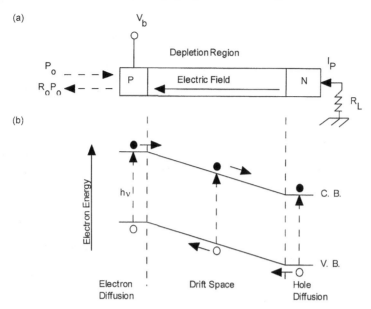

Figure 8.1. A PIN photodiode schematic. (a) Structure: P_o, incident power; V_b, bias voltage; R_o, reflectivity; I_p, photocurrent; R_L, load resistor. (b) Energy band diagram at reverse bias. C. B. stands for conduction band, V. B. for valence band.

The PIN photodiode, consisting of a low-doped (either i or π) region sandwiched between p and n regions, is a popular alternative to the simple p-n photodiode. At a relatively low bias, the intrinsic region becomes fully depleted and the total depletion width, including the p and n regions, remains almost constant as the peak electric field at the junction is increased. A schematic diagram of a PIN photodiode with light incident from the left is shown in Fig. 8.1a; its energy band diagram at reverse bias is depicted in Fig. 8.1b. Typically, for doping concentrations of $\sim 10^{14}$–10^{15} cm^{-3} in the intrinsic region, a bias voltage of less than 5 V is sufficient to deplete several micrometers, and the electron velocity also reaches the saturation value. However, as mentioned earlier, transit time is not the only factor affecting the response speed of the PIN photodiode. Carrier diffusion and the RC time constant can also be important. For p and n regions thinner than one diffusion length, the response time due to diffusion alone is typically 1 ns/μm in p-type silicon and about 100 ps/μm in p-type III–V materials [12]. The corresponding value for n-type III–V materials is several nanoseconds per micrometer due to the lower mobility of holes. To minimize the diffusion effect, one can employ very thin p and n regions [6]; alternatively, one can employ a p or n region transparent to the incident light.

In heterojunction PIN photodiodes, for instance, where the p region consists of materials transparent to the incident photons, only the first term inside the bracket of Eq. (8.3) is significant (W now stands for the intrinsic layer thickness). η can be increased by reducing the surface reflection R_o. One way to achieve this is by depositing an antireflection coating on the input facet of the photodiode. The quantum efficiency in some optimized heterojunction photodiodes can reach a value as high as 95% [13].

As mentioned, for the 1.0–1.55 μm wavelength range suitable for optical fiber communication, germanium and a few III–V compound semiconductor alloys stand out as candidate materials for PIN photodiodes, primarily because of their large absorption coefficients. The absorption edge of germanium is near 1.6 μm at room temperature. Its α_o is almost flat and is in the 10^4 cm^{-1} range over the 1.0–1.55 μm wavelength region [1]. In comparison, silicon's absorption edge is near 1.1 μm. There are problems, however, in using germanium for PIN photodiodes. Due to its small bandgap, the dark current arising from the across-bandgap generation–recombination processes can degrade the signal-to-noise ratio. This is further aggravated by the surface recombination. So far, no satisfactory surface passivation technique has been found for germanium PIN photodiodes, and the surface leakage current tends to be very high and unstable, especially at high ambient temperature.

III–V compound semiconductors have largely replaced germanium as materials for fiber optical compatible detectors [14,15]. By properly selecting the material composition of III–V materials, they can be lattice-matched to each other. For heterostructure PIN photodiodes, one can choose a material composition at the intrinsic region with a bandgap smaller than those of the p and n regions, while at the same time sufficiently smaller than the photon energy. The materials of interest are the ternary alloys $In_{0.53}Ga_{0.47}As$ and the quaternary $In_xGa_{1-x}As_yP_{1-y}$ alloys lattice-matched to InP. Similar to germanium, unpassivated InGaAs surfaces also exhibit an unstable dark current. However, this problem can be lessened by protecting the exposed surface with polyimide or InP. PIN detectors with excellent performances in the 1–1.6 μm wavelength range have been demonstrated in the InGaAs/InP material system. To achieve PIN photodiodes with high yield and high reliability, a planar fabrication process is favored, leakage current less than 25 pA has been achieved in some semi-planar photodiodes [16].

8.2.2 Schottky photodiodes and MSM photodetectors

Schottky photodiodes, also known as metal–semiconductor photodiodes, are majority carrier devices as the slower carriers (holes) generated recombine rapidly at the nearby contact region and thus do not accumulate in the absorption region. The photodiode usually consist of an undoped semiconductor layer (usually n type) on bulk material, and a metal layer deposited on top to form a Schottky barrier. An example is shown in Fig. 8.2. The metal can take the form of a very thin film (~100 Å) so that it becomes semi-transparent to the incident radiation [17]. Due to the strong electric field developed near the metal–semiconductor interface, photogenerated holes and

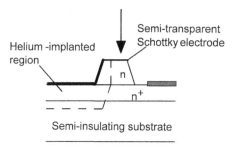

Figure 8.2. Schematic diagram of a Schottky photodiode for high speed operation.

electrons are separated and collected at the metal contact and in the bulk semiconductor materials, respectively. By increasing the reverse bias to the diode, both the peak electric field and the depletion region width inside the undoped layer can be increased, and thus the responsivity can be increased. The complete absorption of optical signals within the high-field region ensures high speed, as absorption outside the high field region generates a slow diffusive current. The photodiode response speed is typically dominated by the electron transit time across the undoped region.

In comparison with PIN photodiodes, Schottky photodiodes are simpler in structure. Unfortunately, while good Schottky contacts are readily made on GaAs, for long wavelength InGaAsP/InP materials, stable Schottky barriers are difficult to obtain. An intermediate approach to stabilize the barrier height is to employ a thin p^+-InGaAs layer between the metal and the detection region (n-InGaAs). By placing a gold-Schottky contact on n-$In_{0.52}Al_{0.48}As$ that is lattice-matched to InP, a wide spectral response range (0.4 to 1.6 μm) Schottky photodiode with a responsivity of 0.34 A/W at 0.8 μm and 0.42 A/W at 1.3 μm has been demonstrated [18].

For comparison, GaAs Schottky photodiodes with a 3-dB bandwidth larger than 100 GHz and operating at less than 4 V reverse bias have been reported [19]. This high speed is attained by restricting the photosensitive area to a very small mesa (5×5 μm^2) to minimize capacitance and by using a semi-transparent platinum Schottky contact such that detection can be achieved with top-side illumination.

The metal–semiconductor–metal (MSM) photodiode is formed by making two Schottky contacts on an undoped (or semi-insulating) layer. The MSM contacts can be single contacts, or interdigitated contacts with light incident through the top or the side of the device (as in the case of the velocity matched waveguide photodiode discussed in Section 8.5.3). The interdigitated MSM photodiode is considered for high speed operation due to its small capacitance and simplicity in layer structure. At zero bias, the dark current of an MSM photodiode on a GaAs substrate is on the order of 1 nA and is similar to that of a PIN photodiode. At higher bias, both the photocurrent and dark current increase rapidly, as results of the photoconductive effect and thermionic emission at the metal–semiconductor interface [20].

8.3 Noise sources in optical receivers

For optical receivers, we generally consider five noise sources: the shot (quantum) noise, the relative intensity laser noise, the photodetector dark current noise, the photodetector excess noise, and the electronic noise. Some of these noise sources are optical intensity dependent, and some are temperature dependent. The conventional model treats these noise sources as independent sources. Since electronic amplifier noise is extensively treated in electronic texts (such as those listed in [3]), readers are referred to them for details.

Both the shot noise and the relative intensity laser noise (RIN) are directly related to the incident optical power, hence they play an important role in the receiver noise figure for analog links operating at high optical power. It is noted that these noise sources have very wide spectral distribution. The power spectral-density of shot (quantum) noise is $i_{sh}^2(\omega) = 2qi_p$ where i_p is photocurrent. This noise arises from the statistical nature of photogenerated carriers flowing through the p-n junction. RIN is attributed to the fact that the output amplitude of the laser is not perfectly constant, as it fluctuates slightly around an average value. The fluctuations are due to temperature, acoustic disturbance, spurious feedback, as well as spontaneous transitions in the laser medium. The amount of noise present in an optical source is represented as RIN defined as the mean of the square noise power divided by the mean power squared. The spectral-density of the relative intensity noise is $i_{rin}^2(\omega) =$ RIN i_p^2. The impact of laser RIN on a receiver can be accounted for by including an additional current-noise generator in parallel with the quantum shot-noise generator. Figure 8.3 illustrates the interplay of these noise sources at the receiver.

Dark current flows whether or not the photodetector is illuminated and it results from the presence of current leakage paths. In photodiodes, leakage currents mainly

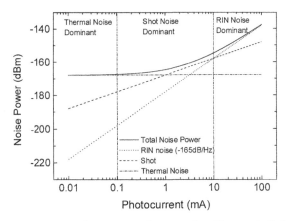

Figure 8.3. Noise power versus photocurrent in an analog fiber optic link as limited by the shot noise, RIN and thermal noise.

come from surface recombination, interface recombination and bulk leakage. Leakage occurs as a result of the small reverse current due to the thermally excited carriers, generation–recombination in the depleted region, and tunneling between the conduction and valence bands. This leakage generally increases with applied bias and temperature. Associated with the dark current i_d is the dark current shot noise generator, $i_{nd}^2(\omega) = 2qi_d$, in parallel with the photocurrent source. There are three main noise contributions intrinsic to the photodiode [21]. At low frequencies, the $1/f$ (flicker) noise dominates in the photodiode noise spectrum (where the noise goes like $1/f^\beta$ with β close to unity), followed by a mid-region of generation–recombination noise due to random fluctuation of the photoconductance in the intrinsic layer. This latter noise has the same cut-off frequency as f_{3dB}. Beyond this frequency the noise is dominated by a thermal noise component that has a broad spectral range. Since a photodiode contains resistive elements in parallel or in series with the photocurrent source, the associated thermal noise sources can be viewed as a current noise generator in parallel with a parallel-resistive element and a voltage noise generator in series with a serial-resistive element [2].

The photodetector excess noise refers to the additional noise power due to the carrier multiplication process. In the case of avalanche photodiodes, for instance, the power spectral-density of shot noise due to multiplication is denoted by $i_M^2 = 2qi_p F(M) M^2$, where $F(M)$ is the excess noise factor of the materials in which the avalanche multiplication occurs. For the uniform electric field case, the excess noise factors for electrons, F_n, and holes, F_p, are given as [4]:

$$F_n = M_n \left[1 - (1-k)\left(\frac{M_n - 1}{M_n}\right)^2 \right] \tag{8.4a}$$

and

$$F_p = M_p \left[1 - \left(1 - \frac{1}{k}\right)\left(\frac{M_p - 1}{M_p}\right)^2 \right], \tag{8.4b}$$

where k is the ratio of the hole to electron ionization coefficients, and M_n and M_p are the dc multiplication factors for pure electron and pure hole injection, respectively.

Amongst these noise sources, the principal ones that add to the noise floor of analog links are the thermal noise, shot noise and laser RIN. As the optical power in the link is increased, the last two factors predominate, as shown in Fig. 8.3. Since the shot noise goes like I_p and the laser RIN goes as I_p^2, the laser RIN dominates at high optical power. The turning point of the two depends on the particular level of the laser RIN. Typically for RIN $\sim$ 170 dBc/Hz, the noise floor will be laser RIN limited at photocurrents on the order of tens of mA. In this case the maximum achievable value for the spurious free dynamic range is the inverse of the RIN (or simply the maximum S/N). The balanced detector approach [22] (see Section 8.5.3) can be used to minimize the effect of the laser RIN by using two identical detectors arranged in such a way that the fundamental RF signals add and the laser RINs are suppressed [23]. With this approach, the noise floor becomes shot noise limited, and the S/N (and thus the SFDR) improves as the optical power is raised.

8.4 Nonlinearity in photodetectors

8.4.1 Carrier transport and circuit element effects

For analog optical links, as mentioned, both the link RF efficiency and spurious free dynamic range can be improved by using a large optical intensity. In links where a PIN photodiode is used at the receiver, the linearity of the photodiode becomes critical in the consideration of the overall link performance. Consequently, it is of interest to obtain high speed photodetectors that can be operated without saturation under a large optical intensity. At saturation, the fundamental signal can be accompanied by a high level of nonlinear distortions [24]. Possible causes of the saturation are absorption saturation, and electric field screening arising from the space charge in the intrinsic layer. The latter can be worsened due to the external circuit effect and the geometric effect.

Severe absorption saturation occurs when there is a significant filling of the conduction and valence band states. For InGaAs materials, for instance, this corresponds to an electron concentration in excess of 2×10^{17} cm^{-3} and a hole concentration in excess of 10^{18} cm^{-3}. However, these concentrations are typically orders of magnitude higher than the carrier concentrations at which the space charge screening effect becomes significant [25]. The screening effect due to the external circuit primarily arises from series resistance outside the junction region of the detector. For example, for load resistance of $50\,\Omega$, a photocurrent of $10\,$mA can lead to a voltage reduction of $0.5\,$V across the intrinsic region. The reduction increases as the photodiode current increases and can become a significant fraction of the applied bias ($5\,$V or so) across the detector.

Both photodiode and photoconductor operate under high electric field for sweeping out the photogenerated carriers. However, under a high flux of photogenerated carriers, the applied electric field can be screened by the local dipolar electric field resulting from the separating electrons and holes. The nonlinear response can arise from a redistribution of the electric field within the intrinsic region, which is accompanied by spatially dependent hole and electron velocity distributions [26]. Consequently, the harmonic distortion levels increase at elevated optical intensities. At even higher intensity, the fundamental RF signal response can be affected. In principle, the screening effect can be neglected when the photodiode is under a large

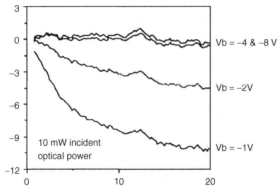

Figure 8.4. Waveguide photodiode frequency response with 10 mW laser power and detector bias voltages of $V_b = -8, -4, -2, -1$ V. The horizontal axis is the frequency in units of GHz, and the vertical axis is relative response in dB.

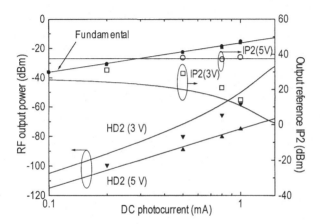

Figure 8.5. Fundamental and second harmonic signals of a waveguide photodiode, at 260 MHz and at -3 V (■, ▼) and -5 V(●, ▲) respectively, versus photocurrent. The dots and solid curves are the measured and theoretical results respectively. The calculated and extracted output referenced IP2 at -3 V and -5 V are included.

reverse bias. Figure 8.4 demonstrates the effect of bias on the frequency response of a waveguide photodetector at high optical flux. Figure 8.5 shows the fundamental and second harmonic signals at -3 and -5 V as a function of photocurrent. An additional nonlinear effect was observed at high bias that has been attributed to absorption at the highly doped contact region of the photodiode [26].

In the literature, nonlinear distortions of the PIN photodiode have been attributed to the nonlinear transport effects induced by the space charge. Numerical models have been developed to describe the nonlinear transport inside the device [27,28]. Basically these solve simultaneously the Poisson equation and the carrier continuity equations under various operating conditions. When the density of photogenerated

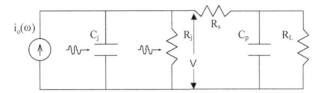

Figure 8.6. The small signal equivalent circuit of a photodiode under illumination.

electron and hole pairs reaches a level high enough to partially screen the biasing electric field, a highly non-uniform carrier velocity profile can result, and the
photodiode is in severe saturation with high nonlinear distortion levels.

Experimentally, the nonlinear distortion is studied by examining the harmonic
levels of the photodiode with the photodiode subject to heterodyned lasers with
wavelength slightly offset to generate an RF tone. Alternatively, beat signals are
generated at the photodiode using three separate optical beams RF-modulated at
frequencies f_1, f_2 and f_3 respectively: the second-order intermodulation distortion
signal at $f_1 + f_2$ and the third-order intermodulation distortion at $f_3 - (f_1 + f_2)$
are measured [29].

There is an alternative experimental approach to access the photodiode's harmonic and intermodulation distortion levels at different frequencies. This approach
relies upon the variation of the photodiode microwave impedance at various optical powers for determining the frequency response and nonlinearity behaviors.
The microwave impedance can be extracted from measured microwave reflection
coefficient (S_{11}) of the photodiode. The distortions can therefore be traced back to
physical parameters of the photodiode. To illustrate this, an equivalent circuit of
the PIN photodiode is shown in Fig. 8.6 where $i_o(\omega)$ represents the photocurrent at
angular frequency ω inside the reverse-biased intrinsic region, C_j is the junction capacitance, R_j is the diode shunt junction resistance, C_p is the parasitic capacitance,
R_s is the series resistance (due to the contacts and the p and n regions of the diode),
and R_L is the load resistor. For this circuit, the output current $i(\omega)$ can be expressed
as $i(\omega) = i_o(\omega) \cdot H(\omega)$, where $H(\omega)$ is the transfer function of the circuit. The form
of $i(\omega)$ suggests that the harmonic distortion can show up in either $i_o(\omega)$ or $H(\omega)$,
or both. As mentioned above, the physical origin of the nonlinearity in $i_0(\omega)$ and
$H(\omega)$ is attributed to the nonlinear carrier transport and the associated impedance
changes induced by the optical input signal. This model can be used to account for
the photodiode distortion even at low optical power.

In the analysis, an increase of junction capacitance C_j of a planar PIN photodiode under optical illumination can be paralleled by an increase in the electric
polarization due to an increase in photogenerated electron–hole pairs, whose density depends on the net electric field and optical illumination level. In general, both
R_j and C_j change with optical illumination. For instance, we consider the effect of

C_j and assume it increases linearly with photocurrent:

$$C_j = C_{dark} + I_{dc}C' \tag{8.5}$$

where C_{dark} is the dark capacitance of the junction, I_{dc} is the dc photocurrent, C' is defined as the differential capacitance with respect to the photocurrent. From the equivalent circuit in Fig. 8.6, the equation for the voltage V across the current source satisfies:

$$\frac{d}{dt}\{(C_{jo} + i_o e^{j\omega t}C')V\} + \frac{V}{Z(\omega)} = i_o e^{j\omega t}, \tag{8.6}$$

where C_{jo} and R_{jo} are the junction capacitance and resistance at a given I_{dc}, $Z(\omega)$ is the equivalent impedance of C_p, R_s and R_L. V can be expressed in a harmonic series of $V_1 e^{j\omega t} + V_2 e^{j2\omega t} + V_3 e^{j3\omega t} + \cdots$, where V_1, V_2, and so on can be determined by comparing the coefficient of the ω^n on both sides of Eq. (8.6). In this model it is observed that a relatively large change in C_j can enhance harmonic levels, while the fundamental power remains largely unchanged. Also, the harmonic distortions have similar RC frequency roll-off as the fundamental signal. At frequencies well below the roll-off frequency, harmonic powers are proportional to ω^2 when the variation of C_j dominates; similar observations hold for higher order harmonics and intermodulation distortions [30].

8.4.2 Geometrical effect

Another issue associated with optical saturation arises from the geometry effect of the photodiode. To illustrate this, consider a surface normal InP/InGaAs/InP PIN diode with a photodetection region of thickness d, and a ring electrode, as shown in Fig. 8.7. As mentioned before, at high power, the photocurrent developed a voltage drop across the photodiode. For the electrode configuration

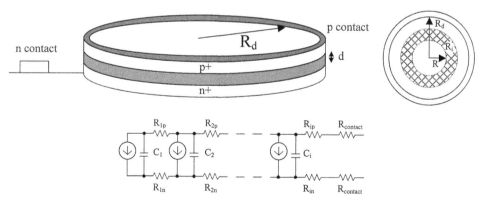

Figure 8.7. Distributed detector model of a surface normal photodiode with a ring electrode. The ring capacitance and resistance are represented in the distributed circuit model.

shown in Fig. 8.7, the lateral charging of the distributed capacitors through the resistors can enhance the debiasing effect. A simple distributed circuit model is shown to estimate the effect of high power optical pulses on the biasing within the photodiode [31]. For the p^+ and n^+ InP regions the corresponding ring resistances R_{ip} and R_{in} are obtained as:

$$R_j = \frac{\rho_s}{2\pi} \ln \frac{R + \Delta R}{R},$$

(8.7)

where the ρ_s is the resistivity of the layer. Similarly, the capacitance between corresponding rings is obtained from the parallel plate capacitance:

$$C_j = \frac{\varepsilon}{d} \pi [(R + \Delta R)^2 - R^2].$$

(8.8)

For simplicity, the incident light on the photodetector is represented by a Gaussian distribution function. The debiasing effect of the high power optical pulse is evident in Fig. 8.8 in which 20 ps, 1 W optical pulses cause a bias larger than 10 V at the photodiode, which is opposite to the bias of the photodiode and thus results in the saturation effect as described in Section 8.4.1. In this model, a high contact resistance can greatly enhance the debiasing effect. To reduce the debiasing of photodiodes, a thick p^+ surface layer and distributed contacts should be used to reduce the internal resistance, or alternatively, one can explore the waveguide-coupled configuration where the lateral charging effect is minimized because of the continuous metal contact along the absorption waveguide. This geometric effect can

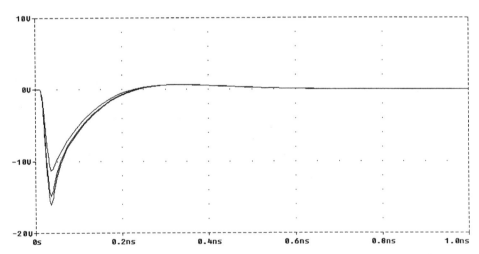

Figure 8.8. Effect of optical debiasing of the photodiode upon high peak power pulse (1 W, 20 ps duration) excitation. The vertical axis is voltage and the horizontal axis is time in ns. The different curves show the cumulative effect from the innermost ring to the outermost ring (lowest curve), as depicted in Fig. 8.7.

be reduced by evenly distributing the photocurrent; an example is the waveguide integrated photodiodes where the optical energy is more evenly distributed and absorbed along the waveguide [32].

8.5 Recent advances in photodiodes

8.5.1 High speed surface normal photodiodes

The PIN photodiodes developed in the early days are surface normal ones where light is coupled perpendicular to the junction. The material and device structure design of these surface normal photodiodes are well documented in the literature. The key parameters for analog applications are speed, responsivity, linearity and saturation power. Since light is incident perpendicular to the absorption layer, there is a trade-off of the layer thickness with respect to speed and responsivity considerations [33]. In addition, there is a trade-off between the carrier transit time at the intrinsic layer and the RC time constant that determines the optimal thickness for a given junction area. In general, the interplay of the transit time and the capacitance consideration sets the limits to the responsivity at a given frequency, with device area constrained to achieve optimal fiber coupling [34]. With respect to device reliability, a planar photodiode, in contrast to the mesa photodiode, has the advantage of minimizing the exposure of the active region. As a compromise, a circular trench can encircle a detection mesa to reduce the junction capacitance, at the same time resulting in small exposure of the active region that helps to reduce the surface recombination.

For high speed, a surface normal photodiode of the mesa type is usually integrated with a coplanar waveguide transmission line, as depicted in Fig. 8.9. Two generic InGaAs/InP PIN photodiode configurations, front-illuminated and backside-illuminated photodiodes, are commonly used. The front-illuminated photodiode has a top window for light coupling; the backside-illuminated photodiode has light

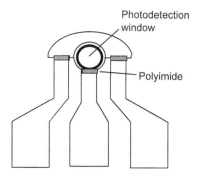

Figure 8.9. Schematic top-view of a surface normal photodiode of the mesa type which is integrated with a co-planar waveguide transmission line. The ring electrode on top of the mesa is connected to the central transmission line on polyimide.

coupled through the thick substrate that is almost transparent to the 1.3 and 1.55 μm light. In the past when liquid phase epitaxy was used, a simple front-illuminated p-InGaAs/i-InGaAs homojunction photodiode could be made with most of the absorption taking place at the i-InGaAs layer. With the advent of metalorganic chemical vapor deposition, one can readily make InP–InGaAs–InP double heterostructure PIN photodiodes [35]. Typically, a layer of $In_{0.53}Ga_{0.47}As$ lattice matched to InP is used as the intrinsic layer. The residual doping level of the layer is in the 5×10^{14}–5×10^{15} cm^{-3} range. After epitaxy, a rounded mesa is etched to define the junction area. Surface protection layers (such as dielectrics and polyimide) and antireflection coating layers are then deposited. Sub-nanoampere dark currents have been obtained for ~50-μm diameter devices. Speeds higher than 20 GHz have been routinely achieved, with device lifetimes $> 10^8$ h. The disadvantage of this front-illuminated structure is that the contact area needed for forming the bonding pad introduced additional capacitance and dark current to the photodiodes. Recently, through careful control of the doping profiles, a photodiode with a similar geometry has been demonstrated up to 50 GHz with responsivity of 0.55 A/W [36].

Back-illumination is possible for material systems whose substrates are transparent to the optical beam. In this case, the active-layer thickness is selected such that at the operating bias, the electric field punches through to n^+-InP material. In principle, the back-illuminated photodiode has the same basic features as the top-illuminated photodiode, however the responsivity can be reduced due to the free-carrier absorption in the relatively thick substrate and some optical design is needed for efficient light coupling. The main advantage of the backside-illumination is that the top surface is not needed for light coupling, therefore, for the same photodetection area, the backside-illuminated diodes potentially have lower capacitance and lower dark current than those of the front-illuminated diodes.

Optimal designs for broadband (> 40 GHz) PIN photodiodes have been obtained by carefully balancing the effects of the transit time, the geometric configuration (including the diode area), and the transmission line characteristics [37].

8.5.2 Waveguide photodiodes

As mentioned, major trade-offs in the design of analog wide bandwidth photodiodes can be related to the intrinsic layer thickness which links directly the quantum efficiency, the carrier transit time and the RC circuit time constant [33]. For conventional surface normal photodiodes, the intrinsic layer thickness needs to be thin for reducing the transit time; however, in doing so, the quantum efficiency is reduced and the capacitance is increased. To reduce the RC time effect, the detection area needs to be reduced and this can reduce the quantum efficiency (and thus the responsivity), as illustrated in Fig. 8.10 [34].

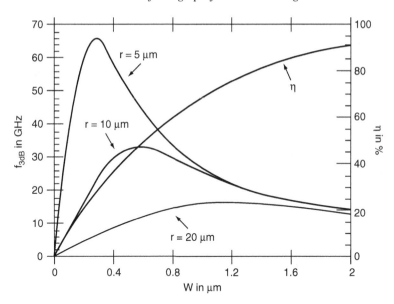

Figure 8.10. The f_{3dB} vs. intrinsic layer thickness for PIN photodiode with surface normal illumination; r is the diode radius.

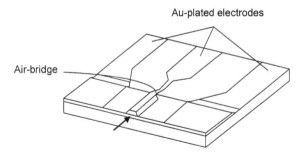

Figure 8.11. Schematic diagram of a waveguide photodiode integrated with a coplanar waveguide transmission line. Light is edge-coupled to the waveguide, the p-electrode on top of the waveguide is connected to the central electrode via an air bridge.

The waveguide photodiode was proposed as a means to ease the trade-off between the speed and responsivity [11]. This type of detector is generally edge-illuminated with light guided parallel to the intrinsic layer and incrementally absorbed. The optical waveguiding structure is similar to that of the double heterostructure lasers. An example of the waveguide photodiode for 1.3 and 1.5 μm detection is shown in Fig. 8.11, where the mesa structure is used to confine the light in the lateral direction. Edge-coupling and optical waveguiding increases the interaction length of the light inside the absorption layer. The responsivity is mainly limited by the coupling efficiency at the fiber-to-waveguide interface, and the scattering loss along the waveguide. In comparison with the surface normal photodiode, the waveguide photodiode typically has a smaller responsivity due to the modal mismatch between the waveguide and the single mode fiber. However, due to the edge-illumination, the responsivity is mainly determined by the waveguide length, and not by absorption layer thickness, therefore the waveguide detector has a larger bandwidth-efficiency performance than the surface normal photodiode. A quantum efficiency as high as 68% has been achieved for a waveguide photodiode with f_{3dB} greater than 50 GHz [11]. In contrast, a conventional PIN photodiode with a quantum efficiency of 68% will be limited to an f_{3dB} of about 30 GHz.

The waveguide photodiode shown in Fig. 8.11 consists of a rib-loaded PIN waveguide integrated with a coplanar waveguide transmission line on a semi-insulating InP substrate. The rib is 5 μm wide at the top and 20 μm long, and the intrinsic InGaAs layer is 0.35 μm thick. For the waveguide photodiode, the f_{3dB} is limited by the transit time across the intrinsic region and the RC time effect. Using a mushroom-mesa structure to reduce the RC time constant (through reducing the contact resistance while keeping the junction area small), a >100 GHz bandwidth waveguide photodiode with a quantum efficiency of 50% has been demonstrated [33].

As noted earlier, the waveguide configuration poses a limitation to the responsivity due to the fiber coupling. The coupling can be improved through the use of lensed fibers, antireflection coating at the waveguide facet, or laterally tapered waveguide section [38].

8.5.3 Traveling wave photodiodes and velocity-matched photodiodes

As lumped element designs, both the surface normal and waveguide photodiodes discussed earlier are limited by the RC circuit time. As a means to overcome the RC bandwidth limitation, the distributed photodiode was first introduced in 1990 in the form of the "traveling wave photodetector" for photoabsorption distributed along a microwave transmission line [39]. In the later monolithic version, the optical waveguide structure of the traveling wave waveguide photodiode is very similar to that of the ordinary wave photodiode described in Section 8.5.2. The device can be viewed as the terminated section of a transmission line with a position dependent photocurrent source distributed along its length. The transit time limitation to the frequency response still persists, and as a transmission line element, the frequency response is related to the matching between the characteristic impedance of the transmission line and termination impedance. The monolithic design where the light is continuously absorbed along the microwave waveguide is commonly referred to as the traveling wave waveguide photodiode [40]. In an alternative design where light is absorbed in a set of discrete photodiodes distributed along the optical waveguide and the transmission line, the photodiode is referred to as the periodic traveling wave photodiode [41].

While the absorption behavior is similar in the two traveling wave photodiode designs, the microwave properties can be quite different. In particular, the microwave (phase) wave is slower than the optical (group) wave in the traveling wave waveguide photodiode and faster in the periodic traveling wave photodiode; this mismatch affects the maximum response bandwidth [33]. The bandwidth can also be reduced due to the effect of reflected microwaves from the termination of the transmission line. For instance, the open termination is bandwidth-limited by the interference between the forward and backward waves; while the impedance-matched termination has larger bandwidth, the available photocurrent can be reduced by a factor of two (and the microwave power by a factor of four). The potential benefit of the traveling wave photodiode, besides the large bandwidth, is the maximum waveguide length allowable (within the constraint placed by the impedance mismatch and the waveguide scattering loss) for maximum absorption, which eases the design for the modal confinement at the absorber and thus is attractive for high optical power handling. Bandwidth as high as 172 GHz has been reported for a GaAs/AlGaAs traveling wave photodiode detecting in the 800–900 nm wavelength range, with an intrinsic GaAs region which is 1 μm wide and 0.17 μm thick [42].

The periodic traveling wave photodiode can take the form of an array of individual photodiodes serially connected by long, passive waveguides and transmission lines. Since the backward microwave seriously disturbs the forward microwave in a long transmission line, a match termination is also required. An advantage of the periodic configuration is the feasibility to perform the velocity matching between the microwave phase velocity and the optical group velocity using an electrical delay line – the resulting structure has been referred to as the velocity-matched waveguide photodiode. The slow-wave line is achieved by periodic capacitive loading using the arrayed photodiodes [41]. The VMPD based on the GaAs-MSM photodiode has reached around 50 GHz and that based on InGaAs has reached 78 GHz (the efficiency was 7.5%) [43].

For externally modulated analog fiber optic links, balanced photodetection can be used to suppress the laser RIN noise and amplified spontaneous noise from the optical amplifier prior to the transmitter. When this happens, the link performance becomes shot noise limited and the spurious free dynamic range and the noise figure can be improved by increasing the optical power. The velocity-matched waveguide photodiode is instrumental for demonstrating this purpose. Figure 8.12 shows a schematic diagram of the distributed velocity-matched waveguide photodiode with twin waveguides coupled to two complementary optical inputs whose RF signals are 180 degrees out of phase [23]. With differential current at the output of each

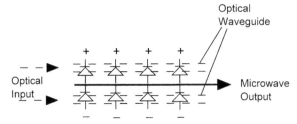

Figure 8.12. Schematic of photodiodes and velocity-matched waveguides connected for balanced detection.

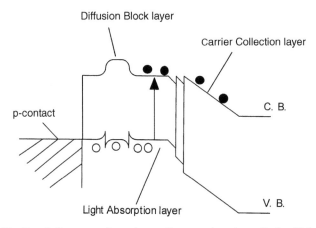

Figure 8.13. Band diagram of a uni-traveling carrier photodiode. (After Ref. 6.)

segment, the detected RF signals are generated and added in a distributive manner along the two waveguides while the noise current associated with the photon fluctuation cancels in the microwave output. Broadband balanced detection from 1–12 GHz has been demonstrated with this technique with more than 24 dB suppression of the laser RIN [23].

8.5.4 Uni-traveling carrier photodiodes

While the waveguide photodiode was introduced to reduce the constraints between the bandwidth and the responsivity, a new means to enhance the f_{3dB} of photodiodes of PIN structures has been demonstrated by Ishibashi [6]. As mentioned, the bandwidth of the conventional surface normal design was largely limited by the trade-off between the transit time and the RC time. In the novel approach, the electron–hole pairs are generated in the low field p region and electrons diffused to the intrinsic region (see Fig. 8.13) are swept out at high velocity while the holes, as majority carriers, responds quickly to the electron flow to maintain quasineutrality. This results in both fast rise and fast fall times for photoresponse. A faster carrier transport eases the layer thickness design for small capacitance. A further advantage is that, since the hole drift current is absent within the intrinsic region, the carrier screening effect is greatly reduced and thus high optical power operation becomes feasible. Optical saturation power as large as 63 mW has been reported up to 40 GHz and is attributed to the space charge effect in the collector layer [44]. In the structure shown in Fig. 8.13 where the p-InGaAs is the absorption layer and the InP is the collection (intrinsic) layer, the bandwidth is limited by the diffusion velocity in the absorption layer. By the use of a diffusion barrier layer, the average diffusion velocity of electrons in the p layer becomes $\sim 2D_n / W_a$, where W_a is the thickness of the absorption layer. Pulsed operation with f_{3dB} as high as 310 GHz has been demonstrated in the uni-traveling carrier photodiode [45].

References

1. R. J. Keyes (ed.), *Optical and Infrared Detectors, Topics in Applied Physics*, Vol. 19, Springer-Verlag, Berlin, 1980.
2. D. P. Schinke, R. G. Smith, and A. R. Hartman, "Photodetectors," in *Semiconductor Devices for Optical Communication*, 2nd Edn., ed. H. Kessel, Topics in Applied Physics, Vol. 39, Springer-Verlag, Heidelberg, 1982.
3. S. B. Alexander, *Optical Communication Receiver Design*, SPIE Press Tutorial Texts in Optical Engineering, Vol. TT22, and IEE Telecommunication Series, Vol. 37, 1997.
4. R. J. McIntyre, *IEEE Trans. Electron. Dev.*, **ED-19**, 703, 1972.
5. A. R. Williams, A. L. Kellner, and P. K. L. Yu, *Electron. Lett.*, **29**, 1298, 1993.
6. T. Ishibashi *et al.*, *Technical Digest, 1999 Microwave Photonics Meeting MWP'99*, p. 75, 1999.
7. K. Kato, *IEEE Trans. Microwave Theory Tech.*, **47**, 1265, 1999.
8. P. L. Liu, K. J. Williams, M. Y. Frankel, and R. D. Esman, *IEEE Trans. Microwave Theory Tech.*, **47**, 1297, 1999.
9. W. Franz, *Z. Naturforsch*, **A13**, 484, 1958; L. V. Keldysh, *Zh. Eskp. Teor. Fiz.*, **34**, 1158, 1958.
10. J. S. Weiner *et al.*, *Appl. Phys. Lett.*, **47**, 1148, 1985.
11. K. Kato, S. Hata, K. Kawano, J. Yoshida, and A. Kozen, *IEEE J. Quantum Electron.*, **28**, 2728, 1992.
12. S. M. Sze, *Physics of Semiconductor Devices*, Wiley, New York, p. 649, 1981.
13. D. Wake, R. H. Walling, S. K. Sargood, and I. D. Henning, *Electron. Lett.*, **23**, 415, 1987.
14. M. A. Washington, R. E. Nahory, and E. D. Beebe, *Appl. Phys. Lett.*, **33**, 854, 1978.
15. T. P. Lee, C. A. Burrus, and A. G. Dentai, *IEEE. J. Quantum Electron.*, **17**, 232, 1981.
16. J. N. Patellon, J. P. Andre, J. P. Chane, J. L. Gentner, B. G. Martin, and G. M. Martin. *Phillips J. Res.*, **44**, 465, 1990.
17. W. Gao, K. Al-Sameen, P. R. Berger, R. G. Hunsperger, G. Zydzik, H. M. O'Bryan, D. Sivco, and A. Y. Cho, *Appl. Phys. Lett.*, **65**, 1930, 1994.
18. K. C. Hwang, S. S. Li, and Y. C. Kao, *Proc. SPIE*, **1371**, 128, 1991.
19. S. Y. Wang and D. M. Bloom, *Electron. Lett.*, **19**, 554, 1983.
20. P. Bhattacharya, *Semiconductor Optoelectronic Device*, Ch. 9, 2nd Ed., Prentice Hall, 1986.
21. A. van der Ziel, *Noise in Solid State Devices and Circuits*, Wiley, 1986.
22. R. J. Deri *et al.*, *IEEE Photon. Technol. Lett.*, **4**, 1238, 1992.

23. M. S. Islam, T. Chau, S. Mathai, T. Itoh. M. C. Wu, D. L. Sivco, and A. Y. Cho, *IEEE Trans. Microwave Theory Tech.*, **47**, 1282, 1999.
24. H. Jiang and P. K. L. Yu, *IEEE Photon. Technol. Lett.*, **10**, 1608, 1998.
25. C. K. Sun, P. K. L. Yu, C. T. Chang, and D. J. Albares, *IEEE Trans. Electron Devices*, **39**, 2240, 1992.
26. K. J. Williams, *Appl. Phys. Lett.*, **65**, 1219, 1994.
27. R. R. Hayes and D. L. Persechini, *IEEE Photon. Technol. Lett.*, **5**, **70**, 1993.
28. J. Harari, G. Jin, J. P. Vilcot, and D. Decoster, *IEEE Trans. Microwave Theory Tech.*, **45**, 4332, 1997.
29. T. Ozeki and E. H. Hara, *Electron. Lett.*, **12**, 80, 1976.
30. H. Jiang, D. S. Shin, G. L. Li, J. T. Zhu, T. A. Vang, D. C. Scott, and P. K. L. Yu, *IEEE Photon. Technol. Lett.*, **12**, 540, 2000.
31. D. Ralston, A. Metzger, Y. Kang, P. Asbeck, and P. K. L. Yu, *Proc. SPIE*, **4112**, 132, 2000.
32. H. Jiang and P. K. L. Yu, IEEE International Microwave Symposium, *IMS-2000 Digest*, Vol. 2, p. 679, 2000.
33. K. Kato, *IEEE Trans. Microwave Theory Tech.*, **37**, 1265, 1999.
34. J. E. Bowers and C. A. Burrus, Jr. *J. Lightwave Technol.*, **5**, 1339, 1987.
35. G. H. Olsen and T. J. Zamesowski, *IEEE J. Quantum Electron.*, **17**, 128, 1981.
36. Catalog, Discovery Semiconductor, Inc., 1999.
37. Z. Zhu, R. Gillon, and A. V. Vorst, *Microwave Opt. Technol. Lett.*, **8**, 8, 1995.
38. L. Giraudet, F. Banfi, S. Demiguel, and G. Herve-Gruyer, *IEEE Photon. Technol. Lett.*, **11**, 111, 1999.
39. H. F. Taylor, O. Eknoyan, C. S. Park, K. N. Choi, and K. Chang, *Proc. SPIE Optoelectronic Signal Processing Phased Array Antenna II*, **1217**, 59, 1990.
40. K. S. Giboney, M. Rodwell, and J. Bowers, *IEEE Photon. Technol. Lett.*, **4**, 1363, 1992.
41. L. Y. Lin, M. C. Wu, T. Itoh, T. A. Vang, R. E. Muller, D. L. Sivco, and A. Y. Cho, *IEEE Trans. Microwave Theory Tech.*, **45**, 1320, 1997.
42. K. S. Giboney, R. Nagarajan, T. Reynolds, S. Allen, R. Mirin, M. Rodwell, and J. E. Bowers, *IEEE Photon. Technol. Lett.*, **7**, 412, 1995.
43. E. Droge, E. H. Bottcher, S. Kollakowski, A. Strittmatter, O. Reimann, R. Steingruber, A. Umbach, and D. Bimberg, *ECOC'98*, vol. 1, p. 20, 1998.
44. N. Shimuzu, Y. Miyamoto, A. Hirano, K. Sato, and T. Ishibashi, *Electron. Lett.*, **36**, 750, 2000.
45. T. Furuta, S. Kodama, T. Ishibashi, *Electron. Lett.*, **36**, 1809, 2000.

9

Opto-electronic oscillators

X. STEVE YAO

General Photonics Corporation

9.1 Introduction

9.1.1 Review of oscillators

Oscillators are devices that convert energy from a continuous source to a periodically varying signal. They represent the physical realization of a fundamental basis of all physics, the harmonic oscillator, and they are perhaps the most widely used devices in modern day society. Today a variety of mechanical [1] (such as the pendulum), electromagnetic (such as LC [2,3] and cavity based [4]), and atomic (such as maser [5] and laser [6]) oscillators provide a diverse range in the approximation to the realization of the ideal harmonic oscillator. The degree of spectral purity and stability of the output signal of the oscillator is the measure of the accuracy of this approximation, and is fundamentally dependent on the energy storage ability of the oscillator, determined by the resistive loss (generally frequency dependent) of the various elements in the oscillator.

An important type of oscillator widely used today is the electronic oscillator. The first such oscillator was invented by L. De Forest [2] in 1912, shortly after the development of the vacuum tube. In this triode based device known as the "van der Pol oscillator [3]" the flux of electrons emitted by the cathode and flowing to the anode is modulated by the potential on the intervening grid. This potential is derived from the feedback of the current in the anode circuit containing an energy storage element (i.e., the frequency selecting LC filter) to the grid, as shown in Fig. 9.1a.

Today the solid state counterparts of these "valve" oscillators based on transistors are pervasive in virtually every application of electronic devices, instruments, and systems. Despite their widespread use, electronic oscillators, whether of the vacuum tube or the solid state variety, are relatively noisy and lack adequate stability for applications where very high stability and spectral purity are required. The

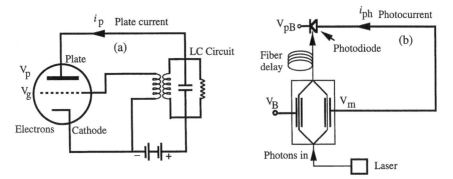

Figure 9.1. Comparison of a van der Pol oscillator (a) with a light induced microwave oscillator (b).

limitation to the performance of electronic oscillators is due to ohmic and dispersive losses in various elements in the oscillator including the LC resonant circuit.

For approximately the past fifty years the practice of reducing the noise in the electronic oscillator by combining it with a high quality factor (Q) resonator has been followed to achieve improved stability and spectral purity. The Q is a figure of merit for the resonator given by $Q = 2\pi f \tau_d$, where τ_d is the energy decay time that measures the energy storage ability of the resonator and f is the resonant frequency. High Q resonators used for stabilization of the electronic oscillator include mechanical resonators, such as quartz crystals [7,8], electromagnetic resonators, such as dielectric cavities [9], and acoustic [10] and electrical delay lines, where the delay time is equivalent to the energy decay time τ_d and determines the achievable Q. This combination with a resonator results in hybrid type oscillators referred to as electromechanical, electromagnetic, or electroacoustic, depending on the particular resonator used with the oscillator circuit. The choice of a particular resonator is generally determined by a variety of factors, but for the highest achievable Qs at room temperatures the crystal quartz is the resonator of choice for the stabilization of the electronic oscillator. However, because quartz resonators have only a few high Q resonant modes at low frequencies [7,8], they have a limited range of frequency tunability and cannot be used to directly generate high frequency signals.

9.1.2 Signal generation for RF photonic systems

As discussed in previous chapters, photonic RF systems [11–13] embed photonic technology into the traditional RF systems. In particular, in a photonic RF system optical waves are used as a carrier to transport RF signals through optical fibers to remote locations. In addition, some of the RF signal processing functions such as signal mixing [14], antenna beam steering [15,16], and signal filtering [17,18] can also be accomplished optically. The photonic technology offers the advantages of low loss, light weight, high frequency, high security, remoting capability, and immunity to electromagnetic interference.

Like in any RF systems, high quality RF oscillators are essential for photonic systems. Traditional RF oscillators cannot meet all the requirements of photonic RF systems. Because photonic RF systems involve RF signals in both optical and electrical domains, an ideal oscillator for the photonic RF systems should be able to generate RF signals in both optical and electrical domains. The optical domain RF signal is essentially an optical wave modulated by a signal at RF frequency, or an optical subcarrier at RF frequency. To fully utilize the advantage of the high frequency optical carrier, the operation frequency for a typical photonic RF system should be high. Therefore the ideal oscillator should have high frequency generation capability. In addition, it should be possible to synchronize or control the oscillator by both electrical and optical references or signals.

Presently, generating a high frequency RF signal in the optical domain is usually done by modulating a diode laser or an external electro-optical (E/O) modulator using a high frequency stable electrical signal from a local oscillator (LO). Such a local oscillator signal is generally obtained by multiplying a low frequency reference (e.g., quartz oscillator) to the required high frequency, say 32 GHz, with several stages of multipliers and amplifiers. Consequently, the resulting system is bulky, complicated, inefficient, and costly. A fundamental shortcoming of this scheme is that the phase noise of the resulting high frequency signal increases by a factor of N^2, where N is the frequency multiplication factor.

An alternative way to generate photonic RF carriers is to mix two lasers with different optical frequencies [19]. However, the resulting bandwidth of the signal is wide (limited by the spectral width of the lasers, typically greater than tens of kilohertz) and the frequency stability of the beat signal is poor, caused by the drift of the optical frequency of the two lasers. It is possible to phase lock the beat frequency to an external RF frequency source; however; such an arrangement still requires a costly high frequency and low noise RF source.

9.1.3 OEO – A new class of oscillators

In this chapter, a new class of oscillators [20–22] that meets the special requirements for photonic RF systems, yet is extremely suitable for conventional RF applications, is introduced and discussed. This oscillator, shown schematically in Fig. 9.1b, is based on converting the continuous light energy from a pump laser to radio frequency (RF) and microwave signals, and thus is given the acronym OEO for opto-electronic oscillator. The OEO is fundamentally similar to the van der Pol oscillator with photons replacing the function of electrons, an electro-optic (E/O) modulator replacing the function of the grid, and a photodetector replacing the function of the anode. The energy storage function of the LC circuit in the van der Pol oscillator is replaced with a long fiber optic delay line in the OEO.

Despite this close similarity, the OEO is characterized by significantly lower noise and very high stability, as well as other functional characteristics which are not achieved with the electronic oscillator. The superior performance of the OEO results from the use of electro-optic and photonic components which are generally characterized with high efficiency, high speed, and low dispersion in the microwave frequency regime. Specifically, currently photodetectors are available with as high as 90% quantum efficiency, which can respond to signals with frequencies as high as 110 GHz [23]. Similarly, E/O modulators with 75 GHz frequency response are also available [24]. Finally, the commercially available optical fiber which has a small loss of 0.2 dB/km for 1550 nm light allows long storage time of the optical energy with negligible dispersive loss (loss dependent on frequency) for the intensity modulations at microwave frequencies.

The OEO may also be considered as a hybrid oscillator in so far as its operation involves both light energy and microwave signals. Nevertheless as a hybrid oscillator, the OEO is unique in that its output may be obtained both directly as a microwave signal, or as intensity modulation of an optical carrier. This property of the OEO is naturally suited for photonic RF applications [22].

The ring configuration consisting of an electro-optic modulator which is fed back with a signal from the detected light at its output has been previously studied by a number of investigators interested in the nonlinear dynamics of bistable optical devices [25–29]. The use of this configuration as a possible oscillator was first suggested by Neyer and Voges [30]. The interest of their investigations however was focused primarily on the nonlinear regime and the chaotic dynamics of the oscillator. This same interest persisted in the work of Aida and Davis [31], who used a fiber waveguide as a delay line in the loop. Our studies, by contrast, are specifically focused on the stable oscillation dynamics and the noise properties of the oscillator. The sustainable quasi-linear dynamics, in both our theoretical and experimental demonstrations, are arrived at by the inclusion of a filter in the feedback loop to eliminate harmonics generated by the nonlinear response of the E/O modulator. This approach yields stable, low noise oscillations and closely supports the analytical formulation presented here.

9.2 Basics of the opto-electronic oscillator

In this section, the opto-electronic oscillator and the physical basis for its operation are described and identified. Then a quasi-linear theory for the oscillator dynamics and for the oscillator noise are developed. Finally, results of the theory are compared with experimental results.

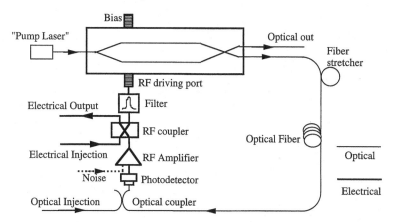

Figure 9.2. Detailed construction of an OEO. Optical injection and RF injection ports are supplied for synchronizing the oscillator with an external reference by either optical injection locking or electrical injection locking. The bias port and the fiber stretcher can be used to fine tune the oscillation frequency. Noise in the oscillator can be viewed as being injected from the input of the amplifier.

9.2.1 Description of the oscillator

The OEO utilizes the transmission characteristics of a modulator together with a fiber optic delay line to convert light energy into stable, spectrally pure RF/microwave reference signals. Detailed construction of the oscillator is shown schematically in Fig. 9.2. In this depiction, light from a laser is introduced into an E/O modulator, the output of which is passed through a long optical fiber, and detected with a photodetector. The output of the photodetector is amplified and filtered and fed back to the electric port of the modulator. This configuration supports self sustained oscillations, at a frequency determined by the fiber delay length, bias setting of the modulator, and the bandpass characteristics of the filter. It also provides for both electric and optical outputs, a feature which is of considerable advantage to photonics applications.

A regenerative feedback approach will be used to analyze the spectral properties of the OEO. Similar methods have been successfully used to analyze lasers [6] and surface acoustic wave oscillators [32]. The conditions for self sustained oscillations include coherent addition of partial waves each way around the loop, and a loop gain exceeding losses for the circulating waves in the loop. The first condition implies that all signals that differ in phase by some multiple of 2π from the fundamental signal may be sustained. Thus the oscillation frequency is limited only by the characteristic frequency response of the modulator, and the setting of the filter, which eliminates all other sustainable oscillations. The second condition implies that with adequate light input power, self sustained oscillations may be obtained, without the need for the RF/microwave amplifier in the loop. These expected characteristics, based on the qualitative analysis of the oscillator dynamics, are mathematically derived in the following sections.

9.2.2 Quasi-linear theory of the OEO

In the following sections a quasi-linear theory to study the dynamics and noise of an OEO is introduced. In the discussion, it is assumed the E/O modulator in the oscillator is of the Mach–Zehnder type. However, the analysis of oscillators made with different E/O modulators follows the same procedure. The flow of the theory is as follows. First the open loop characteristics of a photonic link consisting of a laser, a modulator, a fiber delay, and photodetector are determined. The loop is then closed back into the modulator and a quasi-linear analysis is invoked by including a filter in the loop. This approach leads to a formulation for the amplitude and the frequency of the oscillation. In the next step the influence of the noise in the oscillator is considered, again assisted by the presence of the filter which limits the number of circulating Fourier components. Finally an expression for the spectral density of the OEO is reached, which would be suitable for experimental investigations.

9.2.2.1 Oscillation threshold

The optical power from the E/O modulator's output port that forms the loop is related to an applied voltage $V_{in}(t)$ by

$$P(t) = (\alpha P_o/2)\{1 - \eta \sin \pi [V_{in}(t)/V_\pi + V_B/V_\pi]\}, \qquad (9.1)$$

where α is the fractional insertion loss of the modulator, V_π is its half-wave voltage, V_B is its bias voltage, P_o is the input optical power, and η determines the extinction ratio of the modulator by $(1 + \eta)/(1 - \eta)$.

If the optical signal $P(t)$ is converted to an electric signal by a photodetector, the output electric signal after an RF amplifier is

$$V_{out}(t) = \rho P(t)RG_A = V_{ph}\{1 - \eta \sin \pi [V_{in}(t)/V_\pi + V_B/V_\pi]\}, \qquad (9.2)$$

where ρ is the responsivity of the detector, R is the load impedance of the photodetector, G_A is the amplifier's voltage gain, and V_{ph} is the photovoltage defined as

$$V_{ph} = (\alpha P_o \rho/2)RG_A = I_{ph}RG_A, \qquad (9.3)$$

with $I_{ph} \equiv \alpha P_o \rho/2$ as the photocurrent. The OEO is formed by feeding the signal of Eq. (9.2) back to the RF input port of the E/O modulator. Therefore the small signal open loop gain G_S of the OEO is

$$G_S \frac{dV_{\text{out}}}{dV_{\text{in}}}\bigg|_{V_{\text{in}}=0} = -\frac{\eta\pi V_{\text{ph}}}{V_\pi} \cos\left(\frac{\pi V_B}{V_\pi}\right). \tag{9.4}$$

The highest small signal gain is obtained when the modulator is biased at quadrature, that is when $V_B = 0$ or V_π. From Eq. (9.4) one may see that G_S can be either positive or negative, depending on the bias voltage. The modulator is said to be positively biased if $G_S > 0$, otherwise it is negatively biased. Therefore, when $V_B = 0$, the modulator is biased at negative quadrature, while when $V_B = V_\pi$, the modulator is biased at positive quadrature. Note that in most externally modulated photonic links, the E/O modulators can be biased at either positive or negative quadrature without affecting their performance. However, as will be seen next, the biasing polarity has an important effect on the operation of the OEO.

In order for the OEO to oscillate, the magnitude of the small signal open loop gain must be larger than unity. From Eq. (9.4) we immediately obtain the oscillation threshold of the OEO to be

$$V_{\text{ph}} = V_\pi / [\pi\eta | \cos(\pi V_B / V_\pi)|]. \tag{9.5}$$

For the ideal case in which $\eta = 1$ and $V_B = 0$ or V_π, Eq. (9.5) becomes

$$V_{\text{ph}} = V_\pi / \pi. \tag{9.6}$$

It is important to notice from Eq. (9.3) and Eq. (9.6) that the amplifier in the loop is not a necessary condition for oscillation. So long as $I_{\text{ph}} R \geq V_\pi / \pi$ is satisfied, no amplifier is needed ($G_A = 1$). It is the optical power from the pump laser that actually supplies the necessary energy for the OEO. This property is of practical significance because it enables the OEO to be powered remotely using an optical fiber. Perhaps more significantly, however, the elimination of the amplifier in the loop also eliminates the amplifier noise, resulting in a more stable oscillator. For a modulator with a V_π of 3.14 V and an impedance R of 50 Ω, a photocurrent of 20 mA is required for sustaining the photonic oscillation without an amplifier. This corresponds to an optical power of 25 mW, assuming the responsivity ρ of the photodetector to be 0.8 A/W.

9.2.2.2 Linearization of E/O modulator's response function

In general, Eq. (9.2) is nonlinear. If the electrical input signal $V_{\text{in}}(t)$ to the modulator is a sinusoidal wave with an angular frequency of ω, an amplitude of V_o, and an initial phase β:

$$V_{\text{in}}(t) = V_o \sin(\omega t + \beta), \tag{9.7}$$

the output at the photodetector, $V_{\text{out}}(t)$, can be obtained by substituting Eq. (9.7) in Eq. (9.2) and expanding the right hand side of Eq. (9.2) with Bessel functions

$$
\begin{aligned}
V_{\text{out}}(t) \\
= V_{\text{ph}} \Bigg\{ 1 - \eta \sin\left(\frac{\pi V_{\text{B}}}{V_\pi}\right) \Bigg[J_0\left(\frac{\pi V_0}{V_\pi}\right) + 2 \sum_{m=1}^{\infty} J_{2m}\left(\frac{\pi V_0}{V_\pi}\right) \cos(2m\omega t + 2m\beta) \Bigg] \\
- 2\eta \cos\left(\frac{\pi V_{\text{B}}}{V_\pi}\right) \sum_{m=0}^{\infty} J_{2m+1}\left(\frac{\pi V_0}{V_\pi}\right) \sin[(2m+1)\omega t + (2m+1)\beta] \Bigg\}.
\end{aligned}
\tag{9.8}
$$

It is clear from Eq. (9.8) that the output contains many harmonic components of ω.

The output can be linearized if it passes through an RF filter with a bandwidth sufficiently narrow to block all harmonic components. The linearized output can be easily obtained from Eq. (9.8) to be

$$
V_{\text{out}}(t) = G(V_0)V_{\text{in}}(t),
\tag{9.9a}
$$

where the voltage gain coefficient $G(V_0)$ is defined as

$$
G(V_0) = G_{\text{S}} \frac{2V_\pi}{\pi V_0} J_1\left(\frac{\pi V_0}{V_\pi}\right).
\tag{9.10a}
$$

It can be seen that the voltage gain $G(V_0)$ is a nonlinear function of the input amplitude V_0 and its magnitude decreases monotonically with V_0. However, for a small enough input signal ($V_0 \ll V_\pi$ and $J_1(\pi V_0/V_\pi) = \pi V_0/2V_\pi$), we can recover from Eq. (9.10a) the small signal gain: $G(V_0) = G_{\text{S}}$. If we expand the right hand side of Eq. (9.2) with Taylor series, the gain coefficient can be obtained as

$$
G(V_0) = G_{\text{S}} \Bigg[1 - \frac{1}{2}\left(\frac{\pi V_0}{2V_\pi}\right)^2 + \frac{1}{12}\left(\frac{\pi V_0}{2V_\pi}\right)^4 \Bigg].
\tag{9.10b}
$$

It should be kept in mind that in general, $G(V_0)$ is also a function of the frequency ω of the input signal, because V_{ph} is linearly proportional to the gain of the RF amplifier and the responsivity of the photodetector, which are all frequency dependent. In addition, the V_π of the modulator is also a function of the input RF frequency. Furthermore, the frequency response of the RF filter in the loop can also be lumped into $G(V_0)$. In the discussions below, we will introduce a unitless complex filter function $\tilde{F}(\omega)$ to explicitly account for the combined effect of all frequency dependent components in the loop while treating $G(V_0)$ as frequency independent:

$$
\tilde{F}(\omega) = F(\omega)e^{i\phi(\omega)},
\tag{9.11}
$$

where $\phi(\omega)$ is the frequency dependent phase caused by the dispersive component in the loop and $F(\omega)$ is the real normalized transmission function. Now Eq. (9.9a) can be rewritten in complex form as

$$\tilde{V}_{out}(t) = \tilde{F}(\omega)G(V_o)\tilde{V}_{in}(\omega, t), \tag{9.9b}$$

where $\tilde{V}_{in}(t)$ and $\tilde{V}_{out}(t)$ are complex input and output voltages. Note that although Eq. (9.9b) is linear, the nonlinear effect of the modulator is not lost – it is contained in the nonlinear gain coefficient $G(V_o)$.

9.2.2.3 Oscillation frequency and amplitude

In this section we derive the expressions for the amplitude and frequency of OEO. Like other oscillators, the oscillation of an OEO starts from noise transient, which is then built up and sustained with feedback at the level of the oscillator output signal. We derive the amplitude of the oscillating signal by considering this process mathematically. The noise transient can be viewed as a collection of sine waves with random phases and amplitudes. To simplify our derivation, we use this noise input with the linearized expression Eq. (9.9b) for the loop response. Because Eq. (9.9b) is linear, the superposition principal holds and we can analyze the response of the OEO by first inspecting the influence of a single frequency component of the noise spectrum:

$$\tilde{V}_{in}(\omega, t) = \tilde{V}_{in}(\omega)e^{i\omega t}, \tag{9.12}$$

where $\tilde{V}_{in}(\omega)$ is a complex amplitude of the frequency component.

Once the noise component of Eq. (9.12) is on the oscillator, it will circulate in the loop and the recurrence relation of the fields from Eq. (9.9b) is

$$\tilde{V}_n(\omega, t) = \tilde{F}(\omega)G(V_o)\tilde{V}_{n-1}(\omega, t - \tau'), \tag{9.13}$$

where τ' is the time delay resulting from the physical length of the feedback and n is the number of times the field has circulated around the loop, with $\tilde{V}_{n=0}(\omega, t) = \tilde{V}_{in}(\omega, t)$. In Eq. (9.13), the argument V_o in $G(V_o)$ is the amplitude of the total field (the sum of all circulating fields) in the loop.

The total field at any instant of time is the summation of all circulating fields. Therefore, with the input of Eq. (9.12) injected in the oscillator, the signal measured at the RF input to the modulator for the case that the open loop gain is less than unity can be expressed as

$$\tilde{V}(\omega, t) = G_a\tilde{V}_{in}(\omega) \sum_{n=0}^{\infty} [\tilde{F}(\omega)G(V_o)]^n e^{i\omega(t-n\tau')} = \frac{G_a\tilde{V}_{in}e^{i\omega t}}{1 - \tilde{F}(\omega)G(V_o)e^{-i\omega\tau'}}. \tag{9.14}$$

For loop gain below threshold and with V_o small, $G(V_o)$ is essentially the small signal gain G_S given by Eq. (9.4).

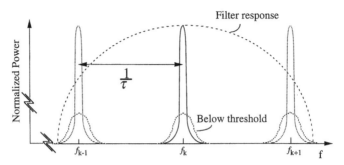

Figure 9.3. Illustration of the oscillator's output spectra below and above the threshold.

The corresponding RF power of the circulating noise at frequency ω is therefore

$$P(\omega) = \frac{|\tilde{V}(\omega, t)|^2}{2R} = \frac{G_A^2 |\tilde{V}_{in}(\omega)|^2 / (2R)}{1 + |F(\omega)G(V_o)|^2 - 2F(\omega)|G(V_o)| \cos[\omega\tau' + \phi(\omega) + \phi_o]},$$

(9.15)

where $\phi_o = 0$ if $G(V_o) > 0$ and $\phi_o = \pi$ if $G(V_o) < 0$.

For a constant $\tilde{V}_{in}(\omega)$, the frequency response of an OEO has equally spaced peaks similar to that of a Fabry–Perot resonator, as shown in Fig. 9.3. These peaks are located at the frequencies determined by

$$\omega_k \tau' + \phi(\omega_k) + \phi_o = 2k\pi \qquad k = 0, 1, 2 \ldots,$$

(9.16)

where k is the mode number. In Fig. 9.3, each peak corresponds to a frequency component resulting from the coherent summation of all circulating fields in the loop at that frequency. As the open loop gain increases; the magnitude of each peak becomes larger and its shape becomes sharper. These peaks are the possible oscillation modes of the OEO. When the open loop gain is larger than unity, each time a noise component at a peak frequency travels around the loop, it is amplified and its amplitude increases geometrically – an oscillation is started from noise.

Because an RF filter is placed in the loop, the gain of only one mode is allowed to be larger than unity, thus selecting the mode that is allowed to oscillate. Because of the nonlinearity of the E/O modulator or the RF amplifier, the amplitude of the oscillation mode cannot increase indefinitely. As the amplitude increases, higher harmonics of the oscillation will be generated by the nonlinear effect of the modulator or the amplifier, at the expense of the oscillation power, and these higher harmonics will be filtered out by the RF filter. Effectively, the gain of the oscillation mode is decreased according to Eq. (9.10) till the gain is, for all practical measures, equal to unity, and the oscillation is stable. As will be shown later, because of the continuous presence of noise, the closed loop gain of an oscillating mode is actually less than unity by a tiny amount on the order of 10^{-10}, which ensures that the summation in Eq. (9.14) converges.

In the discussion that follows only one mode k is allowed to oscillate, and so the oscillation frequency of this mode will be denoted as f_{osc} or $\omega_{\text{osc}}(\omega_{\text{osc}} = 2\pi f_{\text{osc}})$, its oscillation amplitude as V_{osc}, and its oscillation power as $P_{\text{osc}}(P_{\text{osc}} = V_{\text{osc}}^2/2R)$. In this case, the amplitude V_{o} of the total field in Eq. (9.15) is just the oscillation amplitude V_{osc} of the oscillating mode. If we choose the transmission peak of the filter to be at the oscillation frequency ω_{osc} and so $F(\omega_{\text{osc}}) = 1$, the oscillation amplitude can be solved by setting the gain coefficient $|G(V_{\text{osc}})|$ in Eq. (9.15) to unity. From Eq. (9.10a) it leads to

$$\left| J_1\left(\frac{\pi V_{\text{osc}}}{V_\pi} \right) \right| = \frac{1}{2|G_{\text{S}}|} \frac{\pi V_{\text{osc}}}{V_\pi} \tag{9.17a}$$

In deriving Eq. (9.17a), we have assumed that the RF amplifier in the loop is linear enough that the oscillation power is limited by the nonlinearity of the E/O modulator. The amplitude of the oscillation can be obtained by solving Eq. (9.17a) graphically, and the result is shown in Fig. 9.4a. Note that this result is the same as that obtained by Neyer and Voges [30] using a more complicated approach.

If we use Eq. (9.10b), we can obtain the approximated solution of the oscillation amplitude to be

$$V_{\text{osc}} = \frac{2\sqrt{2}V_\pi}{\pi}\sqrt{1 - \frac{1}{|G_{\text{S}}|}} \qquad \text{3rd order expansion,} \tag{9.17b}$$

$$V_{\text{osc}} = \frac{2\sqrt{3}V_\pi}{\pi}\left(1 - \frac{1}{\sqrt{3}}\sqrt{\frac{4}{|G_{\text{S}}|} - 1}\right)^{1/2} \qquad \text{5th order expansion.} \tag{9.17c}$$

The threshold condition of $|G_{\text{S}}| \geq 1$ is clearly indicated in Eq. (9.17b) and Eq. (9.17c). Figure 9.4a shows the normalized oscillation amplitude as a function of $|G_{\text{S}}|$ obtained from Eq. (9.17a), Eq. (9.17b) and Eq. (9.17c). Comparing the three theoretical curves one can see that for $|G_{\text{S}}| \leq 1.5$, the 3rd order expansion result is a good approximation. For $|G_{\text{S}}| \leq 3$, the 5th order expansion result is a good approximation.

The corresponding oscillation frequency $f_{\text{osc}} \equiv f_k = \omega_k/2\pi$ can be obtained from Eq. (9.16) to be

$$f_{\text{osc}} \equiv f_k = (k + 1/2)/\tau \qquad \text{for } G(V_{\text{osc}}) < 0, \tag{9.18a}$$
$$f_{\text{osc}} \equiv f_k = k/\tau \qquad \text{for } G(V_{\text{osc}}) > 0, \tag{9.18b}$$

where τ is the total group delay of the loop, including the physical length delay τ' of the loop and the group delay resulting from dispersive components (such as an

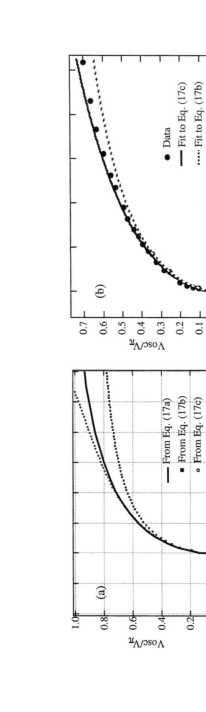

Figure 9.4. Normalized oscillation amplitude of an OEO as a function of small signal gain G_s. (a) Theoretical calculation using Eqs. (9.17a), (9.17b), and (9.17c). (b) Experimental data and curve fitting to Eq. (9.17b) and Eq. (9.17c).

amplifier) in the loop, and it is given by

$$\tau = \tau' + \frac{d\phi(\omega)}{d\omega}\bigg|_{\omega=\omega_{\mathrm{osc}}}. \tag{9.19}$$

For all practical purposes, $J_1(\pi V_{\mathrm{osc}}/V_\pi) \geq 0$ or $V_{\mathrm{osc}}/V_\pi \leq 1.21$ and the sign of $G(V_{\mathrm{osc}})$ is determined by the small signal gain G_S. It is interesting to notice from Eq. (9.18) that the oscillation frequency depends on the biasing polarity of the modulator. For negative biasing ($G_S < 0$), the fundamental frequency is $1/(2\tau)$, while for positive biasing ($G_S > 0$), the fundamental frequency is doubled to $1/\tau$.

9.2.2.4 The spectrum

The fundamental noise in an OEO consists of the thermal noise, the shot noise, and the laser's intensity noise, which for the purpose of analysis can be viewed as all originating from the photodetector. Since the photodetector is directly connected to the amplifier, the noise can be viewed as entering the oscillator at the input of the amplifier, as shown in Fig. 9.2.

We compute the spectrum of the oscillator signal by determining the power spectral density of noise in the oscillator. Let $\rho_N(\omega)$ be the power density of the input noise at frequency ω, we have

$$\rho_N(\omega)\Delta f = \frac{|\tilde{V}_{\mathrm{in}}(\omega)|^2}{2R}, \tag{9.20}$$

where Δf is the frequency bandwidth. Substituting Eq. (9.20) in Eq. (9.15) and letting $F(\omega_{\mathrm{osc}}) = 1$, we obtain the power spectral density of the oscillating mode k to be

$$S_{\mathrm{RF}}(f') = \frac{P(f')}{\Delta f\, P_{\mathrm{osc}}} = \frac{\rho_N G_A^2 / P_{\mathrm{osc}}}{1 + |F(f')G(V_{\mathrm{osc}})|^2 - 2F(f')|G(V_{\mathrm{osc}})|\cos(2\pi f'\tau)}, \tag{9.21}$$

where $f' \equiv (\omega - \omega_{\mathrm{osc}})/2\pi$ is the frequency offset from the oscillation peak f_{osc}. In deriving Eq. (9.21), both Eq. (9.16) and Eq. (9.19) are used.

By using the normalization condition

$$\int_{-\infty}^{\infty} S_{\mathrm{RF}}(f')\,df' \approx \int_{-1/2\tau}^{1/2\tau} S_{\mathrm{RF}}(f')\,df' = 1, \tag{9.22}$$

we obtain

$$1 - |G(V_{\mathrm{osc}})|^2 \approx 2[1 - |G(V_{\mathrm{osc}})|] = \frac{\rho_N G_A^2}{\tau P_{\mathrm{osc}}}. \tag{9.23}$$

Note that in Eq. (9.22), we have assumed that the spectral width of the oscillating mode is much smaller than the mode spacing $1/\tau$ of the oscillator, so that the integration over $1/\tau$ is sufficiently accurate. In addition, in the derivation we have assumed that $|F(f')| \approx 1$ in the frequency band of integration.

Typically, $\rho_N \sim 10^{-17}$ mW/Hz, $P_{osc} \sim 10$ mW, $G_A^2 \sim 100$, and $\tau \sim 10^{-6}$ s. From Eq. (9.23) one can see that the closed loop gain $|G(V_{osc})|$ of the oscillating mode is less than unity by an amount of 10^{-10}. Therefore the equation $|G(V_{osc})| = 1$ is sufficiently accurate for calculating the oscillation amplitude V_{osc}, as proceeded in Eq. (9.17).

Finally, substituting Eq. (9.23) in Eq. (9.21), we obtain the RF spectral density of the OEO to be

$$S_{RF}(f') = \frac{\delta}{(2 - \delta/\tau) - 2\sqrt{1 - \delta/\tau}\cos(2\pi f'\tau)}, \qquad (9.24a)$$

where δ is defined as

$$\delta \equiv \rho_N G_A^2 / P_{osc}. \qquad (9.25)$$

As mentioned before, ρ_N is the equivalent input noise density injected into the oscillator from the input port of the amplifier and P_{osc}/G_A^2 is the total oscillating power measured before the amplifier. Therefore δ is the input noise-to-signal ratio to the oscillator.

For the case where $2\pi f'\tau \ll 1$, we can simplify Eq. (9.24a) by expanding the cosine function in Taylor series:

$$S_{RF}(f') = \frac{\delta}{(\delta/2\tau)^2 + (2\pi)^2(\tau f')^2}. \qquad (9.24b)$$

Equation (9.24b) is a good approximation even for $2\pi f'\tau = 0.7$ at which value the error resulting from neglecting the higher order terms in the Taylor expansion is less than 1%. It can be seen from Eq. (9.24b) that the spectral density of the oscillating mode is a Lorentzian function of frequency. Its full width at half maximum (FWHM) Δf_{FWHM} is

$$\Delta f_{FWHM} = \frac{1}{2\pi}\frac{\delta}{\tau^2} = \frac{1}{2\pi}\frac{G_A^2 \rho_N}{\tau^2 P_{osc}}, \qquad (9.26a)$$

It is evident from Eq. (9.26a) that Δf_{FWHM} is inversely proportional to the square of the loop delay time and linearly proportional to the input noise-to-signal ratio δ. For a typical δ of 10^{-16} /Hz and a loop delay of 100 ns (20 m), the resulting spectral width is sub millihertz. The fractional power contained in Δf_{FWHM} is $\Delta f_{FWHM} S_{RF}(0) = 64\%$.

From Eq. (9.26a) one can also see that for fixed ρ_N and G_A, the spectral width of an OEO is inversely proportional to the oscillation power, similar to the famous Schawlow–Townes formula [33,34] for describing the spectral width $\Delta\nu_{\text{laser}}$ of a laser:

$$\Delta\nu_{\text{laser}} = \frac{1}{2\pi} \frac{\rho_S}{\tau_{\text{laser}}^2 P_{\text{laser}}}, \qquad (9.26b)$$

where $\rho_S = h\nu$ is the spontaneous emission noise density of the laser, P_{laser} is the laser oscillation power, and τ_{laser} is the decay time of the laser cavity. However, as will be shown in Section 2.2.5 both P_{osc} and ρ_N are functions of the photocurrent, the statement that the spectral width of an OEO is inversely proportional to the oscillation power is only valid when thermal noise dominates in the oscillator at low photocurrent levels.

The quality factor Q of the oscillator from Eq. (9.26a) is

$$Q = \frac{f_{\text{osc}}}{\Delta f_{\text{FWHM}}} = Q_D \frac{\tau}{\delta}, \qquad (9.27)$$

where Q_D is the quality factor of the loop delay line and is defined as

$$Q_D = 2\pi f_{\text{osc}} \tau. \qquad (9.28)$$

From Eq. (9.24b) we easily obtain

$$S_{\text{RF}}(f') = \frac{4\tau^2}{\delta} \qquad |f'| \ll \Delta f_{\text{FWHM}}/2, \qquad (9.29a)$$

$$S_{\text{RF}}(f') = \frac{\delta}{(2\pi)^2(\tau f')^2} \qquad |f'| \gg \Delta f_{\text{FWHM}}/2. \qquad (9.29b)$$

It can be shown [35] that for an oscillator with a phase fluctuation much less than unity, its power spectral density is equal to the sum of the single side band phase noise density and the single side band amplitude noise density. In most cases in which the amplitude fluctuation is much less than the phase fluctuation, the power spectral density is just the single side band phase noise. Therefore, it is evident from Eq. (9.29b) that the phase noise of the OEO decreases quadratically with the frequency offset f'. For a fixed f', the phase noise decreases quadratically with the loop delay time. The larger the τ, the smaller the phase noise. However, the phase noise cannot decrease to zero no matter how large τ is, because at large enough τ, Eq. (9.24b) and Eq. (9.29b) are not valid anymore. From Eq. (9.24a), the minimum phase noise is $S_{\text{RF}}^{\text{min}} \approx \delta/4$ at $f' = 1/2\tau$. For the frequency offset f' outside of the passband of the loop filter (where $F(f') = 0$), the phase noise is simply the noise to signal ratio δ, as can be seen from Eq. (9.21).

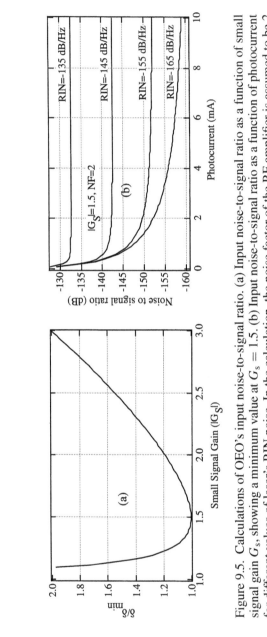

Figure 9.5. Calculations of OEO's input noise-to-signal ratio. (a) Input noise-to-signal ratio as a function of small signal gain G_s, showing a minimum value at $G_s = 1.5$. (b) Input noise-to-signal ratio as a function of photocurrent for different values of laser's RIN noise. In the calculation, the noise factor of the RF amplifier is assumed to be 2 and G_s is fixed at 1.5.

Equation (9.24) and Eq. (9.29b) also indicate that the oscillator's phase noise is independent of the oscillation frequency f_{osc}. This result is significant because it allows the generation of high frequency and low phase noise signals with the OEO. The phase noise of a signal generated using frequency multiplying methods generally increases quadratically with the frequency.

9.2.2.5 The noise-to-signal ratio

As mentioned before, the total noise density input to the oscillator is the sum of the thermal noise $\rho_{\text{thermal}} = 4k_{\text{B}}T(NF)$, the shot noise $\rho_{\text{shot}} = 2eI_{\text{ph}}R$, and the laser's relative intensity noise (RIN) $\rho_{\text{RIN}} = N_{\text{RIN}}I_{\text{ph}}^2R$ densities [36,37]:

$$\rho_{\text{N}} = 4k_{\text{B}}T(NF) + 2eI_{\text{ph}}R + N_{\text{RIN}}I_{\text{ph}}^2R, \qquad (9.30)$$

where k_{B} is the Boltzman constant, T is the ambient temperature, NF is the noise factor of the RF amplifier, e is the electron charge, I_{ph} is the photocurrent across the load resistor of the photodetector, and N_{RIN} is the RIN of the pump laser.

From Eq. (9.25) and Eq. (9.30) one can see that if the thermal noise is dominant, then δ is inversely proportional to the oscillating power P_{osc} of the oscillator. In general, P_{osc} is a function of photocurrent I_{ph} and amplifier gain G_A, as determined by Eq. (9.17), and the noise-to-signal ratio from Eq. (9.25) is thus

$$\delta = \frac{|G_{\text{S}}|^2}{1 - 1/|G_{\text{S}}|} \frac{4kT(NF) + 2eI_{\text{ph}}R + N_{\text{RIN}}I_{\text{ph}}^2R}{4\eta^2 \cos^2(\pi V_{\text{B}}/V_{\pi})|_{\text{ph}}^2R}. \qquad (9.31)$$

In deriving Eq. (9.31), Eq. (9.4) and Eq. (9.17b) are used. From Eq. (9.31) one can see that δ is a nonlinear function of the small signal gain of the oscillator. As shown in Fig. 9.5a, it reaches the minimum value at $|G_{\text{S}}| = 3/2$:

$$\delta_{\text{min}} = \frac{4kT(NF) + 2eI_{\text{ph}}R + N_{\text{RIN}}I_{\text{ph}}^2R}{(16/27)I_{\text{ph}}^2R}, \qquad (9.32)$$

where $\eta = 1$ and $\cos(\pi V_{\text{B}}/V_{\pi}) = 1$ are assumed. The oscillation amplitude at $|G_{\text{S}}| = 3/2$ can be obtained from Eq. (9.17b) to be

$$V_{\text{osc}} = 2\sqrt{2}V_{\pi}/(\pi\sqrt{3}) \approx 0.52V_{\pi}, \qquad (9.33a)$$

and the corresponding RF power is

$$P_{\text{osc}} = 4V_{\pi}^2/(3\pi^2R) = 10P_{\text{m}}^{1\text{dB}}/3, \qquad (9.33b)$$

where $P_{\text{m}}^{1\text{dB}}$ is the input 1 dB compression power of the E/O modulator [37,38]. From Eq. (9.33b) one can conclude that in order to have minimum noise, the oscillation power measured at the input of the E/O modulator should be 5 dB above the 1 dB compression power of the modulator. Equation (9.33a) indicates that the noise of the oscillator is minimum when the oscillating amplitude is roughly half V_π or the voltage in the oscillator is varying between the peak and the trough of the sinusoidal transmission curve of the E/O modulator. This makes sense because the modulator has its minimum sensitivity to voltage variations at the maximum and the minimum of the transmission curve, and the most likely cause of voltage variations in an OEO is the noise in the loop.

It is evident from Eq. (9.32) that the higher the photocurrent, the less the noise-to-signal ratio of the oscillator until it flattens out at the laser's RIN level. Therefore, the ultimate noise to signal ratio of an OEO is limited by the pump laser's RIN. If the RIN of the pump laser in an OEO is -160 dB/Hz, the ultimate noise-to-signal ratio of the oscillator is also -160 dB/Hz and the signal-to-noise ratio is 160 dB/Hz. Figure 9.5b shows the noise-to-signal ratio δ as a function of photocurrent I_{ph} for different RIN levels. In the plot, the small signal gain G_S is chosen to be a constant of 1.5, which implies that when I_{ph} is increased, the amplifier gain G_A must be decreased to keep G_S constant. From the figure one can easily see that δ decreases quadratically with I_{ph} at small I_{ph} and flattens out at the RIN level at large I_{ph}.

9.2.2.6 Effects of amplifier's nonlinearity

In the discussions above, we have assumed that the nonlinear distortion of a signal from the E/O modulator is more severe than from the amplifier (if any) used in the oscillator so that the oscillation amplitude or power is limited by the nonlinear response of the E/O modulator. Using an engineering term, this simply means that the output 1 dB compression power of the amplifier is much larger than the input 1 dB compression power of the E/O modulator [37].

For cases in which the output 1 dB compression power of the amplifier is less than the input 1 dB compression power of the modulator, the nonlinearity of the amplifier will limit the oscillation amplitude V_{osc} (or power P_{osc}) of the oscillator, resulting in an oscillation amplitude less than that given by Eq. (9.17). The exact relation between the oscillation amplitude (or power) and the small signal gain G_S can be determined using the same linearization procedure as that for obtaining Eq. (9.17) if the nonlinear response function of the amplifier is known. However, all the equations in Section 9.2.2.4 for describing the spectrum of the oscillator are still valid, provided that the oscillation power in those equations is determined by the nonlinearity of the amplifier. For a high enough small signal gain G_S, the oscillation power is approximately a few dB above the output 1 dB compression power of the amplifier.

It is important to note that the analysis performed here was for the specific case of the OEO with a Mach–Zehnder electro-optic modulator. Other modulation schemes such as those with electroabsorptive modulators or the direct modulation of semiconductor lasers [38] will also lead to signals with characteristics similar to those obtained in this work. For these cases the theoretical approach developed above is still applicable after suitable modifications. The major change required in the analysis is the replacement of Eq. (9.1) which describes the transmission characteristics of a Mach–Zehnder modulator with the appropriate equation for the specific modulation scheme. All other equations can then be derived in the same way as described in the theory.

9.2.3 Experimental verification

In all the experiments discussed below, the output 1 dB compression power of the amplifiers chosen is much larger than the input 1 dB compression power of the modulator so that the oscillation power is limited by the modulator

9.2.3.1 Amplitude vs. open loop gain

In the experiments described below, a highly stable diode-pumped Nd:YAG ring laser [34] with a built-in RIN reduction circuit [39] to pump the OEO was used. The experimental setup for measuring the oscillation amplitude as a function of the open loop gain is shown in Fig. 9.6a. Here an RF switch was used to open and close the loop. While the loop was open, an RF signal from a signal generator with the same frequency as the oscillator was injected into the E/O modulator. The amplitudes of the injected signal and the output signal from the loop were measured with an oscilloscope to obtain the open loop gain which was the ratio of the output amplitude to the injected signal amplitude. The open loop gain was varied by changing the bias voltage of the E/O modulator, or by attenuating the optical power of the loop, or by using a variable RF attenuator after the photodetector, as indicated by Eq. (9.4). When closing the loop, the amplitude of the oscillation was conveniently measured using the same oscilloscope. We measured the oscillation amplitudes of the OEO for different open loop gains at an oscillation frequency of 100 MHz, and the data obtained is plotted in Fig. 9.4b. It is evident that the experimental data agrees well with our theoretical predictions.

9.2.3.2 The phase noise measurement setup

The phase noise of the OEO was measured using the frequency discriminator method [40] and the experimental setup is shown in Fig. 9.6b. The advantage of this method is that it does not require a frequency reference and hence can be used to measure an oscillator of any frequency. Using a microwave mixer in the experiment the phase of a signal from the electrical output port of the OEO was compared with its delayed replica from the optical output port. The length of the delay line is important because the longer the delay line the lower the frequency

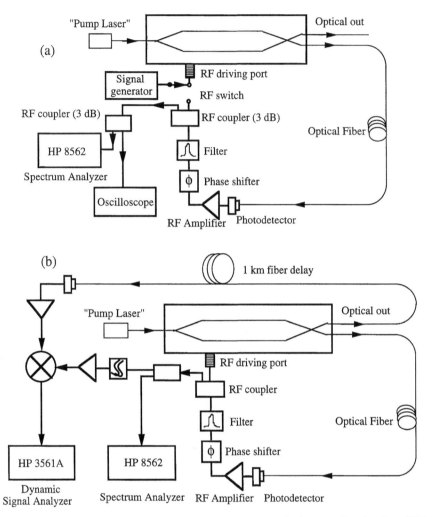

Figure 9.6. Experimental setups. (a) For measuring the oscillation amplitude of an OEO as a function of the small signal gain. (b) For measuring the phase noise of an OEO using the frequency discrimination method.

offset at which the phase noise can be accurately measured. On the other hand, if the delay line is too long, the accuracy of the phase noise at higher frequency offset will suffer. The length of delay used in the experiment is 1 km or 5 μs. Because of this delay, any frequency fluctuation of the OEO will cause a voltage fluctuation at the output of the mixer. The spectrum of this voltage fluctuation was then measured with a high dynamic range spectrum analyzer and the spectral data was transferred to a computer. Finally, this information was converted into the phase noise spectrum of the OEO according to the procedures given in Ref. 30. In these experiments, the noise figure of the RF amplifier was 7 dB.

9.2.3.3 Phase noise as a function of offset frequency and loop delay

Figure 9.7a is the log vs. log scale plot of the measured phase noise as a function of the frequency offset f'. Each curve corresponds to a different loop delay time. Clearly, the phase noise has a 20 dB per decade dependence on the frequency offset, in excellent agreement with the theoretical prediction of Eq. (9.29b).

Figure 9.7b is the measured phase noise at 30 kHz from the center frequency as a function of the loop delay time, extracted from the different curves of Fig. 9.7a. Because the loop delay is increased by adding more fiber segments, the open loop gains of the oscillator with longer loops decrease as more segments are connected, causing the corresponding oscillation power to decrease. From the results of Fig. 9.9 below, the phase noise of the OEO decreases linearly with the oscillation power. To extrapolate the dependence of the phase noise on the loop delay only from Fig. 9.7a, each data point in Fig. 9.7b is calibrated using the linear dependence of Fig. 9.9, while keeping the oscillation power for all data points at 16.33 mW. Again, the experimental data agrees well with the theoretical prediction.

9.2.3.4 Phase noise independence of oscillation frequency

To confirm the prediction that the phase noise of the OEO is independent of the oscillation frequency, the phase noise spectrum as a function of the oscillation frequency was measured and the result is shown in Fig. 9.8a. In the experiment, the loop delay was kept at 0.28 μs, and the oscillation frequency was varied by changing the RF filter in the loop. The frequency was fine tuned using an RF line stretcher. It is evident from Fig. 9.8a that all phase noise curves at frequencies 100 MHz, 300 MHz, 700 MHz, and 800 MHz overlap with one another, indicating a good agreement with the theory. Figure 9.8b is a plot of the phase noise data at 10 kHz as a function of the frequency. As predicted, it is a flat line, in contrast with the case when a frequency multiplier is used to obtain higher frequencies. This result is significant because it confirms that the OEO can he used to generate high frequency signals up to 75 GHz with a much lower phase noise than can be attained with frequency multiplying techniques.

9.2.3.5 Phase noise as a function of oscillation power

The phase noise spectrum of the OEO as a function of oscillation power with the results is shown in Fig. 9.9. In this experiment, the loop delay of the OEO was 0.06 μs, the noise figure of the RF amplifier was 7 dB, and the oscillation power was varied by changing the photocurrent I_{ph} according to Eqs. (9.3), (9.4), and (9.17). With this amplifier and the photocurrent level (1.8–2.7 mA), the thermal noise in the oscillator dominates. Recall that in Eq. (9.24) and Eq. (9.25), the phase noise of an OEO is shown to be inversely proportional to the oscillation power. This is true

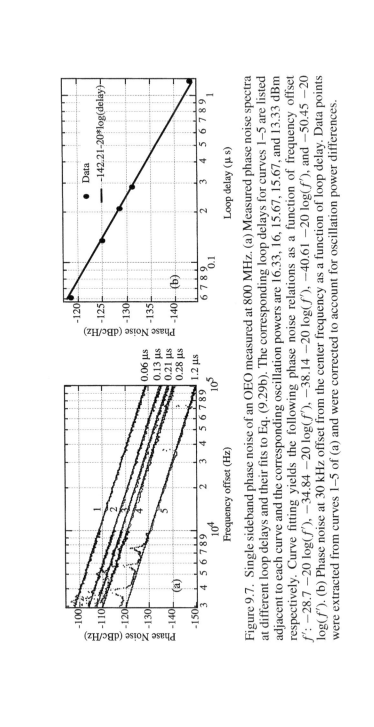

Figure 9.7. Single sideband phase noise of an OEO measured at 800 MHz. (a) Measured phase noise spectra at different loop delays and their fits to Eq. (9.29b). The corresponding loop delays for curves 1–5 are listed adjacent to each curve and the corresponding oscillation powers are 16.33, 16, 15.67, 15.67, and 13.33 dBm respectively. Curve fitting yields the following phase noise relations as a function of frequency offset f': $-28.7 - 20\log(f')$, $-34.84 - 20\log(f')$, $-38.14 - 20\log(f')$, $-40.61 - 20\log(f')$, and $-50.45 - 20\log(f')$. (b) Phase noise at 30 kHz offset from the center frequency as a function of loop delay. Data points were extracted from curves 1–5 of (a) and were corrected to account for oscillation power differences.

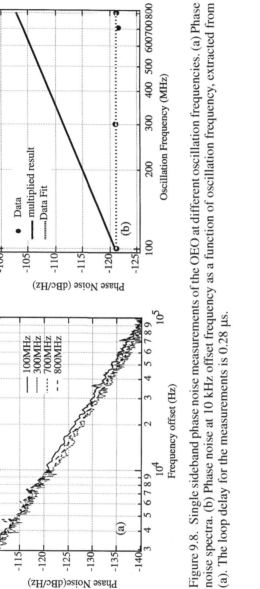

Figure 9.8. Single sideband phase noise measurements of the OEO at different oscillation frequencies. (a) Phase noise spectra. (b) Phase noise at 10 kHz offset frequency as a function of oscillation frequency, extracted from (a). The loop delay for the measurements is 0.28 μs.

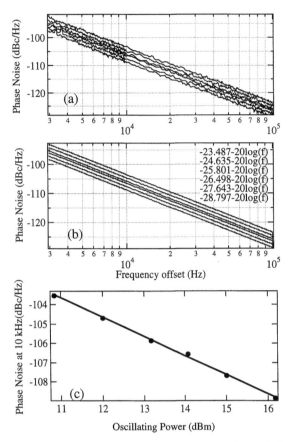

Figure 9.9. Single sideband phase noise spectra as a function of oscillation power measured at 800 MHz. (a) Experimental data. (b) The fit to Eq. (9.29b). (c) Phase noise at 10 kHz offset as a function of oscillation power extracted from (b).

if the gain of the amplifier is kept constant and the photocurrent is low enough to ensure that the thermal noise is the dominant noise term. In Fig. 9.9a, each curve is the measurement data of the phase noise spectrum corresponding to an oscillating power, and the curves in Fig. 9.9b are the fits of the data to Eq. (9.29b). Figure 9.9c is the phase noise of the OEO at 10 kHz as a function of the oscillation power, extracted from the data of Fig. 9.9b. The resulting linear dependence $-92.605 - P_{\mathrm{osc}}(\mathrm{dBm})$ agrees well with the theoretical prediction of Eq. (9.29b).

9.2.4 Compact OEO with integrated DFB laser/modulator module

In the experiments described in the previous section, the OEO was constructed with expensive and bulky diode-pumped YAG lasers, and LiNbO$_3$ modulators. For communication, radar, and space applications, compact and low cost OEOs are

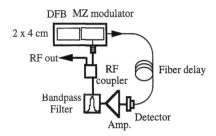

Figure 9.10. The configuration of a compact OEO constructed with an integrated module of DFB laser and modulator.

preferred. In this section, a compact and high performance OEO constructed with an integrated module consisting of a DFB laser and a semiconductor modulator is described. The experimental results demonstrate the feasibility of using low cost semiconductor lasers and modulators to construct high performance and compact opto-electronic oscillators.

Figure 9.10 illustrates the configuration of a compact 10 GHz OEO. The key component in the oscillator is a commercial module consisting of a 1550 nm DFB laser integrated with a GaAs Mach–Zehnder modulator that has a 3-dB modulation bandwidth of 8.6 GHz. The size of the module is only $2(W) \times 4(L) \times 1(H) \, cm^3$ and the output power of the module is about 2 mW. The measured RIN (relative intensity noise) of the laser output is -110 dBc/Hz at 10 Hz and -135 dBc/Hz at 10 KHz, which sets the limitation of the phase noise of the OEO. The oscillation frequency of the OEO is determined by a bandpass filter centered at 9.56 GHz with a bandwidth of 2 MHz. The filter was constructed from a small dielectric cylindrical disc (8.4 mm × 4 mm) that has a dielectric constant of 35 and an exceptionally low temperature coefficient (down to ±1 ppm/°C at room temperature).

The phase noises of this compact OEO with 2-km, 4-km and 6-km loop lengths were measured with the frequency discriminator method described in the last section and the results are shown in Fig. 9.11. It is evident that exceptionally low phase noises of -50 dBc/Hz at 10 Hz and -125 dBc/Hz at 10 kHz away from the 9.56 GHz carrier were achieved. As expected, the phase noise decreases with the increase of the loop length. However, the rate of decrease as a function of loop length at higher offset frequencies is slower than that at lower offset frequencies, indicating that the phase noise of the OEO at higher offset frequencies is limited by the RIN noise level of the DFB laser. It should be noticed that phase noise as a function of frequency and length deviates from that of Eq. (9.24). This is because when deriving Eq. (9.24), the noise sources in the OEO were assumed to be white noise. However, in practice both amplifier phase noise and multiplied RIN noise [37] at close-to-carrier frequencies are not white noise anymore. Modification of

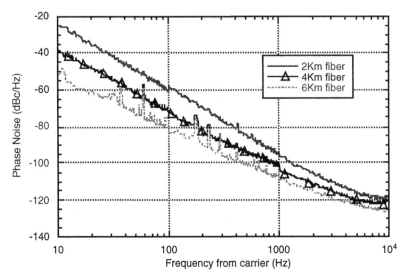

Figure 9.11. Measured phase noise of the compact OEO with 2 km, 4 km, and 6 km loop lengths.

the theory is required to take into account the $1/f$ noise features at close-to-carrier frequencies.

9.3 Multi-loop optoelectronic oscillator

9.3.1 Single mode selection

It was shown in Section 9.2 that the length of the fiber determines the phase noise of the OEO (in particular, the phase noise is inversely proportional to the loop length squared), while the operation frequency is limited only by the characteristics of the modulator.

The highest spectral purity signals with the OEO are achieved with the longest fiber length. Nevertheless the length of the fiber introduces a practical difficulty when the OEO operates at frequencies above a few GHz. This is because the OEO is essentially a multimode device, with its mode spacing determined by the length of the fiber. For a 1 km fiber length, the mode spacing is 200 kHz, requiring a filter with narrow enough bandwidth to select a mode for operation at a single frequency. Narrow-band filters centered at 10 GHz or higher frequencies are difficult to realize, and if they are realized, the filter bandwidth prohibits tuning the oscillator frequency.

In this section, a new configuration is introduced to solve the difficulty associated with the RF filter. The approach is based on the implementation of a second fiber loop to function as the filter. As discussed below, the two fiber loops essentially act as the short and long cavities in a laser to select a single operation mode [41], while preserving tunability. As a bonus, the second loop in this dual-loop OEO also reduces the oscillation threshold to reduce the gain requirement of the RF amplifier.

9.3.2 Analysis

Two configurations of dual-loop OEO are shown in Fig. 9.12. Unlike a single loop OEO with only one feedback loop, in a double loop OEO there are two fiber loops of different lengths. The condition for self sustained oscillation is that the open loop gain of its feedback loop must be greater than one. In the dual-loop oscillator, the open loop gain of each feedback loop may be less than unity, as long as the combined open loop gain of both loops is larger than unity. In addition, the possible oscillation frequencies, f_{osc}, must add up in phase after each round trip around both loops. For the configurations of Fig. 9.12, the oscillation frequency must satisfy

$$f_{osc} = (k + 1/2)/\tau_1 = m/\tau_2 \qquad (9.34)$$

where k and m are integers, and τ_1 and τ_2 are the loop delays of loop 1 and loop 2 respectively. The 1/2 in Eq. (9.34) arises because the modulator is biased such that the optical power in loop 1 decreases with the increase of applied voltage (see Eq. (9.18a)). The mode spacing is then determined by the shorter loop: $\Delta f = 1/\tau_2$. On the other hand, the phase noise of the oscillator is determined by the longer loop, resulting in an oscillator having large mode spacing and low phase noise.

The RF spectrum of the dual-loop OEO can be analyzed using the same quasi-linear theory developed for the single loop OEO. Similar to the single loop OEO,

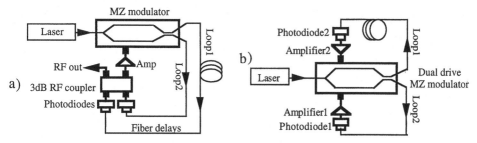

Figure 9.12. Two examples of the dual-loop OEO. (a) Configured with a single RF drive Mach–Zehnder electro-optic modulator. When oscillating, most power at the 3 dB coupler (hybrid or directional coupler) goes to the amplifier due to the interference of the two input RF signals at the coupler. (b) Configured with a dual RF drive Mach–Zehnder modulator. In both cases, the modulator should be biased at quadrature for maximum modulation efficiency.

the recursive relation can be expressed as:

$$\tilde{V}_j(\omega) = (g_1 e^{i\omega\tau_1} + g_2 e^{i\omega\tau_2})\tilde{V}_{j-1}(\omega), \tag{9.35}$$

where $\tilde{V}_j(\omega)$ is the complex amplitude of the circulating field after a round trip, g_1 is the complex gain of loop 1 and g_2 is the complex gain of loop 2. The total field of all circulating fields is thus

$$\tilde{V}_{out}(\omega) = \sum_{j=0}^{\infty}(g_1 e^{i\omega\tau_1} + g_2 e^{i\omega\tau_2})^j \tilde{V}_j(\omega)$$

$$= \frac{\tilde{V}_o}{1 - (g_1 e^{i\omega\tau_1} + g_2 e^{i\omega\tau_2})}. \tag{9.36}$$

The corresponding RF power $P(\omega) \equiv |\tilde{V}_{out}(\omega)|^2/2R$ is therefore

$$P(\omega)\frac{|V_o|^2/2R}{1 + |g_1|^2 + |g_2|^2 + 2|g_1||g_2|\cos(\Phi_1 - \Phi_2) - 2(|g_1|\cos\Phi_1 + |g_1|\cos\Phi_1)}, \tag{9.37}$$

where

$$\Phi_i(\omega) = \omega\tau_i + \phi_i \qquad i = 1, 2. \tag{9.38}$$

In Eq. (9.38), ϕ_i is the phase factor of the complex gain g_i.

If the gain of each loop is less than unity, no oscillation may start independently in either loop. However, for the frequency components satisfying Eq. (9.34), oscillations can start collectively in the two loops. When Eq. (9.34) is satisfied, we have

$$\Phi_1(\omega) = 2k\pi, \tag{9.39a}$$

$$\Phi_2(\omega_o) = 2m\pi, \tag{9.39b}$$

$$\Phi_1(\omega_o) - \Phi_2(\omega_o) = 2(k - m)\pi, \tag{9.39c}$$

where ω_o is the oscillation frequency. Substituting Eq. (9.39) in Eq. (9.37) yields

$$P(\omega) = \frac{|V_o|^2/2R}{1 + |g_1|^2 + |g_2|^2 + 2|g_1||g_2| - 2|g_1| - 2|g_1|}. \tag{9.40}$$

In order for the oscillation to start from noise, we must have

$$1 + |g_1|^2 + |g_2|^2 + 2|g_1||g_2| - 2|g_1| - 2|g_1| = 0. \tag{9.41}$$

For $|g_1| = |g_2|$, we obtain

$$|g_1| = |g_2| = 0.5. \tag{9.42}$$

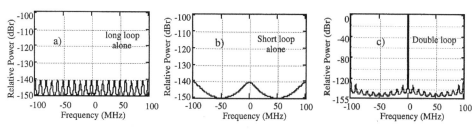

Figure 9.13. Calculated spectra of a double loop OEO. (a) Long loop alone closed. (b) Short loop alone closed. (c) Both loops closed.

This is the oscillation threshold for the double loop OEO. If initially the small signal gain in each loop is larger than 0.5, the nonlinearity of the E/O modulator or the amplifier may bring the gain to 0.5 after the oscillation is started and stabilized.

Figure 9.13 shows the calculated RF spectrum of the dual-loop OEO from Eq. (9.37). In the calculation, $|g_1| = |g_2| = 0.5 - 10^{-8}$, $|V_o|^2/2R = 1$, $\tau_1 = 0.1$ µs and $\tau_2 = 10\tau_1 = 1$ µs were chosen. Figure 9.13a shows the spectrum of an OEO of loop 2 alone, Fig. 9.13b is the spectrum of loop 1 alone, while Fig. 9.13c is the spectrum of the OEO with both loops present. The relative power unit on the vertical axis is the calculated power of each case divided by the calculated peak power of the double loop OEO. Note that the relative powers for Fig. 9.13a and Fig. 9.13c are extremely low, due to the fact that the loop gain is too small for each loop to oscillate individually. For the double loop case of Fig. 9.13c, a strong oscillation at selected frequencies is present, with the mode spacing determined by the shorter loop and the spectral width determined by the long loop, in agreement with our expectation.

Because of the relatively large mode spacing, single mode operation of the OEO can easily be achieved by including an RF filter of wide bandwidth. Since $\Delta f/f = -\Delta L/L$, where f is the oscillation frequency, L is the loop length, and Δf is the frequency change caused by the loop length change ΔL, oscillation frequency of the dual-loop OEO can be more sensitively tuned by changing the loop length of the shorter loop. The tuning range is determined by the mode spacing of the shorter loop and the tuning resolution is determined by the mode spacing of the long loop. For example, if the short loop is 20 cm and the long loop is 10 km, the tuning range will be 1 GHz and the tuning resolution will be 20 kHz. In practice, due to the mode-pulling effect, the tuning is rather continuous before hopping to a different mode of the long loop.

For a larger range, a tunable RF filter can be used. However, the tuning of the center frequency of the filter should be synchronized with that of the loop length, which tunes the oscillation frequency of the OEO. The combination of the double loop OEO and the tunable RF filter produces a novel frequency synthesizer

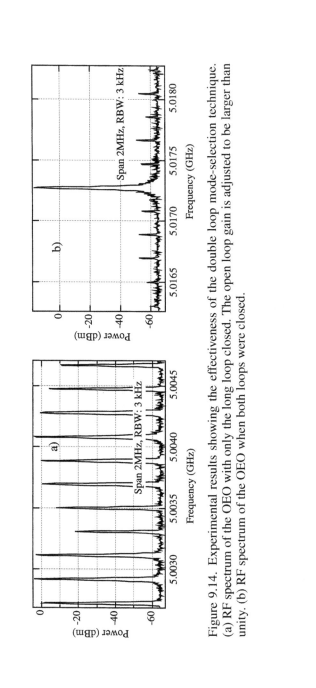

Figure 9.14. Experimental results showing the effectiveness of the double loop mode-selection technique. (a) RF spectrum of the OEO with only the long loop closed. The open loop gain is adjusted to be larger than unity. (b) RF spectrum of the OEO when both loops were closed.

scheme with wide frequency tuning range (tens of GHz), low phase noise, and high frequency resolution.

9.3.3 Experiment

The configuration depicted in Fig. 9.12b was used in the experimental setup, where a dual-drive Mach–Zehnder modulator and a diode pumped YAG laser at 1319 nm were used. The shorter loop ($\sim$ few meters) includes a photodetector and an RF amplifier. The longer loop ($\sim$1 km in length includes another photodetector, a filter centered at 5 GHz, and another RF amplifier. Figure 9.14a shows the multimode oscillation of a single loop OEO in which the shorter loop is disconnected, however the longer loop is closed and its open loop gain is larger than unity. Fig. 14b shows the single mode oscillation of a double loop OEO in which both loops are closed and the gain of each loop was adjusted by inserting attenuators of proper values to keep the gain below the oscillation threshold. However, as required by the analysis above, the combined loop gain for a particular mode was larger than unity. It is evident that the presence of the shorter loop effectively suppresses other modes of the longer loop and only leaves one mode to oscillate. In this configuration the mode suppression ratio is more than 60 dB. We find that even when the gain of each loop is larger than unity (each loop may oscillate by itself), the dual loop configuration is still effective in single mode selection. The side mode suppression ratio of more than 70 dB has been achieved.

It should be noted that an added advantage of the double loop configuration is that for a fixed optical pump power, the second loop increases the open loop gain by 6 dB. Hence the oscillation threshold is lowered by 6 dB, making it easier to realize an OEO without employing an RF amplifier. In fact, with this configuration, we have succeeded in achieving an oscillation at 30 MHz (a frequency for which the modulator is much more efficient) without any RF amplifiers in any loops. Operating the OEO without using amplifiers is important because it reduces the cost and power consumption of the device. In addition, it allows the OEO to be remotely powered with an optical fiber. Perhaps more significantly, however, the elimination of the amplifier in the loops also eliminates amplifier $1/f$ noise, resulting in an oscillator with higher spectral purity.

Figure 9.15 illustrates another dual loop OEO constructed with a 2 km thermally stabilized fiber in the long loop and a filter centered at 10 GHz with a bandwidth of 40 MHz in each loop. The spectra of the OEO output are shown in Fig. 9.16 and are compared with the spectra of an HP high performance RF synthesizer (model HP8617B) for different spectrum analyzer settings. It is evident that the spectrum purity of the OEO is significantly better than the HP synthesizer.

The phase noise of the double loop OEO can be evaluated using the frequency discriminator method and the experimental setup is shown in Fig. 9.15. In the setup,

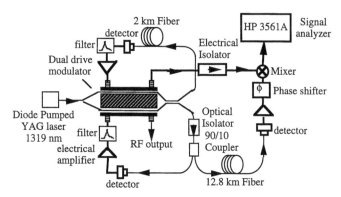

Figure 9.15. Experimental setup of a 10 GHz dual-loop OEO and the frequency discriminator arrangement for measuring the phase noise of the dual-loop OEO. The travelling wave Mach–Zehnder modulator used in the experiments has two input ports and two corresponding output RF ports.

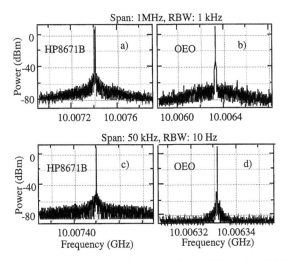

Figure 9.16. Spectral purity comparison of a double loop OEO and an HP8671B synthesizer. The spectra were measured with an HP8563E spectrum analyzer. (a) & (b) The analyzer's frequency span and resolution bandwidth (RBW) settings were 1 MHz and 1 kHz respectively. (c) & (d) The analyzer's frequency span and resolution bandwidth (RBW) settings were 50 kHz and 10 Hz respectively.

the reference fiber has a length of 12.8 km, significantly longer than the longer loop of the OEO. The RF phase shifter was used to adjust the relative phase of the delayed and undelayed signals so that they were in quadrature at the mixer. Any frequency fluctuation of the oscillator was converted to a voltage fluctuation at the output of the mixer and detected by a HP3561A signal analyzer. The measured voltage fluctuation was then converted into phase noise using a simple calibration procedure detailed in Ref. 40. Both the loop fiber and the reference fiber are acoustically

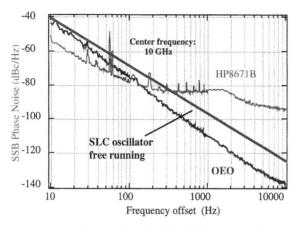

Figure 9.17. Phase noise comparison of the 10 GHz OEO free running at room temperature with a sapphire cavity loaded (SCL) oscillator (also free running at room temperature) and an HP8671B synthesizer. The superior performance of the OEO is evident.

isolated in a box padded with lead-backed foam. Figure 9.17 shows the measured phase noise as a function of offset frequency. It should be noticed that at 10 kHz from the 10 GHz carrier, the phase noise of the double loop OEO is -140 dBc/Hz. As a comparison, the measured phase noise of a high performance HP frequency synthesizer (HP8617B) using the same measurement setup is shown in Fig. 9.15. In this measurement the 2 km loop of the OEO at the RF input port to the modulator was disconnected. Instead, the modulator input was connected to the output of the frequency synthesizer. The result indicates that the synthesizer's phase noise at 10 kHz away from the 10 GHz carrier is about -96 dBc/Hz, more than 40 dB higher than that of the double loop OEO. For the purpose of comparison the phase noise of a free running (no carrier suppression noise reduction scheme was employed) sapphire loaded cavity (SLC) oscillator [42] is also presented in the figure.

Figure 9.17 also shows that the slope of the phase noise of the OEO as a function of the frequency offset is 30 dB/decade, about 10 dB/decade higher than predicted in Section 9.2. This is an indication that a $1/f$ noise source is present in the loop and must be removed for better performance. It is believed that the RF amplifiers used in the loop are responsible for the observed $1/f$ noise; thus the noise can be removed with the carrier suppression technique first employed in SLC oscillators for further phase noise reduction [42,43].

9.4 Summary and future directions

The opto-electronic oscillator is a new class of oscillators that bridges the photonics and the traditional RF worlds, and its potential is yet to be fully realized. Despite its young age, OEO has proven itself as a high performance oscillator for high frequency (tens of GHz) signal generation. A free running OEO has achieved an unsurpassed low phase noise of -140 dBc/Hz at 10 kHz away from a 10 GHz carrier, compared with other free running oscillators. Such a low phase noise oscillator can find wide applications in radar systems, microwave photonic systems [22], mm-wave radio systems, fiber optic communication systems [44], and high speed analog-to-digital conversion systems.

One area for OEO to have a great impact is in high performance mm-wave generation due to the inherent low-loss advantage of the photonic delay lines at high frequencies. Although Gunn oscillators and IMPATT oscillators are available for directly generating mm-wave signals, their phase noise is too high to be useful in many systems. The OEO is expected to outperform a phase locked dielectric resonator oscillator (DRO) by at least 20 dB in phase noise specification.

Continuously improving the phase noise performance of the OEO is another area of intensive research. The author has implemented a carrier suppression noise reduction method [45] to further reduce the close-to-carrier phase noise of an OEO by more than 20 dB. This carrier suppression method is capable of removing the contributions of the $1/f$ phase noise of the microwave components in the OEO loop and the contribution of the RIN noise of the pump laser. Further phase noise improvement may result from using low phase noise photodetectors and compensating for the fiber dispersion. The fiber dispersion converts the laser frequency fluctuation into signal delay fluctuation and thus degrades OEO phase noise.

Another fruitful area of OEO research will be in device miniaturization, including board level integration and chip level integration. The board level integration will involve putting DFB laser, electroabsorption modulator, photodetector, RF filter, RF amplifier, and compact fiber coil on a same circuit board. To reduce the size of the fiber coil, a fiber resonator [46] with a loop length of tens of meters may be used to replace a fiber length of a few kilometers. An 8 cm $\times$ 8 cm $\times$ 5 cm or smaller total package size may be expected from the board level integration.

On the other hand, the chip level integration will involve integrating laser, electroabsorption (E/A) modulator, microsphere, and photodetector on the same semiconductor chip, as shown in Fig. 9.18. The optical microsphere [47,48] is used both as an energy storage element to replace the fiber delay line and as an RF filter for mode selection. When activated by a current source, the active waveguides provide gain for the laser system. Electroabsorption modulators are fabricated at the ends of the waveguides by etching out an insolating gap to separate the electrodes of the gain section and the modulator section. The upper electroabsorption modulator is coated with high reflection coating to induce pulse colliding in the modulator and thus enhance the mode-locking capability [49,50]. At the lower section, the gap between the electroabsorption modulator and the waveguide is etched deeper to induce optical separation. The gap interface acts as a partial mirror to reflect light

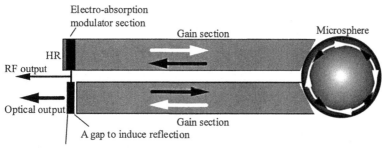

Figure 9.18. A conceptual COEO chip that integrates semiconductor laser, electroabsorption modulator, microsphere, and photodetector on a semiconductor substrate. Note that one of the electroabsorption (E/A) modulators is used as a photodetector (it is strongly biased). Because the impedances of the detector and E/A modulator are close (~ 1 kΩ), efficient RF power transfer to the modulator is expected. The microsphere is located inside the laser cavity for both laser mode selection and RF mode selection. HR: high reflection coating.

back into the lower waveguide and form a laser cavity with the high reflection mirror in the upper section. The lower electroabsorption modulator is strongly biased so that it acts as a photodetector. The output from the photodetector is connected to the upper electroabsorption modulator via a simple matching circuit to induce microwave oscillation. Because the photodetector and the electroabsorption modulator are essentially the same device, they have similar impedance on the order of a few kΩ and thus are essentially impedance matched. Taking typical values of 2 V modulator switching voltage, 1 kΩ of modulator and photodetector impedance, and 0.5 A/W of photodetector responsivity, we estimate that the optical power required for the sustained RF oscillation is only 1.28 mW. Such an optical power is easily attainable in semiconductor lasers. Therefore, the power hungry RF amplifier can be eliminated with the proposed OEO design. The chip level integration will result in an OEO with a size of 1 cm $\times$ 1 cm $\times$ 0.5 cm or smaller.

The coupled opto-electronic oscillator (COEO) [51,52] is another important area of continued research. The COEO is a variant of the OEO which converts light energy to microwave oscillation. However, unlike the original OEO in which the laser oscillation is isolated from the opto-electronic oscillation, in a COEO, the optical oscillation is coupled with the opto-electronic oscillation and therefore the generated RF influences the optical oscillation. The optical and electrical feedback loops of a COEO form a coupled pair of oscillators, one producing microwave oscillation, and the other short optical pulses at the rate of the oscillation frequency in the microwave loop. Because of this coupling, the jitter in the optical pulses will be determined by the spectral characteristics of the microwave signal. Since the signal in the microwave loop can have extremely high spectral purity, as already demonstrated with the "conventional" OEO, optical pulses with jitter in a few femtosecond range may be readily generated. Such a device can clearly have important applications in a number of fields where a high level of synchronization and low jitter can improve the system performance, such as in photonic analog-to-digital conversion systems.

Acknowledgements

The author thanks L. Maleki, Y. Ji, and V. S. Ilchenko for providing relevant materials used in this chapter, and M. Calhoun for the careful review of the manuscript. The work described in this chapter was supported by the Jet Propulsion Laboratory under contracts with the National Aeronautics and Space Administration and by U.S. Air Force Rome Laboratories.

References

1. J. Marion and W. Hornyak, *Physics for Science and Engineering*, Saunders College Publishing, Philadelphia, 1982, Ch. 15.
2. B. van der Pol, "A theory of the amplitude of free and forced triode vibrations," *Radio Review*, **7**, 701–54, 1920.
3. B. van der Pol, "The nonlinear theory of electric oscillations," *Proc. Inst. Radio Eng.*, **22** (9), 1051–86, 1934.
4. O. Ishihara *et al.*, "A highly stabilized GaAsFED oscillator using a dielectric resonator feedback circuit in 9–14 GHz," *IEEE Trans. Microwave Theory Tech.* **MTT-28** (8), 817–24, 1980.
5. A. Siegman, *Microwave Solid State Masers*, McGraw-Hill, 1964.
6. A. Siegman, *Lasers*, University Science Books, Mill Valley, 1986, Ch. 11.
7. A. Ballato, "Piezoelectric resonators," in *Design of Crystal and Other Harmonic Oscillators*, ed., B. Parzen, John Wiley and Sons, 1983, pp. 66–122.
8. W. L. Smith, "Precision oscillators," in *Precision Frequency Control*, Vol. 2, eds. E. A. Gerber and A. Ballato, Academic Press, 1985, pp. 45–98.
9. J. K. Plourde and C. R. Ren, "Application of dielectric resonators," *IEEE Trans. Microwave Theory Tech.*, **MTT-29** (8), 754–69, 1981.
10. M. W. Lawrence, "Surface acoustic wave oscillators," *Wave Electronics*, **2**, 199–218, 1976.
11. H. Ogawa, D. Polifko, and S. Banba, "Millimeter-wave fiber optics systems for personal radio communication," *IEEE Trans. Microwave Theory Tech.*, **40** (12), 2285–93, 1992.
12. P. Herczfeld and A. Daryoush, "Fiber optic feed network for large aperture phased array antennas," *Microwave J.*, 160–6, 1987.
13. X. S. Yao and L. Maleki, "Field demonstration of X-band photonic antenna remoting in the deep space network," TDA Progress Report 42–117, Jet Propulsion Laboratory, pp. 29–34, 1994.
14. G. K. Gopalakrishnan, W. K. Burns, and C. H. Bulmer, "Microwave-optical mixing in $LiNbO_3$ modulators," *IEEE Trans. Microwave Theory Tech.*, **41** (12), 2383–91, 1993.
15. E. Toughlian and H. Zmuda, "A photonic variable RF delay line for phased array antennas," *J. Lightwave Technol.*, **8**, 1824–8, 1990.

16. X. Steve Yao and L. Maleki, "A novel 2-D programmable photonic time-delay device for millimeter-wave signal processing applications," *IEEE Photon. Technol. Lett.*, **6** (12), 1463–5, 1994.

17. D. Nortton, S. Johns, and R. Soref, "Tunable wideband microwave transversal filter using high dispersive fiber delay lines," *Proc. 4th Biennial Department of Defense Fiber Optics and Photonics Conference*, Mclean, Virginia, 1994, pp. 297–301.

18. B. Moslehi, K. Chau, and J. Goodman, "Fiber-optic signal processors with optical gain and reconfigurable weights," *ibid.* pp. 303–9.

19. Lightwave Electronics Corp., "Introduction to diode-pumped solid state lasers," Technical Information No. 1, 1993.

20. X. S. Yao and L. Maleki, "High frequency optical subcarrier generator," *Electron. Lett.* **30** (18), 1525–6, 1994.

21. X. S. Yao and L. Maleki, "Opto-electronic oscillator," *J. Opt. Soc. Am. B*, **13**, (8), 1725–35, 1996.

22. X. S. Yao and and L. Maleki, "Opto-electronic oscillator for photonic systems", *IEEE J. Quantum Electron.*, **32** (7), 1141–9, 1996.

23. M. Rodwell, J. E. Bowers, R. Pullela, K. Gilboney, J. Pusl, and D. Nguyen, "Electronic and optoelectronic components for fiber transmission at bandwidths approaching 100 GHz," in *The LEOS Summer Topical Meetings*, 1995 Digest of the LEOS Summer Topical Meetings (Institute of Electrical and Electronics Engineering, Piscataway, NJ. IEEE catalog number 95TH8031), RF Optoelectronics pp. 21–2.

24. K. Noguchi, H. Miyazawa, and O. Mitomi, "75 GHz broadband Ti:LiNbO$_3$ optical modulator with ridge structure," *Electron. Lett.* **30**, (12), 949–51, 1994.

25. A. Neyer and E. Voges, "Nonlinear electrooptic oscillator using an integrated interferometer," *Opt. Commun.*, **37**, 169–74, 1980.

26. A. Neyer and E. Voges, "Dynamics of electrooptic bistable devices with delayed feedback," *IEEE J. Quantum Electron.*, **QE-18** (12), 2009–15, 1982.

27. H. F. Schlaak and R. Th. Kersten, "Integrated optical oscillators and their applications to optical communication systems," *Opt. Commun.*, **36**, (3), 186–8, 1981.

28. H. M. Gibbs, F. A. Hopf, D. L. Kaplan, M. W. Derstine, R. L. Shoemaker, "Periodic oscillation and chaos in optical bistability: possible guided wave all optical square-wave oscillators," *Proc. SPIE*, **317**, *Integrated Optics and Millimeter and Microwave Integrated Circuits*, pp. 297–304.

29. E. Garmire, J. H. Marburger, S. D. Allen, and H. G. Winful, "Transient response of hybrid bistable optical devices," *Appl. Phys. Lett.*, **34**, (6), 374–6, 1979.

30. A. Neyer and L. Voges, "High-frequency electro-optic oscillator using an integrated interferometer," *Appl. Phys. Lett.*, **40**, (1), 6–8, 1982.

31. T. Aida and P. Davis, "Applicability of bifurcation to chaos: Experimental demonstration of methods for switching among multistable modes in a nonlinear resonator," in *OSA Proc. Nonlinear Dynamics in Optical Systems*, eds. N. B. Abraham, E. Garmire, and P. Mandel, Vol. 7, pp. 540–4, Optical Society of America, 1990.

32. M. F. Lewis, "Some aspects of saw oscillators," *Proc. 1973 Ultrasonics Symposium*, IEEE, 1973, pp. 344–7.

33. A. L. Schawlow and C. H. Townes, "Infared and optical masers," *Phys. Rev.* **112**, (6), 1940–9, 1958.

34. R. L. Byer, "Diode laser-pumped solid-state lasers," *Science*, **239**, 742–7, 1988.

35. L. S. Culter and C. L. Searle, "Some aspects of the theory and measurement of frequency fluctuations in frequency standards," *Proc. IEEE*, **54**, (2), 136–54, 1966.

36. A. Yariv, *Introduction to Optical Electronics*, 2nd Edn., Holt, Rinehart and Winston, New York, 1976, Ch. 10.

37. X. S. Yao and L. Maleki, "Influence of an externally modulated photonic link on a microwave communications system," The Telecommunication and Data Acquisition Progress Report 42–117, Vol. January–March, Jet Propulsion Laboratory, Pasadena, California, pp 16–28 (May 15, 1994), obtainable at *http://tmo.jpl.nasa.gov/progress_ report.*

38. M.F. Lewis, "Novel RF oscillator using optical components," *Electron. Lett.* **28**, (1), 31–2, 1992.

39. T. J. Kane, "Intensity noise in diode-pumped single-frequency Nd:YAG lasers and its control by electronic feedback," *IEEE Photon. Technol. Lett.*, **2**, (4), 244–5, 1990.

40. Hewlett-Packard Co., "Phase noise characterization of microwave oscillators - Frequency discriminator method," Product note 11729C-2.

41. P. W. Smith, "Mode selection in lasers," *Proc. IEEE*, **60** (4), 422–40, 1972.

42. E. N. Ivannov, M. E. Tobar, and R. A. Woode, "Advanced phase noise suppression technique for next generation of ultra low noise microwave oscillator," *Proc. 1995 IEEE International Frequency Control Symposium*, pp. 314–20, 1995.

43. J. Dick and D. Santiago, "Microwave frequency discriminator with a cryogenic sapphire resonator for ultra-low phase noise," *Proc. 6th European Frequency and Time Forum*, held at ESTEC, Noordwijk, NJ, 17–19 March 1992, pp. 35–39.

44. X. S. Yao and G. Lutes, "A high speed photonic clock and carrier regenerator," *IEEE Photon. Technol. Lett.*, **8**, (5), 688–90, 1996.

45. X. S. Yao, L. Maleki, and J. Dick, "Opto-electronic oscillator incorporating carrier suppression noise reduction technique," *1999 IEEE Frequency Control Symposium*, France, 1999. X. S. Yao, L. Maleki, and J. Dick, "Multiloop opto-electronic oscillator," *IEEE J. Quantum Electron.*, **36** (1), 79–84, 2000.

46. X. S. Yao, "Optical resonator based opto-electronic oscillators," New Technology Report, #NPO-20547, Jet Propulsion Laboratory, Pasadena, California, 1999.

47. V. S. Ilchenko, X. S. Yao, L. Maleki, "High-Q microsphere cavity for laser stabilization and optoelectronic microwave oscillator", *Proc. SPIE*, **3611**, 190–8, 1999.

48. V. S. Ilchenko, X. S. Yao, L. Maleki, "Pigtailing the high-Q microsphere cavity: a simple fiber coupler for optical whispering-gallery modes," *Opt. Lett.*, **24**, (11), 723–5, 1999.

49. Y. K. Chen and M. Wu, "Monolithic colliding-pulse mode-locked quantum well lasers," *IEEE J. Quantum Electron.*, **28** (10), 2176–85, 1992.

50. M. C. Wu, T.K. Chen, T.Tanbun-Ek, and R.A. Logan, *Monolithic CPM Diode Lasers*, Springer Series on Chemical Physics, Vol. 55, Ultrafast Phenomena VIII, eds. J. L. Martin, A. Migus, G. A. Mourou, and A. G. Zewail, pp. 211–5, Springer-Verlag, Berlin Heidelberg, 1993.

51. X. S. Yao and L. Maleki, "Dual microwave and optical oscillator," *Opt. Lett.*, **22**, (24), 1867–9, 1997.

52. X. S. Yao, L. Maleki, C. Wu, L. Davis, and S. Forouhar, "Recent results with coupled optoelectronic oscillator," *Proc. SPIE*, **3463**, 237–45, 1998.

10

Photonic link techniques for microwave frequency conversion

STEPHEN A. PAPPERT[1]
ROGER HELKEY[2]
RONALD T. LOGAN JR.[3]

[1]*Lightwave Solutions, Inc., San Diego*
[2]*MIT Lincoln Laboratory (presently with Calient Networks)*
[3]*JDS Uniphase (presently with Phasebridge, Inc.)*

10.1 Introduction

Microwave frequency conversion techniques using analog photonic link technology are reviewed in this chapter. Opto-electronic or photonic radio-frequency (RF) signal mixing refers to converting intensity modulation of an optical carrier at one modulation frequency to intensity modulation or an electrical output at a different frequency. Frequency conversion optical links integrate the functions of electrical frequency mixing, traditionally provided by electronics, together with the transport of the RF carrier by the optical link. Photonic RF signal mixing using fiber optic link technology has recently become a topic of interest for reducing front-end hardware complexity of antenna systems and efficiently extending link frequency coverage into the millimeter-wave (MMW, 30–300 GHz) range. As commercial and military systems push to higher operating frequencies, microwave optical transmission and signal conversion techniques offer attractive benefits to designers of RF systems for communications, radar and electronic warfare applications.

The frequency converting link diagram displayed in Fig. 10.1 can be used to introduce the concept of photonic link signal mixing as well as to introduce some nomenclature that will be used throughout the chapter. Here, an example antenna remoting configuration is shown that incorporates both optical RF up-conversion for transmit operation and optical RF down-conversion for receive operation. Referring to the transmit-mode up-conversion path of Fig. 10.1, the photonic frequency conversion occurs by multiplying the MMW optical local oscillator (LO) signal at f_{LO1}, with the lower frequency RF input or information bearing signal at f_{IF1} in the integrated optical modulator (IOM). The photodetection of this complex optical signal produces MMW frequency up-converted signals at frequencies $f_{LO1} \pm f_{IF1}$. In this way, the photonic frequency converting link acts as a distributed mixer, where the frequency conversion occurs both in the modulator (RF–optical) and in the photodetector (optical–RF). In a similar manner, frequency down-conversion

293

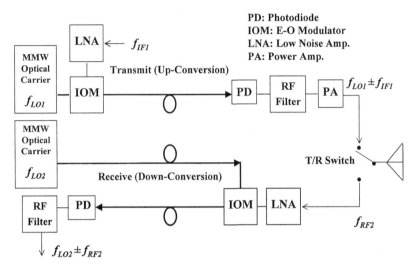

Figure 10.1. Example fiber optic antenna remoting configuration incorporating optical RF up-conversion for transmit-mode operation and optical RF down-conversion for receive-mode operation.

from f_{RF2} to $f_{LO2} - f_{RF2}$ is obtained from the receive-mode path of Fig. 10.1. In this case, f_{LO2} can be chosen close in frequency to f_{RF2} allowing direct, single-stage down-conversion to a convenient processing frequency. Other opto-electronic RF mixing schemes and configurations are possible and will also be discussed in this chapter. Depending on system demands, photonic microwave signal conversion can be tailored and configured for either frequency up-conversion or down-conversion using either high-side or low-side LO signals.

For many of the RF frequency conversion approaches to be discussed, a MMW optical carrier or optical LO signal is required. The MMW optical LO signal can and has been generated using a number of optical techniques in addition to direct electro-optic modulation at f_{LO}. In this chapter, two general approaches for high spectral purity MMW signal generation are discussed; two-mode heterodyning techniques and modulator sideband generation. One LO signal generation approach is based on heterodyning two optical sources that are offset by the desired LO frequency from each other, referred to here as sinusoidal optical carrier modulation. An interesting version of this approach discussed later in the chapter uses two distributed-feedback (DFB) semiconductor diode lasers that are optically injection-locked to different optical modes of an optical comb generated by a mode-locked laser or a frequency-modulated laser. A second optical LO approach discussed in some detail is based on using the MZ optical modulator as an efficient frequency multiplier to generate high quality signals into the MMW regime. These approaches for microwave LO signal generation have their unique advantages and drawbacks that can serve to illustrate the important issues relating to this technology. This is discussed in Section 10.2.

Developments in the area of frequency conversion optical links are reviewed in Section 10.3, including frequency conversion configurations with opto-electronic mixing in lasers, modulators, interferometers, photodetectors, and optical fiber. Implementation issues such as passive and active impedance matching, optical amplification, selective sideband amplification, differential noise suppression, single-sideband up-conversion, and image rejection are discussed. The dynamic range performance of frequency conversion links is also discussed. A discussion of the commercial and military applications of this technology close out Section 10.3 and a brief summary is given in Section 10.4. Prior to discussing the details of frequency converting links, some background information is provided to help identify the need and benefits of this important RF photonic topic.

10.1.1 RF system frequency allocation and requirements

The demand for increased information capacity and larger instantaneous signal bandwidths have driven commercial and military RF system architects well beyond UHF (300–3000 MHz) frequencies for satisfying their communication, radar, and electronic warfare antenna system requirements. To illustrate this point for the case of satellite communications (SATCOM), Table 10.1 shows the present frequency spectrum and how it is allocated amongst different users. Commercial and military

Table 10.1. *Commercial and military SATCOM frequency bands (courtesy of Dr. Roy Axford, U.S. Navy SPAWAR System Center, San Diego)*

Band designation	System(s)	Earth-to-space (Uplink)	Space-to-Earth (Downlink)
VHF	ORBCOMM	148.000–150.050 MHz	137.000–138.000 MHz
UHF	UHF SATCOM	292.850–317.325 MHz	243.855–269.950 MHz
L	INMARSAT-2	1626.5–1649.5 MHz	1530.0–1548.0 MHz
L	INMARSAT-3	1626.5–1660.5 MHz	1525.0–1559.0 MHz
L	Iridium Handsets	1616.0–1626.5 MHz	
L	METOC	N/A	~1694 MHz
C	Challenge Athena	5925–6425 MHz	3700–4200 MHz
X	SHF SATCOM (DSCS)	7.90–8.40 GHz	7.25–7.75 GHz
Ku			10.95–11.45 GHz
Ku			11.20–11.70 GHz
Ku	Predator UAV	14.00–14.50 GHz	11.45–11.95 GHz
Ku	PrimeStar (GE 2)	14.00–14.50 GHz	11.70–12.20 GHz
Ku	DirecTV, EchoStar	17.30–17.80 GHz	12.20–12.70 GHz
Ku			12.25–12.75 GHz
K/Ka	GBS, ACTS, Future Commercial	27.50–31.00 GHz	17.70–21.20 GHz
K/Q	MILSTAR	43.50–45.50 GHz	20.20–21.20 GHz

systems share the precious bandwidth that spans the VHF (30–300 MHz) to EHF (30–300 GHz) frequency range. In addition, wireless communication systems have frequency requirements from 900 MHz to beyond 60 GHz as well as specialized military applications spanning the full microwave spectrum. With these enormous system frequency ranges comes the need to efficiently transmit mirowave/MMW signals over both short-haul and long-haul distances. These commercial and military antenna systems have fueled interest in optically transmitting high frequency microwave signals from base stations to remote antenna sites.

Microwave/MMW signal antenna remoting using analog photonic links becomes attractive as low loss, high dynamic range links are made available. The RF performance and cost of high frequency optical links have steadily improved over time making them more practical for general antenna remoting, and ultimately enabling extended link functions such as MMW frequency conversion. The loss and linearity of an optimized frequency-converting link are minimally degraded from that of a standard non-converting link. Hence, in applications where microwave transmission using fiber optic links is practical, photonic RF frequency conversion should also be practical.

10.1.2 Benefits of frequency converting photonic links

The chief advantage of using photonic link technology for generating and mixing microwave signals is the ability to transport the signals to and from remote locations, and to minimize the electronic hardware needed at the remote location. Conventional electronic microwave/MMW receiver front-end designs incorporate LO generation and down-conversion. Front-end amplifiers and multiple-stage down-converting mixers with associated LO oscillators have limitations associated with them. Size, weight, and power requirements, reduced dynamic range due to multiple-stage down-conversion, and limited mixer port-to-port isolation head the list of drawbacks. These deficiencies are magnified as wider bandwidth systems are developed. Photonic link signal mixing offers a unique solution for some of the most demanding, wideband antenna receiver applications. Photonic RF mixing allows the ability to efficiently up-convert or down-convert directly to IF frequencies with excellent port-to-port isolation. Achieving high LO–RF isolation is difficult in conventional electronic mixers but is extremely important in surveillance and electronic warfare (EW) systems. In a photonic down-converting link, the LO–RF isolation is essentially infinite since there is no leakage path from the optical carrier to the RF electrodes on the modulator, and it comes with no added effort. Single-stage conversion from MMW frequencies to baseband with low conversion loss is possible using photonic link technology. Many choices of optically modulated LO signal are available depending on phase noise, IF bandwidth, and conversion efficiency requirements. In all cases, the need for electronic mixers is eliminated and this allows the use of lower bandwidth optical modulators upon up-conversion, and higher efficiency low frequency photodetectors upon down-conversion. The problem of obtaining high efficiency modulators at MMW frequencies [1,2] to achieve low RF link loss and noise figure [3] is mitigated somewhat if reduced bandwidth can be traded for increased efficiency using resonant tuning techniques [4].

To summarize this last important point, improved transmit-mode antenna remoting of narrowband high center frequency microwave/MMW signals is possible using photonic signal up-conversion due to the use of a lower bandwidth modulator with low half-wave voltage (V_π) to introduce the information signal. Improved receive-mode antenna remoting of microwave/MMW signals is also possible using photonic signal down-conversion due to the use of a higher responsivity, high optical power handling photodetector, and the potential for using a modulator with improved efficiency over a narrow bandwidth at MMW frequencies.

10.2 Optical local oscillator signal generation

To achieve efficient microwave frequency conversion using photonic link technology configured as in Fig. 10.1 for example, a tailored MMW optical carrier or optical LO signal must be used. In fact, the conversion loss associated with the photonic link signal mixer is determined in large measure by the optical LO waveform. This can be understood by examining the time domain representation of the optical LO signal intensity, $P_{LO}(t)$, which can be represented generally by its Fourier expansion as

$$P_{LO}(t) = a_0 + a_1 \sin(\omega_{LO}t + \theta_1) + a_2 \sin(2\omega_{LO}t + \theta_2)$$
$$+ a_3 \sin(3\omega_{LO}t + \theta_3) + \cdots, \tag{10.1}$$

where a_n represent the Fourier coefficients of the optical LO signal, $\omega_{LO} = 2\pi f_{LO}$ is the LO microwave radian frequency, and phase factors θ_n are included for completeness. Assuming linear optical modulation at the signal frequency f_{RF} (i.e., proportional to $\sin(2\pi f_{RF})$), the RF power from the general frequency converting link at $nf_{LO} \pm f_{RF}$ is proportional to $a_n^2/4$, whereas the RF power at f_{RF} from the standard link is proportional to a_0^2. An important objective of the frequency converting link is to maximize the converted RF power out of the link, hence maximizing a_n for conversion to $nf_{LO} \pm f_{RF}$.

For the purposes of comparison, a mixer conversion gain will be defined here as the ratio of the detected power at $nf_{LO} \pm f_{RF}$ to the detected electrical power at f_{RF}, which is given by $\Gamma_n = a_n^2/4a_0^2$. This will be called the *differential conversion gain* of the frequency converting link. In the case where an IOM is used to generate the optical LO, the value of a_0 is obtained at the quadrature bias point for the LO modulator, the point of maximum link gain at f_{RF}. By evaluating a_0 in this manner for both even and odd n guarantees that minimizing the *differential conversion loss* simultaneously implies maximizing the converted electrical power. Frequency conversion link optimization analysis sometimes uses this definition of differential conversion gain [5–7] as it normalizes away factors that do not directly depend on the frequency conversion process. The differential conversion gain definition does not include the optical insertion loss resulting from additional components used for frequency mixing [8]. The frequency conversion definition conventionally used to characterize RF components is *power conversion gain*, the ratio of output electrical IF power to input RF electrical power [9]. This definition includes the RF link loss, and is affected by electronic or optical amplifiers in the transmission path. Both these mixer conversion efficiency definitions are useful and will be used later in the chapter when various opto-electronic microwave mixing approaches are compared.

As mentioned earlier, the optical LO can be generated using a number of different techniques. A few approaches are introduced here including laser heterodyne techniques and modulator sideband generation, with many others cited in the references.

10.2.1 Heterodyned laser techniques

Heterodyning two laser modes that are offset from each other by ω_{LO} is the most commonly used and, on the surface, the simplest method for efficiently producing an optical LO signal. The modes can originate from the same laser cavity [10–15] or come from two independent optical sources that are either electrically/optically phase locked [16–26] or injection locked [27–33] together. In either case, the optical LO signal intensity can be generally expressed in the form given in Eq. (10.1) as

$$P_{LO}(t) = a_0 + a_1 \sin(\omega_{LO}t + \theta), \tag{10.2}$$

where $a_0 = (P_1 + P_2)/2$ is the average optical intensity of the two combined modes, $a_1 = (P_1 P_2)^{1/2}$, and ω_{LO} is the offset frequency of the two laser modes. For simplicity, this expression assumes the polarization of the combined laser field is aligned and that a 3 dB loss is incurred in the optical combining process (ideal conditions). In the extreme case of orthogonally polarized light beams, $a_1 = 0$ and there is no field interaction and consequently no optically modulated signal at ω_{LO}. In this case, a polarizer oriented at 45 degrees can be used to mix the two polarizations and produce the beat frequency as given by Eq. (10.2). Ideally, the electric field vectors for the individual laser modes are perfectly aligned, the intensities of each mode are equal, and the offset frequency is well stabilized. In this case, 100% optical modulation is achieved, $a_0 = a_1$, and the differential conversion gain obtainable from a photonic RF mixer based on this LO source is $\Gamma_1 = -6$ dB, quite a good result compared with conventional electronic mixers. It is important to remember though, that the link loss at the input RF signal frequency is not included in this result. Nonetheless, heterodyning laser modes represents a very efficient approach for generating optical LO signals for frequency conversion. Both semiconductor laser diodes and solid-state lasers have been used with this technique to generate high quality microwave signals.

An important feature of this laser heterodyne technique is that $a_n = 0$ for $n > 1$, so that there is no unintentional mixing with higher frequency components. This results in a very clean microwave frequency spectrum, which is important for multi-octave frequency converting applications where deleterious spurious signals must be avoided. Additionally, this technique can provide large frequency tunability [18] that is also important and sometimes required in certain applications. On the downside, an active phase-locked-loop (PLL) with an electronic microwave frequency source is typically required to stabilize the resulting microwave frequency. The performance of the PLL is extremely important in achieving low phase noise signals [34–37]. Using a dual-mode laser, low phase noise signals can be more easily attained due to the common-mode noise cancellation between the interfering modes. To illustrate this, a dual-mode Nd:YAG laser has been used to generate a microwave signal at 8.6 GHz [10]. The 8.6 GHz detected microwave beat of the free-running two-mode laser is shown in Fig. 10.2. The single-sideband phase noise power spectral density as read from this plot is approximately −90 dBc/Hz at 1 kHz offset frequency, but this measurement is at the level of the spectrum analyzer phase noise. For this measurement, the laser was not locked to any external RF frequency reference in order to illustrate its inherent stability and demonstrate the common-mode noise rejection resulting from the use of a single optical source to generate the two heterodyned modes.

Combining the attractive signal generation features of heterodyned lasers with the attractive phase-locking features of mode-locked lasers, interesting MMW signal generation approaches can be obtained. Passive and active mode-locking techniques together with optical injection-locking techniques have resulted in a number of high purity microwave signal generation demonstrations [38–47]. The "photonic synthesizer" approach that is highlighted here consists of two CW DFB slave laser diodes that are injection-locked to different modes of an actively mode-locked external-cavity laser diode [43,44]. A block diagram of the photonic synthesizer is shown in Fig. 10.3. When a strong electrical modulation signal is applied to the mode-locked laser at a frequency equal to the mode spacing or a harmonic thereof, active mode-locking can be achieved [48–49]. When mode-locked, the phase fluctuations of the individual longitudinal modes of the mode-locked laser become highly correlated. In the time domain, the output is a train of short pulses with repetition rate equal to the inverse of the mode-locking frequency and pulsewidth determined by the number of modes locked. Semiconductor mode-locked lasers

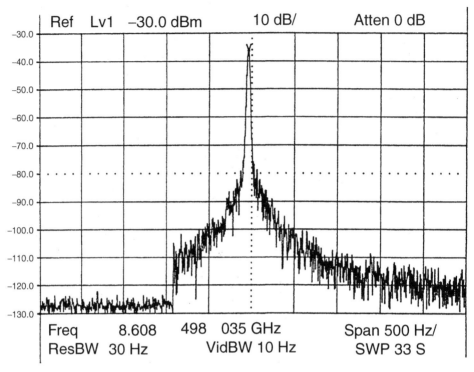

Figure 10.2. Detected microwave beat frequency of the free-running two-mode Nd:YAG laser at 8.6 GHz. Photocurrent 0.4 mA, no post-amplifiers.

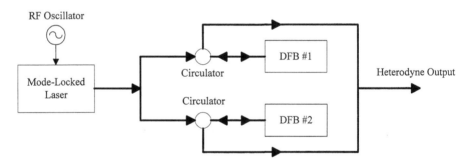

Figure 10.3. Block diagram of RF photonic synthesizer using a mode-locked laser to injection lock two tunable slave laser diodes.

can produce pulses in the sub-picosecond regime, implying that the locked modes span a frequency range in excess of 1000 GHz.

The injection-locked DFB lasers are then tuned so that they become injection-locked by the two desired modes of the mode-locked laser comb. Tuning is accomplished by varying chip temperature, injection current, or using a multi-electrode configuration in which the laser cavity refractive index can be varied to modulate the output frequency. The injection-locked laser outputs are then combined to produe a high-stability, low-phase-noise heterodyne optical signal with frequency separation in the 1 GHz through MMW frequency range. The phase fluctuations of the hetero-dyne output signal are minimized due to the high degree of correlation between the phases of the mode-locked laser frequency comb. As in electronic oscillator injection locking, the injection-locking process causes the slave laser phase to become coherent with the injection signal. Thus, the relative coherence of the modes in the reference comb is conferred to the two slave lasers, so that the heterodyne output frequency phase noise is determined by the relative phase stability of the modes in the comb. The heterodyne output frequency is tunable across the extent of the frequency comb, and low-phase noise operation to 94 GHz is achievable using a mode-locked laser with a pulsewidth in the 5–10 ps range.

By moving the frequency multiplication into the optical domain, this technique can provide for generation of an extremely broad range of frequencies in a common apparatus, with no need for electronic high-frequency phased-locked loops (PLLs) and reference sources operating at the MMW output frequency. It also avoids the conventional mode-locked laser problem of limited available individual mode power and extraneous comb lines. Other types of frequency comb generation techniques have also been used, such as direct current modulation of a laser diode to produce AM and FM, or a modulated resonant optical cavity. However, the mode-locked laser technique is attractive due to the extremely wide bandwidth comb that can be generated with a smooth mode envelope.

To illustrate the signal quality obtainable with this technique, Fig. 10.4 plots the phase noise of the photonic synthesizer, commercial electronic synthesizer, mode-locked laser, and ideally multiplied RF drive signal (calculated) for frequencies from 1 GHz to 50 GHz. The optical heterodyne LO generator has single-sideband phase noise of -93 dBc at 10 kHz frequency offset from $f_{LO} = 50$ GHz which is competitive with the best available electronic synthesizers at these frequencies. Frequencies to 94 GHz have been generated and measured with this photonic synthesizer technique.

Optical modulation depths of 100% are possible using this approach which yields a differential conversion gain of $\Gamma_1 = -6$ dB when applied to a frequency converting link. The other advantages of heterodyned lasers for MMW signal generation are equally applicable to this photonic synthesizer approach.

10.2.2 Harmonic carrier generation using integrated optical modulators

Efficient generation of optical LO signals can also be obtained using a single laser source and a MZ amplitude or optical phase modulator [50–54], using modulation sidebands of an externally modulated fiber optic link. Analog photonic link

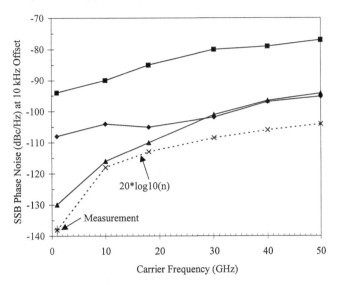

Figure 10.4. Plots of single-sideband RF phase noise at 10 kHz offset from carrier for output frequencies from 1 GHz to 50 GHz for the optical heterodyne LO output (♦), a state-of-the-art commercial microwave synthesizer (■) (Hewlett-Packard HP87350L), the mode-locked laser used in the optical LO generation scheme (▲), and the RF source oscillator used to mode lock the laser (×). The RF source phase noise was measured at 1 GHz; the other points were calculated from the ideal multiplication law of 20 log(n).

technology using an electrically overdriven MZ modulator can be exploited to efficiently generate optical LO signals for frequency conversion [55]. The generation of LO signals directly at f_{LO} or harmonic signals at nf_{LO} can be obtained. By adjusting the modulator bias point and RF drive power to the IOM introducing the low frequency reference signal at f_{LO}, efficient frequency multiplication to generate MMW signals upon detection can be achieved. Using sinusoidal modulation applied to a MZ IOM to introduce the LO signal, the time domain Fourier series representation of the optical intensity can be expressed as

$$P_{LO}(t) = t_m P_0\{1 + \cos\{\pi[A\sin(\omega_{LO}t) + V_b]/V_\pi\}\}$$
$$= t_m P_0\{[1 + \cos(\pi V_b/V_\pi)J_0(m)] - 2[\sin(\pi V_b/V_\pi)J_1(m)]\sin(\omega_{LO}t)$$
$$+ 2[\cos(\pi V_b/V_\pi)J_2(m)]\sin(2\omega_{LO}t + \pi/2)$$
$$+ 2[\sin(\pi V_b/V_\pi)J_3(m)]\sin(3\omega_{LO}t) + \cdots\}, \tag{10.3}$$

where $m = \pi A/V_\pi$ is the modulation index, $2P_0$ is the laser power incident on the modulator, t_m is the modulator transmission factor (<1), V_b is the modulator bias voltage in units of V_π, A is the modulator RF driving amplitude, and $J_n(x)$ is the Bessel function of the first kind of order n. The reason a laser power of $2P_0$ has been chosen will become clear in a moment. Due to the periodic cosine nature

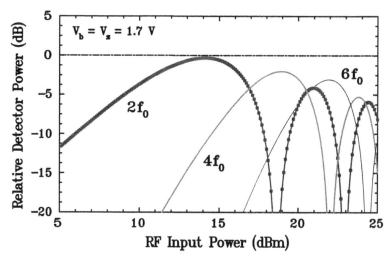

Figure 10.5. Simulated MZ modulator photonic link response versus input RF drive power for 2nd, 4th, and 6th harmonics of f_{LO} driving signal with modulator biased for maximum or minimum transmission.

of the MZ modulator transfer function, given by $2t_m P_0 \cos^2[\pi(V_b + V(t))/2V_\pi]$, where $V(t) = A\sin(\omega_{LO}t)$, a modulator with optimum drive power and bias can produce either an efficient fundamental or a specified harmonic LO signal. With the modulator biased for either minimum or maximum transmission (i.e., $V_b = 0, V_\pi, 2V_\pi, 3V_\pi, \ldots$), the calculated harmonic content of the detector output as a function of the input RF drive power at f_{LO} is shown in Fig. 10.5 for the 2nd, 4th, and 6th harmonics. This simulation shows efficient frequency conversion to any of the even harmonics for specified RF drive powers. For comparison, the 0 dB output level in Fig. 10.5 corresponds to the electrical output obtainable at f_{LO} from 100% modulation depth heterodyned lasers with optical frequency offset f_{LO}. For this comparison, total available laser power was fixed at $2P_0$ in both cases, and an ideal optical insertion loss IOM of 0 dB (i.e., $t_m = 1.0$) has been assumed. Using a more realistic IOM insertion loss of 4.5 dB (i.e., $t_m = 0.35$), the 4th harmonic is down 11 dB from the maximum that can be obtained from heterodyning two independent lasers to directly produce the RF signal.

Quadrature IOM biasing (i.e., $V_b = V_\pi/2, 3V_\pi/2, 5V_\pi/2, \ldots$) produces efficient optical LO generation directly at the fundamental or at a specified odd harmonic of f_{LO}, depending on the RF drive power. Table 10.2 summarizes the detector output power relative to the heterodyned laser case and the associated modulator RF drive voltage requirements (in units of V_π) to 10th order. In generating this table, equal total laser power has again been assumed along with a modulator optical insertion loss of 4.5 dB. This assumption will be discussed in Section 10.2.3. The RF power delivered to the detector output load at the fundamental frequency is 7.7 dB weaker

Table 10.2. *Calculated link relative electrical output power at* nf_{LO} *using a MZ modulator and harmonic carrier generation with respective optimum modulator RF drive voltages (peak amplitude) and bias conditions. The 0 dB output level corresponds to the electrical output obtainable at* f_{LO} *from 100% modulation depth heterodyned lasers. A 4.5 dB optical insertion loss* $(t_m = 0.35)$ *is assumed for the LO modulator*

IOM Bias	Quadrature					Max. or min. transmission				
$n =$	1	3	5	7	9	2	4	6	8	10
Drive voltage (V_π)	0.59	1.3	2.1	2.7	3.4	0.97	1.7	2.4	3.1	3.7
Rel. power (dB)	−7.7	−10.2	−11.5	−12.4	−14.1	−9.3	−11.0	−12.0	−12.8	−13.4

than the 100% modulated heterodyned laser case. The 10th harmonic is only down 13.4 dB relative to the fundamental heterodyned laser output implying that this technique is useful for efficiently generating signals well beyond the modulation limit of present modulators. Certainly MMW signals to 100 GHz are possible, as efficient detectors become available at these high frequencies. The case of the 4th harmonic is especially interesting in that the resulting spectrum is clean out to the desired output frequency. The fundamental and the 3rd harmonic are nulled due to the modulator bias position and the 2nd harmonic experiences a null near the 4th order peak, as is evident in Fig. 10.5. The 6th harmonic has an 8.8 dB suppression relative the 4th order output at the 2nd order null point and must be filtered.

Using the nth harmonic as an LO signal, the differential conversion gain using cascaded IOMs can be expressed as:

$$[\sin(\pi V_b/V_\pi)J_n(m)]^2 \qquad (\text{for } n = 1, 3, 5, \ldots)$$
$$[\cos(\pi V_b/V_\pi)J_n(m)]^2 \qquad (\text{for } n = 2, 4, 6, \ldots) \qquad (10.4)$$

Either quadrature modulator bias (fundamental/odd harmonics) or maximum/ minimum transmission modulator bias (even harmonics) is used to minimize the frequency conversion loss. The differential conversion gain for the $n = 1$ fundamental signal $(f_{LO} \pm f_{RF})$ is $\Gamma_1 = -4.7$ dB. This result is 1.3 dB better than that for heterodyned lasers assuming equal average photodetector powers, however, in practice LO modulator optical insertion loss must be considered which decreases the available power at the detector. Interestingly, the 4th harmonic differential conversion gain to $(4 f_{LO} \pm f_{RF})$ is only $\Gamma_4 = -8$ dB, which is down only 3.3 dB from the fundamental $n = 1$ conversion gain $(f_{LO} \pm f_{RF})$. As pointed out earlier, the 4th harmonic is particularly interesting since the 1st through 3rd harmonics are nulled out in this case. This approach conveniently allows one to extend the frequency range coverage of analog fiber optic links into the EHF frequency range. It can be useful out to 10 times the modualtion bandwidth of present MZ modulators.

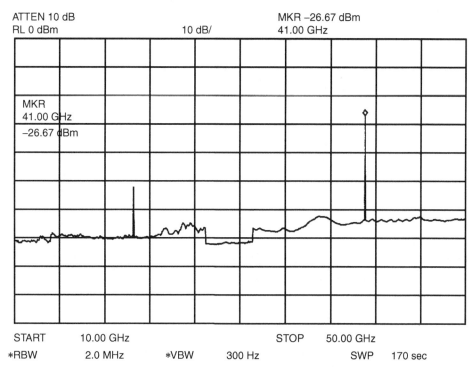

ATTEN 10 dB　　　　　　　　　　　　　　MKR −26.67 dBm
RL 0 dBm　　　　　　　　　10 dB/　　　　41.00 GHz

MKR
41.00 GHz
−26.67 dBm

START　　　10.00 GHz　　　　　　　　　STOP　　50.00 GHz
∗RBW　　　2.0 MHz　　　∗VBW　　300 Hz　　　SWP　　170 sec

Figure 10.6. Spectrum analyzer trace in the 10–50 GHz frequency range for optimized $n = 4$ optical multiplier stage using MZ modulator and a microwave frequency reference of $f_{LO} = 10.24$ GHz.

To demonstrate the utility of this harmonic carrier signal generation technique, an $n = 4$ optical multiplier stage has been used with a microwave frequency reference of $f_{LO} = 10.24$ GHz to generate the output spectrum analyzer trace shown in Fig. 10.6. The full frequency range from 10 GHz to 50 GHz including spurious signals is displayed. The 4th harmonic at 41 GHz has a −26.7 dBm output power. Using the spectrum analyzer directly to measure spectral purity, an absolute single-sideband phase noise of −110 dB/Hz at 10 kHz offset frequency is obtained which is limited by the electronic frequency source at 10.24 GHz. As expected, an n^2 phase noise dependence with n times multiplication is observed with this technique [56]. Approximately 30 dB suppression of the 1st, and >40 dB suppression of the 2nd and 3rd harmonics are realized and close to 50% of an octave frequency tuning range has been demonstrated [53] with this technique. The suppression of the unwanted harmonics and spurious signals of the optical frequency multiplier is ultimately limited by a combination of the modulator bias point accuracy, the modulator drive power accuracy, the symmetry of the modulator optical transmission curve, and the RF filter rejection. As is evidenced by the above full spectrum plot, all of these limiting error contributions can be controlled to a level where a relatively clean output spectrum is obtainable.

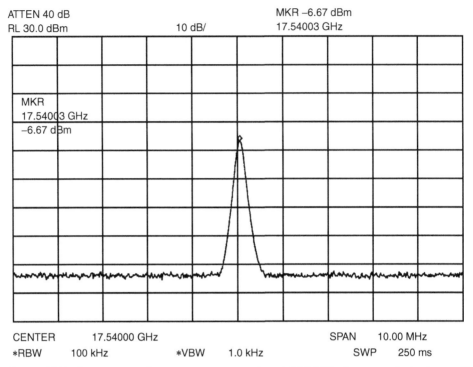

Figure 10.7. Spectrum analyzer output centered at 17.54 GHz for an optimized $n = 2$ optical multiplier stage using MZ biased at minimum transmission.

Output MMW electrical power obtainable from the optical frequency multiplier is currently limited by the optical power handling capabilities of high-speed detectors. This is the case shown in Fig. 10.6 near 41 GHz. At lower frequencies, higher electrical output powers are available due to higher power, higher responsivity optical detectors. Shown in Fig. 10.7, an output electrical power of -6.7 dBm at 17.5 GHz has been demonstrated for an $n = 2$ modulator-based optical multiplier [53]. The absolute phase noise for this high-quality signal is -96 dBc/Hz at 1 kHz offset frequency, again limited by the electronic source. Finally, as lower V_π optical modulators and higher power optical detectors become available at high frequencies, this frequency multiplication technique becomes more attractive due to the reduced RF drive power requirements, higher output power levels, and higher MMW frequencies.

10.2.3 Optical local oscillator generation comparison

In addition to the optical LO signal generation approaches discussed and cited above, other approaches have been investigated. The direct use of a mode-locked laser is one example. Using a conventional mode-locked laser for MMW signal generation, the available optical power is spread among many harmonics of the fundamental mode-locking frequency. In practice, due to the mode power distribution and rather limited high-speed photodetector saturation power, limited RF power is obtainable from any of the isolated harmonics in the mode-locked laser frequency comb. This typically limits the value of employing a conventional mode-locking approach for MMW signal generation and frequency conversion. Other LO signal generation approaches including the opto-electronic oscillator (OEO) [57], gain-coupled DFB lasers [58], and fiber dispersion induced signal generation [59] have been proposed and demonstrated. The OEO approach was covered in detail in Chapter 9 and has its unique advantages. Gain-coupled DFB lasers have also been demonstrated with large frequency tunability available. Detailed comparisons of the different microwave/MMW signal generation approaches have appeared in the literature [60–63]. For microwave frequency conversion applications, optical LO signal generation with large modulation index is required. This leads us to the heterodyned laser (using either an optical phase-locked loop or injection locking with a mode-locked laser) and the overdriven MZ modulator approaches.

It is interesting to compare the MMW signal power and the differential conversion gain for the optimally overdriven modulator and heterodyned laser approaches. The differential conversion gain of the $n = 1$ optimally overdriven modulator is 1.3 dB higher than that of the 100% modulated heterodyned lasers. This can be understood if the modulated optical signals for the two cases are analyzed in the time domain as shown in Fig. 10.8. Considering conversion from f_{RF} to $f_{LO} \pm f_{RF}$, the ratio of the Fourier coefficient at the fundamenetal frequency to the DC component, a_1/a_0,

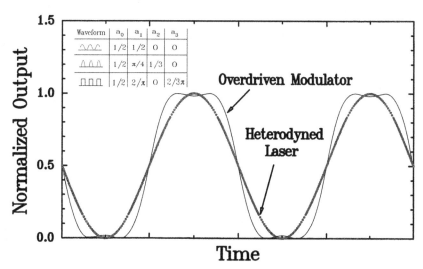

Figure 10.8. Comparison of sinusoidal modulation waveform and the waveform from an over-driven MZ modulator biased at quadrature [8] © *1996 IEEE.*

is larger for the overdriven modulator approach (pseudo square-wave) than for the heterodyned laser approach (raised sine-wave). Maximizing this ratio yields the highest RF mixer differential conversion gain. The first few Fourier coefficients and a_1/a_0 value for the raised sine-wave, square-wave, and half-rectified sine-wave are given in the inset to Fig. 10.8. The raised sine-wave time domain signal has $a_1/a_0 = 1$ resulting in an attainable differential conversion gain of -6 dB and the square-wave has $a_1/a_0 = 4/\pi$ resulting in a conversion gain of -3.9 dB. A half-rectified sine-wave time domain signal with $a_1/a_0 = \pi/2$ can be used to improve the conversion gain to only -2.1 dB. By low biasing the M–Z modulator with the optimal RF drive voltage, conversion losses approaching 3 dB are predicted. In this case, the time domain signal more closely approaches the half-rectified sine-wave. However, in this analysis, we have assumed equal average optical power at the detector (constant a_0) which is not necessarily the case.

Comparing the throughput optical power for the heterodyned laser and overdriven modulator cases is important. In practice, two identical lasers for heterodyning are used versus a single laser for the overdriven modulator. Combining the two laser outputs in the heterodyne case results in a 3 dB optical loss. However, even if we assume $+3$ dB optical power from the single laser compared to the two identical sources, the optical insertion loss of the LO modulator will reduce the optical power obtainable at the detector for the overdriven modulator case, typically in the 6–10 dB range (t_m^2). Hence, the heterodyned laser approach will in practice provide larger average detector powers than the optimally overdriven modulator case, resulting in higher available MMW signal powers. For source power limited systems, this translates into a higher mixer power conversion gain for the heterodyned lasers. This is a case where the differential conversion gain definition does not tell the entire story. Differential conversion gain is most important in detector power limited systems [8]. In practice, both source power limited and detector power limited operation can occur. This mixer conversion efficiency issue will be discussed further in Section 10.3.

The suppression of the unwanted harmonics and spurious signals of the optical LO signal is extremely important for wideband applications. For the heterodyned lasers, extremely clean microwave spectrums are obtained. For the optimally over-driven modulator, the spectral purity is limited by a combination of the modulator bias point accuracy, the modulator drive power accuracy, the symmetry of the modulator optical transmission curve, and the RF filter rejection. As is evidenced by the full spectrum plots shown in Section 10.2.2, all of these limiting error contributions can be controlled to a level where a relatively clean output spectrum is obtainable. However, the spectrum will never be as clean as the heterodyned laser case.

Phase noise performance is also a major concern for most microwave applications. The overdriven modulator approach can achieve electronic source limited phase noise performance over the entire range of offset frequencies. The phase noise of the heterodyned laser is limited by either the PLL or injection-locking performance. Excellent absolute phase noise performance is achievable at offset frequencies beyond approximately 1 kHz. In the PLL case, high-gain feedback loops with low delay time are required to achieve premium performance. LO tuning speed is also important for some applications and must be considered along with phase noise performance. There exists a tradeoff between premium phase noise performance and frequency tuning speed. Current tuned laser diodes can exhibit large frequency tuning in the sub-microsecond regime. The same is true for the overdriven modulator approach whose tuning time is limited only by that of the electronic source.

To summarize, efficient MMW signal generation can be achieved using a number of approaches. Heterodyned lasers and photonic link harmonic carrier generation using MZ optical modulators have been highlighted here. The periodic transfer curve of the MZ modulator provides a unique freqency multiplication appraoch with high efficiency. Nearly complete conversion from f_{LO} to $4f_{LO}$ with low spurious signals can be obtained. An absolute phase noise of -86 dBc/Hz at 1 kHz from a 40.96 GHz LO signal has been measured using the photonic link frequency multiplier. Heterodyned laser diodes or solid-state lasers can be used to generate high-quality frequency tunable microwave signals to beyond 60 GHz. An absolute phase noise of -93 dBc/Hz at 10 kHz from a 50 GHz carrier has been measured using the heterodyne laser with mode-locked-laser injection locking. Both these optical LO signal generation approaches can provide attractive signal synthesis alternatives for MMW frequency converting applications where signal quality is paramount.

10.3 Microwave frequency converting photonic links

A conventional optical link with electrical frequency conversion is shown in Fig. 10.9a. As mentioned earlier, the complexity of the receiver following the link can be reduced by including opto-electronic mixing in the optical link, as the first frequency conversion stage is eliminated and an all-digital receiver operating at the link output frequency may be possible. A number of opto-electronic mixing configurations have been investigated to achieve efficient frequency conversion and some of these are displayed in Fig. 10.9. These optical approaches to microwave frequency conversion are now discussed.

10.3.1 Frequency conversion configurations

Opto-electronic mixing has been performed using the nonlinearity of a photodiode optical detector by intensity modulating an optical signal with the RF input

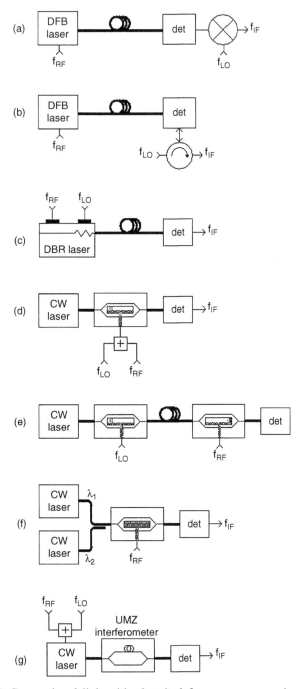

Figure 10.9. (a) Conventional link with electrical frequency conversion, and frequency conversion configurations with (b) photodetector nonlinearity beating, (c) laser modulation, (d) single MZ modulator, (e) series MZ modulator, (f) optical mode beating, and (g) interferometric detection of FM modulation.

signal and electrically coupling the LO signal to the photodetector [5,6,64–70], with an example shown in Fig. 10.9b. The generation of opto-electronic mixing in the photodiode has been achieved using both the capacitance–voltage nonlinearity [64] and the nonlinearity of the current–voltage dependence and photodiode responsivity [6,67,69,71]. At high frequencies, opto-electronic mixing can occur due to modulation of the carrier drift velocity by the applied electric field [70]. Frequency mixing can also take place in phototransistors [72,73]. Using this general technique, frequency down-conversion can be performed by modulating the laser with an input RF signal and generating a low intermediate frequency (IF) mixing product. Alternatively, frequency up-conversion can be performed by modulating the optical signal with a low IF signal, and generating a higher frequency RF mixing product at the photodetector.

The LO signal can also generate mixing products by modulating the gain of an avalanche photodiode [5,65,66], allowing harmonic frequency conversion [5]. Because the avalanche photodiode output power decreases with increasing frequency, the signal- to-noise ratio can be increased using opto-electronic mixing for weak signals where thermal noise from the post-amplifier is larger than shot noise [65]. Opto-electronic mixing has also been accomplished using an intensity modulated optical LO signal to gate an electrical signal in an optical photoconductive switch [74], with bandwidths of up to 4.5 GHz using a fast device [75]. With an interdigitated photoconductive switch, uniform optical illumination is needed to avoid generating optical mixing intermodulation products [76].

Opto-electronic mixing has also been performed in a semiconductor laser. In one configuration, a semiconductor laser is modulated with the RF and LO signals and the intensity modulation is detected with a photodetector [77–80]. The frequency conversion takes place due to nonlinearity in the laser light–current characteristic. This technique has been demonstrated with a passively Q-switched laser, where the LO signal was internally generated by self-oscillations in the device [79]. Up-conversion and down-conversion have been demonstrated using mirror loss modulation in a distributed Bragg reflector (DBR) semiconductor laser [80] as shown in Fig. 10.9c, where the RF input singal is applied to the gain region of the laser, and the LO signal modulates the Bragg grating frequency. The frequency range for fundamental frequency conversion is limited by the frequency response of the laser. Higher frequency operation can be obtained using harmonic generation in the laser [71,81].

The original demonstration of modulator-based opto-electronic mixing links for RF applications by Kolner and Dolfi exploited the nonlinearity in a single MZ modulator as shown in Fig. 10.9d [82]. Frequency conversion links using optical modulators have become popular due to the strong well-controlled nonlinearity of the modulator. Although previous links using MZ modulators had been operated in the linear regime, the MZ sinusoidal transfer function is highly nonlinear for large input signals. By simultaneously injecting an RF signal and a large LO signal into the same modulator, mixing products between the RF input signal and the local oscillator are generated at the photodetector. Either down-conversion or up-conversion products can be generated with a suitable choice of frequencies at the link input and electrical filter at the link output.

Later, the configuration shown in Fig. 10.9e using two MZ optical modulators in series was adopted [9], where the order of the RF and LO modulators can be reversed from that shown in the figure. This series modulator configuration gives extremely high isolation between the RF and LO signals. It also has an advantage over the single modulator configuration when the input RF signal and LO signal are physically separated, for example in antenna remoting applications. This configuration does not rely on the nonlinear transfer function of the MZ modulator for opto-electronic mixing and provides a direct multiplication of the two RF signal inputs.

In addition to applying a sinusoidal drive to an optical modulator, there are a variety of techniques to produce the optical modulation for frequency conversion. Sinusoidal modulation can be achieved directly by beating together two optical modes as was discussed in Section 10.2 on heterodyning. These optical modes can be obtained by combining the outputs of two single-mode lasers as shown in Fig. 10.9f [10], from a laser designed for dual optical mode operation [83], or from the heterodyne lasers injection-locked to a frequency comb as discussed in Section 10.2. The LO modulation occurs at the difference in optical frequency between source wavelengths λ_1 and λ_2. The frequency noises of two modes from the same optical source are correlated, producing lower beat noise than for two separate optical sources [10,84]. Direct modulation of a conventional semiconductor laser can also produce sinusoidal modulation for frequency conversion in a modulator. Alternately, the RF input signal can directly modulate a laser, with the output of this laser externally modulated at the LO frequency.

A laser followed by an unbalanced path-length MZ (UMZ) interferometer also can be used for frequency conversion [6], by modulating the source laser and using the interferometer to detect the resulting frequency modulation (FM) as shown in Fig. 10.9g. Frequency conversion in this configuration is based on the nonlinearity inherent in FM detection with an interferometer [81], and does not rely on any nonlinearity in the laser itself. Performance of this configuration is similar to injecting the RF and LO signals into a conventional MZ modulator. An advantage of this configuration is that using a large difference in the interferometer optical path lengths is equivalent to opto-electronic mixing occurring in a modulator with a small V_π, so the frequency conversion gain can be quite large and the optimum LO drive level can be quite small. A disadvantage of this configuration is that the unbalanced path-length interferometer also converts the laser FM noise to amplitude noise, so that a narrow linewidth laser with large FM efficiency is needed to achieve a low link noise figure.

MZ modulators and electroabsorption modulators can have very high bandwidth; frequency conversion of an RF signal has been realized to 60 GHz using both MZ [85] and electroabsorption [86] modulators. For high frequency modulation, optical fiber dispersion can disrupt the phase between modulation sidebands, leading to periodic variations in modulation amplitude as a function of fiber length. A single-sideband generation technique shown in Fig. 10.10a has been developed that eliminates dispersion-induced gain variations by suppressing one modulation sideband [87]. In this configuration, the LO modulation is applied separately to a phase modulator in each arm of an MZ modulator, with a 90° phase shift between the modulation in the two arms. One difficulty with this technique is tuning the LO frequency over a broad bandwidth while maintaining the 90° phase shift.

There is an analogous frequency down-conversion configuration corresponding to this single-sideband up-conversion configuration. In addition to the desired input RF frequency band, down-conversion links have an image-frequency band at the input with the opposite frequency offset from the LO signal as the modulation signal. This image band produces the same output intermediate frequency (IF) band as the desired modulation band. Noise at this image-band frequency can cause 3 dB of degradation in system noise figure. Spurious signals and noise in the image band are usually rejected with an image-reject filter at the input of the frequency conversion link. This image-reject filter becomes difficult to implement for low IF frequencies where the frequency separation between the input band and the image band becomes small.

The usual techniques to allow output at a low frequency are multiple down-conversion stages with an image-reject filter per conversion stage, or a single image-reject mixer [88]. An image-rejection frequency conversion link configuration similar to that shown in Fig. 10.10b has been demonstrated for opto-electronic mixing [78,89], which allows a single-stage conversion to low frequency. This configuration consists of two frequency down-conversion links where either the RF signals or LO signals have a 90° phase shift. The IF outputs of the two links are combined with a 90° phase shift, resulting in suppression of mixing products from the image band.

The image-reject frequency conversion link configuration allows the image frequency to be suppressed even for mixing down to very low output frequencies. However, laser intensity noise gives an additional constraint to achieving a low-frequency IF output directly from the frequency conversion link [90]. Even low-noise solid state lasers have significant low-frequency intensity noise, which for current commercial devices can cause the noise figure to increase for IF frequencies below ~10 MHz.

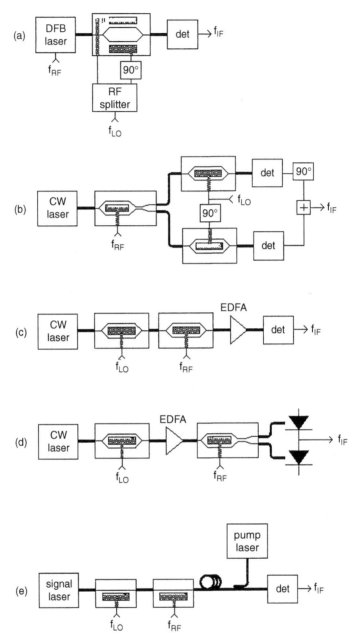

Figure 10.10. Frequency conversion configurations with (a) single-sideband generation, (b) image-band rejection, (c) series modulator with optical amplification, (d) optical amplification with differential noise suppression amplifier, and (e) Brillouin selective sideband amplification.

The gain of the link can be increased using an erbium-doped fiber amplifier (EDFA) before the photodetector [91] as shown in Fig. 10.10c. A limit to increasing the frequency conversion gain with optical amplification is photodetector saturation [92]. Using a large detector active area can increase the optical power limit of the photodetector [93], although this limits the maximum frequency of operation. Several high-frequency photodetectors have been used in parallel in order to increase the maximum optical power while minimizing the reduction in frequency response [91]. Traveling wave photodetectors have demonstrated high saturation power at high frequency [94]. For detector-power limited systems, amplifying the modulation sidebands with respect to the carrier power can increase the gain. One technique to realize this increase in gain is by operating the signal modulator at a low bias to decrease the average optical power by more than the decrease in modulation sideband power, and then increasing the average optical power using an optical amplifier [95].

Adding an optical amplifier just before the photodetector [91] increased the link conversion gain, but also added a large amount of intensity noise. The noise from an optical amplifier is similar to using a laser with an effectively higher RIN [91]. Putting the optical amplifier after both modulators minimizes the pump power required for the optical amplifier. However, putting gain stages before loss stages minimizes noise. In addition, this intensity noise can be cancelled with a differential external modulation link configuration if the optical amplifier is moved to between the two modulators [96] as shown in Fig. 10.10d. In this configuration, the intensity noise from the two complementary optical outputs is the same, while the mixing products from the two outputs are 180° out-of-phase. The two optical outputs are subtracted using photodetectors with opposite polarities, which cancels the intensity noise while enhancing the mixing products. Intensity noise cancellation of 22 dB has been demonstrated using this approach [96].

Another technique that allows amplifying a modulation sideband while suppressing the carrier is Brillouin selective sideband amplification [97] shown in Fig. 10.10e, with a counter-propagating pump laser near the signal laser wavelength. Brillouin amplification results from the high-power pump laser, which sets up an acoustic grating that produces a backward-scattered pump beam. By tuning the optical frequency of the pump laser to overlap with a modulation sideband of the signal, the backscattered pump signal adds up in phase with the signal sideband and amplifies the signal. The two phase modulators shown in Fig. 10.10e produce mixing sidebands on the optical carrier. Selective amplification of one modulation sideband converts phase modulation mixing products into amplitude modulation mixing products. Harmonic up-conversion can also be achieved using this selective sideband amplification technique [97].

10.3.2 Link gain and noise suppression

Gain, noise figure, and dynamic range usually specify the performance of analog components. The gain and noise figure of the opto-electronic mixers will be discussed in this section, and dynamic range in the next section.

For opto-electronic mixing in a photodiode detector, the optimum bias for conventional photodetectors is near 0 V [6,69] where the nonlinearity in responsivity is the largest. Opto-electronic mixing with avalanche photodetectors allows an additional degree of freedom for optimizing performance by adjusting the multiplication factor through the bias voltage. Experimental performance for avalanche photodiodes is shown in Fig. 10.11. Here, the optimum differential conversion gain ranges from −3 dB for fundamental mixing to −15 dB for 5th harmonic mixing, comparing favorably to a system using a conventional optical link and a separate Schottky diode mixer [5]. Differential conversion gain >1 can be achieved with photodetector mixing at the appropriate LO power [6]. A differential conversion gain of 10.4 dB has been achieved using opto-electronic mixing in a transistor [73], where the transistor was operated in the active mode, and using the exponential dependence of the transistor current gain on the base-emitter voltage V_{BE}.

For mixing in a conventional diode laser, the conversion efficiency drops off rapidly with increasing modulation frequency. Up-conversion is found both

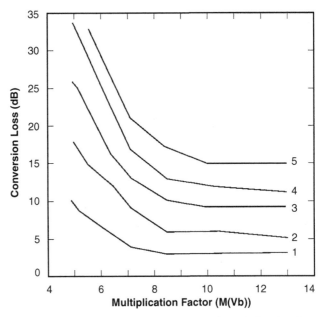

Figure 10.11. Frequency conversion loss as a function of multiplication factor for harmonic mixing in a silicon avalanche photodiode for fundamental mixing ($n = 1$) and harmonic mixing ($n = 2$ to $n = 5$) [6] © *1986 IEEE*.

theoretically and experimentally to be more efficient than down-conversion by an order of magnitude at low frequency [79], but down-conversion is found to vary more slowly with frequency [80]. If both the RF input signal and the LO signal provide current injection into a single-section laser, the mixing products decrease very rapidly with increasing LO frequency [79]. More efficient frequency conversion can be accomplished by directly modulating the loss of the laser [79], resulting in less frequency dependence. The frequency dependence with loss modulation is of order $(1/f_{LO})^3$ instead of $(1/f_{LO})^4$ for up-conversion, and is of order $(1/f_{LO})^2$ instead of $(1/f_{LO})^3$ for down-conversion [79, 80]. Experimental results are shown in Fig. 10.12 using mirror loss modulation in a distributed Bragg reflector (DBR) semiconductor laser by injecting one signal into the grating tuning section [80] as shown in Fig. 10.9c. Up-conversion results are shown in Fig. 10.12a, with higher frequency conversion gain than for the down-conversion results shown in Fig. 12.12b. The up-conversion results are fit to a curve with a frequency dependence of order $1/\omega^{3.17}$, compared to a theoretical frequency dependence of order $1/\omega^3$. The down-conversion results are fit to a curve with a frequency dependence of order $1/\omega^{2.25}$, compared to a theoretical frequency dependence of order $1/\omega^2$.

For opto-electronic mixing in an optical modulator with the RF signal applied to the electrical input, the differential conversion gain for sinusoidal optical modulation at the LO frequency is $G_{\text{diff,sin}} = -6$ dB, independent of the link configuration or optical power. The reason for this was explained in Section 10.2. The differential down-conversion gain for the series modulator configuration with the LO signal applied to one MZ modulator and the RF signal applied to another modulator is $G_{\text{diff,series}} = -4.7$ dB [7]. As discussed earlier, the gain improvement of the series MZ configuration over sinusoidal modulation is due to the square-wave like time-domain optical LO signal, which has a larger Fourier amplitude coefficient relative to the DC term than a pure sine-wave with the same peak–peak amplitude. However, power conversion gain must be considered for source power limited systems.

The MZ LO modulator can instead be biased to optimize harmonic frequency conversion [43,98]. Up-conversion and down-conversion using second order and fourth order harmonics of the LO signal have been demonstrated [43], with a differential conversion gain $G_{\text{diff,series}}$ of -3.2 dB experimentally achieved for mixing with the second harmonic of the LO frequency [99]. Harmonic up-conversion can allow higher link gain and lower LO power by operating the optical modulator at a lower input frequency where the modulator half-wave voltage V_π is lower [43].

Harmonic down-conversion has been used to reduce the frequency of a 192 GHz optical pulse stream in order to measure the timing jitter of a mode-locked laser [100]. Here, the laser repetition frequency was stabilized using the saturable absorber to tune the repetition rate of a mode-locked laser in an optical phase-locked loop [101,102]. A reduction of chromatic dispersion effects in optical fiber has

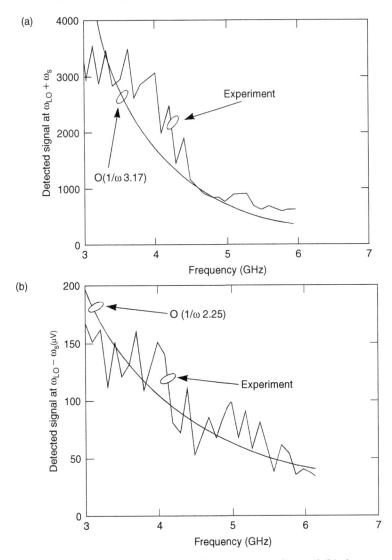

Figure 10.12. Experimental demonstration of (a) up-conversion and (b) down-conversion using modulation of the gain region and mirror region in a DBR laser, plotted as a function of local oscillator frequency [79] © *1986 IEEE.*

also been reported when using harmonic up-conversion [103,104], as compared to up-conversion using mixing products with the LO fundamental frequency.

The power conversion gain G_{series} for the series modulation configuration (shown in Fig. 10.9e) and G_{sin} for quadrature-biased optical modulators with sinusoidal modulation at the optical input (shown in Fig. 10.9f) is [8]

$$G_{\text{series}} = \left(\pi J_1^2(m) I_{\text{ave}} / V_\pi\right)^2 R_{\text{s}} R_{\text{L}}, \tag{10.5}$$

$$G_{\text{sin}} = (\pi I_{\text{ave}} / 2 V_\pi)^2 R_{\text{s}} R_{\text{L}}, \tag{10.6}$$

where R_L is the output load impedance, R_s is the input source impedance, V_π is the modulator RF half-wave voltage, I_{ave} is the detector DC photocurrent delivered to the load, and $J_n(x)$ is the Bessel function of the first kind of order n. In Eq. (10.5), $J_1(m_{opt}) = 0.5819$ when the LO modulator is driven by a sinusoidal signal with the optimum amplitude. Equation (10.6) assumes sinusoidal modulation with a 100% modulation index. Frequency conversion links using optical modulators follow the same scaling laws as conventional external modulation links [105]. Power conversion gain is maximized by using a low V_π modulator, by impedance matching the modulator and photodetector, and by using as high an optical pump power as possible.

For detector-power limited systems, I_{ave} is the same in both Eq. (10.5) and Eq. (10.6). Consequently, the sinusoidal modulation configuration has 1.3 dB lower frequency conversion gain G_{sin} than the series modulator configuration G_{series}, because the overdriven modulator produces more efficient modulation for frequency conversion as discussed earlier. However, for source-power limited systems or modulator-power limited systems, the series modulator configuration increases the optical insertion loss, and sinusoidal modulation can allow higher I_{ave} and higher gain [8]. This is important in practice, since systems are often source-power limited. Calculated power conversion gain is shown in Fig. 10.13 as a function of modulator V_π for several different configurations. A number of the configurations shown in Fig. 10.9 are included for comparison: the series modulator case of Fig. 10.9e (monolithically integrated modulators and two separate fiber pigtailed modulators);

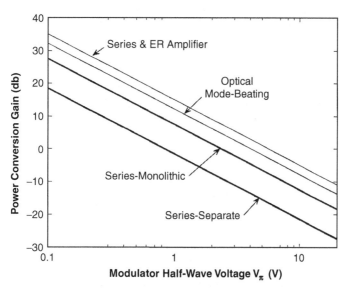

Figure 10.13. Calculation of power conversion gain for several modulator-based frequency conversion configurations.

the optical mode beating configuration of Fig. 10.9f: and a series modulator case with an optical amplifier between both modulators as shown in Fig. 10.10d. In the last case, only a single optical output from the second modulator is included. These calculations assume a 1.3 μm laser or 1.55 μm optical amplifier with 400 mW of optical power, and source impedance of 50 Ω. The assumed fiber coupling insertion loss is 4.5 dB for either a simple M–Z modulator or for a monolithically integrated series MZ modulator pair. The series-separate configuration would have 9 dB of total optical insertion loss for the two separate modulators. The dashed line shows the value of V_π needed for unity frequency conversion gain with a 50 Ω load.

The optical mode-beating configuration produces higher optical gain in this example than the configuration using a second optical modulator to generate the optoelectronic mixing, as the quadrature-bias loss and excess optical insertion loss of the second optical modulator is avoided for the optical mode-beating case. The series modulator configuration with an optical amplifier has much higher gain than either series modulator configuration without an amplifier, because it overcomes the optical loss of the optical mixing modulator. This optical amplifier configuration also has somewhat higher gain than the optical mode-beating configuration because the overdriven modulator waveform has a larger fundamental Fourier component than for pure sinusoidal modulation [7]. However, the series modulator configuration produces harmonics of the LO signal, which can produce undesired mixing products.

The series modulator configuration cannot have differential conversion gain >1 [7], but can have power conversion gain >1 [8]. In the UHF band, the V_π of a LiNbO$_3$ modulator can be on the order of 1 V using reactive impedance matching of the modulator electrode. Power conversion gain from a down-conversion link with a 50 Ω detector impedance [8] is illustrated in Fig. 10.14, with $f_{IF} = 10$ MHz (so $f_{LO} = f_{RF} - 10$ MHz). The noise figure measured with a 50 Ω photodetector impedance was 17.4 dB without electrical pre-amplification, which is in good agreement with theoretical predictions. Even for down-conversion links with high input frequency, the output frequency can be quite low and link gain can be improved by ~15 dB using photodetector impedance matching [106]. Narrowband impedance matching is also demonstrated in Fig. 10.14 with $f_{LO} = 430$ MHz, resulting in the same 15 dB passive impedance matching gain improvement in this experiment as well [8]. With the LO frequency fixed, the power conversion efficiency graph mirrors the frequency response of the photodetector impedance matching circuit. Active photodetector impedance matching has also been used to increase the link gain [107], using a common source FET stage with a high input impedance to match the high photodetector output impedance.

The half-wave voltage V_π of MZ modulators is much larger at high microwave frequencies (~10 V), leading to large link loss. As shown in Eq. (10.5) and

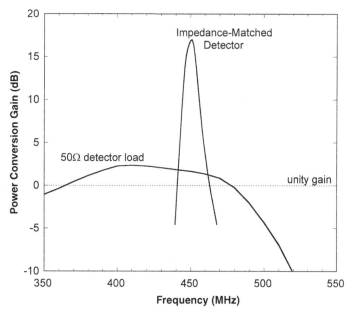

Figure 10.14. Experimental power conversion gain using a monolithic series-modulator configuration *without* photodetector impedance matching ($f_{IF} = 10$ MHz) and *with* photodetector impedance matching ($f_{LO} = 430$ MHz) [9] © *1997 IEEE.*

Eq. (10.6), the frequency conversion gain increases with large average photocurrent I_{ave}, which can be maximized by high laser power or with an optical amplifier. A decrease in power conversion loss to 23 dB was demonstrated at 9 GHz [91], using an optical amplifier between the second optical modulator and the photodetector as shown in Fig. 10.10c. The experimental data is shown in Fig. 10.15.

The intensity noise from the optical amplifier can be cancelled in the differential configuration of Fig. 10.10d. Experimental equivalent input noise of a down-converting link is shown in Fig. 10.16, with an RF frequency of 8 GHz and LO frequency of 8.16 GHz. The theoretical curves show the equivalent input noise as a function of laser RIN. Using a single-output link, the effect of optical amplifier noise is equivalent to using a laser with RIN of approximately −145 dB/Hz. The shot-noise limit in the absence of laser RIN is a straight line labeled "No RIN", while any intensity noise that is not cancelled by the differential configuration would cause a floor to the input noise as the photocurrent is increased. The differential configuration is demonstrated to suppress the intensity noise close to this limit set by shot-noise, with no observable noise floor [96].

Brillouin selective sideband amplification has been used to generate power gain of more than 20 dB at 5.5 GHz for a conventional link without frequency conversion [108] and a differential conversion gain of more than 10 dB at 10 GHz for a frequency conversion link [109]. An interesting aspect of the Brillouin selective

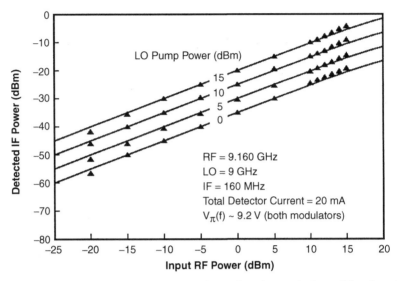

Figure 10.15. Series modulator power conversion gain using optical amplification after the second modulator [89] © *1995 IEEE.*

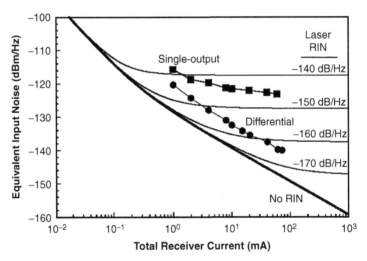

Figure 10.16. Equivalent input noise for frequency conversion link using differential noise suppression, as a function of total photocurrent in the optical receiver with different values of laser RIN [93] © *1996 IEEE.*

sideband amplification scheme is that phase modulators can be used to replace the amplitude modulators in a photonic link. Phase modulation produces two equal amplitude optical sidebands with opposite phase, so that amplitude modulation generated by the two sidebands cancels. Brillouin amplification is very frequency selective, and can convert optical phase modulation to amplitude modulation by

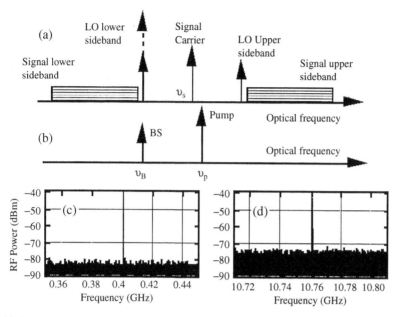

Figure 10.17. Frequency conversion using selective Brillouin amplification. (a) Input optical spectrum to Brillouin amplifier, (b) pump and backscattered gain region, (c) measured spectrum of down-converted signal, and (d) measured spectrum of up-converted signal [106] © *1998 IEEE.*

changing the relative amplitude of the two optical modulation sidebands. This phase modulation to amplitude modulation conversion allows frequency conversion products to be generated from cascaded modulators. In addition to the gain from Brillouin amplification, opto-electronic mixing using phase modulators as shown in Fig. 10.10e has the additional advantage of reducing the optical insertion loss that occurs in quadrature-biased amplitude modulators. Placing the backscattered optical gain at one of the LO sideband wavelengths as shown in Figs. 10.17a and 10.17b allows broadband amplification of the frequency mixing products [109], even though the Brillouin amplification is inherently narrowband. Experimental results of frequency conversion using selective Brillouin amplification are shown in Figs. 10.17c and 12.17d, for an LO frequency of 5.18 GHz and an RF frequency of 5.5873 GHz. Due to the narrowband nature of this approach, care must be taken in setting and maintaining the Brillouin amplification frequency with respect to the optical modulation sidebands.

10.3.3 Dynamic range

Dynamic range is usually the most stringent parameter for analog optical links, as gain and noise figure can typically be improved by suitable electrical amplification.

An SFDR of 93 dB $Hz^{2/3}$ has been measured for photodetector mixing at 66 GHz [70]. For modulator mixing, the sinusoidal transfer function of an MZ modulator produces intermodulation products. Second order distortion is minimal for frequency conversion links with modulators biased at quadrature. For a shot-noise limited optical source and quadrature-biased modulators, the SFDR for the two mixer configurations is [8,10]

$$SFDR_{series} = 10\log_{10}\left[\left(2J_1^2(m)I_{ave}/q\right)\right]^{2/3} \text{dB Hz}^{2/3} \qquad (10.7)$$

$$SFDR_{sin} = 10\log_{10}[(I_{ave}/2q)]^{2/3} \text{dB Hz}^{2/3} \qquad (10.8)$$

where q is the electronic charge. Here, $SFDR_{series}$ is the dynamic range for the series modulator configuration with the LO modulation amplitude set for maximum gain, and $SFDR_{sin}$ is the dynamic range for frequency conversion using 100% sinusoidal optical modulation at the modulator optical input. Unlike noise figure, dynamic range does not depend on the modulator V_π, and the dynamic range is not frequency dependent for a shot-noise limited optical source. A dynamic range of 113 dB $Hz^{2/3}$ was demonstrated using the monolithic series modulator configuration [8].

The dynamic range can be improved by using a linearized modulator for the RF signal. Using an electroabsorption modulator, a dynamic range of 110 dB $Hz^{4/5}$ was achieved with the modulator biased for third order linearization [7], where the dependence on the measurement noise bandwidth has a 4/5 exponent resulting from a dynamic range dominated by fifth order distortion. A dynamic range of 128 dB $Hz^{4/5}$ is projected [7,110] if the power-handling capability and optical insertion loss of these modulators can be improved.

High optical power can be used with a $LiNbO_3$ modulator, leading to a larger demonstrated dynamic range. Using a half-coupler linearized reflective modulator [111], a dynamic range for opto-electronic frequency conversion of 122 dB $Hz^{4/5}$ was obtained [112]. Performance using the reflective half-coupler modulator was affected by the delay between forward and reverse propagating optical waves due to the offset of the reflective modulator electrode from the mirror.

The propagation delays between optical and electrical paths can be equalized in a series MZ linearized modulator configuration [113]. A dynamic range of 130 dB $Hz^{4/5}$ was achieved using a monolithic series MZ linearized modulator for the RF signal together with a separate MZ modulator for the local oscillator signal [114]. The experimental configuration is shown in Fig. 10.18a, where an RF splitter is used to apply the modulation signal to the two inputs of the linearized series MZ modulator. The powers of the fundamental tones and intermodulation products are shown in Fig. 10.18b. For bias point I, the modulator bias is set to minimize third order nonlinearity, resulting in third order intermodulation products with a fifth order dependence on input RF power. For bias point II, the modulator

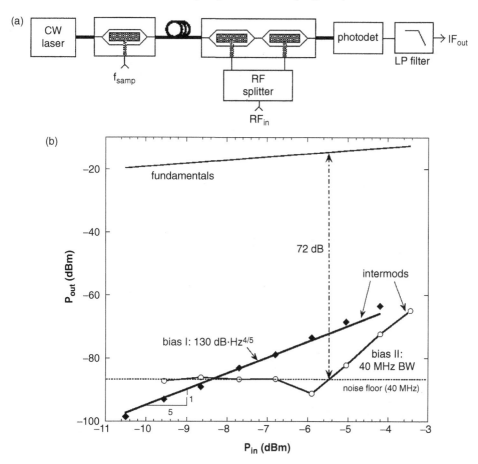

Figure 10.18. (a) Configuration for frequency mixing with a series MZ linearized modulator and (b) experimental results. The bias I data point line corresponds to a dynamic range of 130 dB Hz$^{4/5}$. The dynamic range with the bias II data points is 72 dB in a 40 MHz instantaneous bandwidth.

bias point is set to maximize the dynamic range in a 40 MHz instantaneous bandwidth by a cancellation between third order and fifth order distortion, which occurs at an input power of −6 dBm. These high dynamic range frequency converting link demonstrations clearly indicate their utility in demanding RF transmission system applications. Some of the near-term applications of this enabling RF photonic technology will now be identified.

10.3.4 Applications

The rapid commercial deployment of point-to-multipoint MMW radio systems for voice, video, and data distribution has resulted in much interest in developing high performance MMW photonic links [115–119]. Wireless access to broadband networks providing high data rates to the user demand operation in the MMW range due to the available bandwidth, small antenna, and well-defined cell boundaries. The frequency bands around 30 and 60 GHz have been allocated for applications like interactive multimedia services and multicast distribution services. The frequency converting photonic links are ideally suited for these frequency bands and enable compact antenna systems. The radio-over-fiber application areas represent a large commercial growth area for frequency converting photonic links. As information capacity and bandwidth become more of an issue, these fiber-to-the-air techniques for commercial MMW antenna applications will become more coveted.

On the military side, high-quality tunable microwave and MMW synthesizers are essential for many broadband radar, communication, and electronic warfare systems. The low-phase-noise optical signal generation approaches discussed in this chapter are ideally suited for these applications. One obvious use is as a source of high-frequency LO signals for distribution over an optical fiber network to multiple locations in a microwave or MMW system for phase detection applications or in a phased array antenna [120,121]. All military microwave and MMW systems require up-conversion and down-conversion stages to allow digital signal processing techniques to be used. Hence, remoting antenna signals and photonic link signal mixing aboard military platforms is applicable to most microwave and MMW antenna systems including multiband SATCOM arrays and high performance radars [122,123]. As photonic link performance continues to improve, more military penetration of this technology will occur.

10.4 Summary

Advances in frequency conversion optical links have been reviewed in this chapter. In addition, microwave signal generation techniques most applicable to photonic link signal mixing have been presented. Commercial and military application areas of frequency converting link technology have been identified. The references contained herein provide a good foundation for assessing past work in this area and providing insight into remaining work to be performed.

Summarizing the technical development, significant improvement in the optoelectronic mixing performance over the first demonstration that used a single MZ modulator has been obtained. These improvements include the series modulator configuration, a number of linearized modulator configurations, reactive impedance matching of the modulator and detector, active detector impedance matching, a balanced receiver with an optical amplifier, signle-sideband up-conversion, Brillouin selective sideband amplification, harmonic mixing, and image rejection. A frequency conversion link dynamic range of 130 dB $Hz^{4/5}$ has been demonstrated using a linearized optical modulator in series with a conventional MZ modulator. As photonic link technology continues to improve in terms of cost and performance, frequency conversion links will play an increasingly important role in meeting the most demanding commercial and military MMW transmission system requirements.

Acknowledgements

The authors would like to thank E. Ackerman, C. Cox, G. Eisenstein, H. Roussell, A. Seeds, J. Twichell, K. J. Williams, W. K. Burns, C. K. Sun, E. Gertel and S. Yao for their assistance and collaboration.

References

1. G. K. Gopalakrishnan, C. H. Bulmer, W. K. Burns, R. W. McElhanon, and A. S. Greenblatt, "40 GHz, low half-wave voltage Ti:LiNbO$_3$ intensity modulator," *Electron. Lett.*, **28**, 826–7, 1992.
2. D. Noguchi, H. Miyazawa, and O. Mitomi, "75 GHz broadband Ti:LiNbO$_3$ optical modulator with ridge structure," *Electron. Lett.*, **30**, 949–51, 1994.
3. C. Cox, E. Ackerman, R. Helkey, and G. Betts, "Techniques and performance of intensity-modulation direct-detection analog optical links," *IEEE Trans. Microwave Theory Tech.*, **45**, 1375–83, 1997.
4. N. Mineo, K. Yamada, K. Nakamura, S. Sakai, and T. Ushikobo, "60 GHz band electroabsorption modulator module," Optical Fiber Communication Conference, Vol. 2 of *1998 OSA Technical Digest Series*, Optical Society of America, Washington, D.C., pp. 287–8.
5. A. J. Seeds and B. Lenoir, "Avalanche diode harmonic optoelectronic mixer," *IEE Proc.*, **133**, pt. J, 353–7, 1986.
6. G. Maury, A. Hilt, T. Berceli, B. Cabon, and A. Vilcot, "Microwave-frequency conversion methods by optical interferometer and photodiode," *IEEE Trans. Microwave Theory Tech.*, **45**, 1481–5, 1997.
7. C. K. Sun, R. J. Orazi, and S. A. Pappert, "Efficient microwave frequency conversion using photonic link signal mixing," *IEEE Photon. Technol. Lett.*, **8**, 154–6, 1996.
8. R. Helkey, J. Twichell, and C. Cox, "A down-conversion optical link with RF gain," *J. Lightwave Technol.*, **15**, 956–61, 1997.
9. G. K. Gopalakrishnan, W. K. Burns, and C. H. Bulmer, "Microwave-optical mixing in LiNbO$_3$ modulators," *IEEE Trans. Microwave Theory Tech.*, **41**, 2383–91, 1993.
10. R. T. Logan and E. Gertel, "Millimeter-wave photonic downconvertors: theory and demonstrations," *Proc. SPIE*, **2560**, 58–65, 1995.
11. C. R. Lima, D. Wake, and P. A. Davies, "Compact optical millimeter-wave source using a dual-mode semiconductor laser," *Electron. Lett.*, **31**, 364–6, 1995.
12. D. Wake, C. R. Lima, and P. A. Davies, "Optical generation of millimeter-wave signals for fiber-radio systems using a dual-mode DFB semiconductor laser," *IEEE Trans. Microwave Theory Tech.*, **43**, 2270–6, 1995.
13. W. H. Loh, J. P. de Sandro, G. J. Cowle, B. N. Samson, and A. D. Ellis, "40 GHz optical-millimetre wave generation with a dual polarization distributed feedback fibre laser," *Electron. Lett.*, **33**, 594–5, 1996.

14. S. Pajarola, G. Guekos, and H. Kawaguchi, "Frequency tunable beat note from a dual-polarization emitting external cavity diode laser," *Opt. Quantum Electron.*, **29**, 489–99, 1997.

15. S. Pajarola, G. Guekos, P. Nizzola, and H. Kawaguchi, "Dual-polarization external-cavity diode laser transmitter for fiber-optic antenna remote feeding," *IEEE Trans. Microwave Theory Tech.*, **47**, 1234–40, 1999.

16. R. C. Steele, "Optical phase-locked loop using semiconductor laser diodes," *Electron. Lett.*, **19**, 69–70, 1983.

17. J. Harrison and A. Mooradian, "Linewidth and offset frequency locking of GaAlAs lasers," *IEEE J. Quantum Electron.*, **25**, 1152–5, 1989.

18. K. J. Williams, L. Goldberg, R. D. Esman, M. Dagenais, and J. F. Weller, "6–34 GHz offset phase-locking of Nd:YAG 1319 nm nonplanar ring lasers," *Electron. Lett.*, **25**, 1242–3, 1989.

19. G. J. Simonis and K. G. Purchase, "Optical generation, distribution, and control of microwaves using laser heterodyne," *IEEE Trans. Microwave Theory Tech.*, **38**, 667–9, 1990.

20. M. J. Wale and M. G. Holliday, "Microwave signal generation using optical phased locked loops," *21st European Microwave Conference Workshop*, pp. 78–82, 1991.

21. R. T. Ramos and A. J. Seeds, "Fast heterodyne optical phase lock loop using double quantum well laser diodes," *Electron. Lett.*, **28**, 82–3, 1992.

22. U. Gliese, T. N. Nielsen, M. Bruun, E. L. Christensen, K. E. Stubkjaer, S. Lindgren, and B. Broberg, "A wideband heterodyne optical phase-locked loop for generation of 3–18 GHz microwave carriers," *IEEE Photon. Technol. Lett.*, **4**, 936–8, 1992.

23. G. Santarelli, A. Clairon, S. N. Lea, and G. M. Tino, "Heterodyne optical phase-locking of extended-cavity semiconductor lasers at 9 GHz," *Opt. Commun.*, **104**, 339–44, 1994.

24. R. P. Braun, G. Grosskopf, D. Rohde, and F. Schmidt, "Optical millimetre-wave generation and transmission experiments for mobile 60 GHz band communications," *Electron. Lett.*, **32**, 626–8, 1996.

25. Z. F. Fan, P. J. S. Heim, and M. Dagenais, "Highly coherent RF signal generation by heterodyne optical phase locking of external cavity semiconductor lasers," *IEEE Photon. Technol. Lett.*, **10**, 719–21, 1998.

26. A. C. Davidson, F. W. Wise, and R. C. Compton, "Low phase noise 33–40 GHz signal generation using multilaser phase-locked loops," *IEEE Photon. Technol. Lett.*, **10**, 1304–6, 1998.

27. L. Goldberg, H. F. Taylor, J. F. Weller, and D. M. Bloom, "Microwave signal generation with injection locked laser diodes," *Electron. Lett.*, **19**, 491–3, 1983.

28. L. Goldberg, A. M. Yurek, H. F. Taylor, and J. F. Weller, "35 GHz microwave signal generation with an injection-locked laser diode," *Electron. Lett.*, **21**, 814–15, 1985.

29. L. Goldberg, R. D. Esman, and K. J. Williams, "Generation and control of microwave signals by optical techniques," *IEE Proc.*, **139**, 288–95, 1992.

30. R. P. Braun, G. Grosskopf, R. Meschenmoser, D. Rohde, F. Schmidt, and G. Villino, "Microwave generation for bidirectional broadband mobile communications using optical sideband injection locking," *Electron. Lett.*, **33**, 1395–6, 1997.

31. R. P. Braun, G. Grosskopf, D. Rohde, and F. Schmidt, "Low-phase-noise millimeter-wave generation at 64 GHz and data transmission using optical sideband injection locking," *IEEE Photon. Technol. Lett.*, **10**, 728–30, 1998.

32. A. C. Bordonalli, C. Walton, and A. J. Seeds, "High-performance phase locking of wide linewidth semiconductor lasers by combined use of optical injection locking and optical phase-lock loop," *J. Lightwave Technol.*, **17**, 328–42, 1999.

33. C. Laperle, M. Svilans, M. Poirier, and M. Tetu, "Frequency multiplication of microwave signals by sideband optical injection locking using a monolithic dual-wavelength DFB laser device," *IEEE Trans. Microwave Theory Tech.*, **47**, 1219–24, 1999.

34. R. T. Ramos and A. J. Seeds, "Delay, linewidth and bandwidth limitations in optical phase-locked loop design," *Electron. Lett.*, **26**, 389–90, 1990.

35. U. Gliese, E. L. Christensen, and K. E. Stubkjaer, "Laser linewidth requirements and improvements for coherent optical beam forming networks in satellites," *J. Lightwave Technol.*, **9**, 779–90, 1991.

36. L. N. Langley, C. Edge, M. J. Wale, U. Gliese, A. J. Seeds, C. Walton, J. Wright, and L. Coryell, "Optical phase locked loop signal sources for phased array communications antennas," *Proc. SPIE*, **3160**, 142–53, 1997.

37. L. N. Langley, M. D. Elkin, C. Edge, M. J. Wale, U. Gliese, X. Huang, and A. J. Seeds, "Packaged semiconductor laser optical phase-locked loop (OPLL) for photonic generation, processing and transmission of microwave signals," *IEEE Trans. Microwave Theory Tech.*, **47**, 1257–64, 1999.

38. M. Margalit, M. Orenstein, G. Eisenstein, and V. Mikhailshvili, "Injection locking of an actively mode-locked semiconductor laser," *Opt. Lett.*, **19**, 2125–7, 1994.

39. D. Y. Kim, M. Pelusi, Z. Ahmed, D. Novak, H. F. Liu, and Y. Ogawa, "Ultrastable millimeter-wave signal generation using hybrid modelocking of a monolithic DBR laser," *Electron. Lett.*, **31**, 733–4, 1995.

40. Z. Ahmed, H. F. Liu, D. Novak, M. Pelusi, Y. Ogawa, and D. Y. Kim, "Low phase noise millimetre-wave signal generation using a passively modelocked monolithic DBR laser injection locked by an optical DSBSC signal," *Electron. Lett.*, **31**, 1254–5, 1995.

41. D. Novak, Z. Ahmed, R. B. Waterhouse, and R. S. Tucker, "Signal generation using pulsed semiconductor lasers for application in millimeter-wave wireless links," *IEEE Trans. Microwave Theory Tech.*, **43**, 2257–62, 1995.

42. A. C. Bordonalli, B Cai, A. J. Seeds, and P. J. Williams, "Generation of microwave signals by active mode locking in a gain bandwidth restricted laser structure," *IEEE Photon. Technol. Lett.*, **8**, 151–3, 1996.

43. R. T. Logan, "Photonic radio-frequency synthesizer," *Proc. SPIE*, **2844**, 312–17, 1996.

44. R. T. Logan, R. D. Li, and R. Perusse, "Demonstration of a 1–94 GHz photonic synthesizer," *9th Annual DARPA Photonic Systems for Antenna Applications Conference Proc.*, 1999.

45. Z. Ahmed, H. F. Liu, D. Novak, Y. Ogawa, M. Pelusi, and D. Y. Kim, "Locking characteristics of a passively mode-locked monolithic DBR laser stabilized by optical injection," *IEEE Photon. Technol. Lett.*, **8**, 37–9, 1996.

46. A. Takada and W. Imajuku, "Linewidth narrowing and optical phase control of mode-locked semiconductor ring laser employing optical injection locking," *IEEE Photon. Technol. Lett.*, **9**, 1328–30, 1997.

47. T. Jung, J. L. Shen, D. T. K. Tong, S. Murthy, M. C. Wu, T. Tanbun-Ek, W. Wang, R. Lodenkamper, R. Davis, L. J. Lembo, and J. C. Brock, "CW injection locking of a mode-locked semiconductor laser as a local oscillator comb for channelizing broad-band RF signals," *IEEE Trans. Microwave Theory Tech.*, **47**, 1225–33, 1999.

48. J. E. Bowers, P. A. Morton, A. Mar, and S. W. Corzine, "Actively mode-locked semiconductor lasers," *IEEE J. Quantum Electron.*, **25**, 1426–39, 1989.

49. A. Mar, D. Derickson, R. Helkey, J. E. Bowers, R. T. Huang, and D. Wolf, "Actively mode-locked external-cavity semiconductor lasers with transform-limited single-pulse output," *Opt. Lett.*, **17**, 868–70, 1992.

50. J. J. O'Reilly, P. M. Lane, R. Heidelmann, and R. Hofstetter, "Optical generation of very narrow linewidth millimetre wave signals," *Electron. Lett.*, **28**, 2309–11, 1992.

51. J. J. O'Reilly and P. M. Lane, "Fiber-supported optical generation and delivery of 60 GHz signals," *Electron. Lett.*, **30**, 1329–30, 1994.

52. J. J. O'Reilly and P. M. Lane, "Remote delivery of video services using mm-waves and optics," *J. Lightwave Technol.*, **12**, 369–75, 1994.

53. S. A. Pappert, C. K. Sun, and R. J. Orazi, "Tunable RF optical source using optical harmonic carrier generation," *Proc. SPIE*, **3038**, 89–96, 1997.

54. J. Menders and E. Miles, "Agile MM-wave generation by sideband filtering," *Proc. SPIE*, **3795**, 272–8, 1999.

55. C. K. Sun, R. J. Orazi, S. A. Pappert, and W. K. Burns, "A photonic-link millimeter-wave mixer using cascaded optical modulators and harmonic carrier generation," *IEEE Photon. Technol. Lett.*, **8**, 1166–8, 1996.

56. W. P. Robins, *Phase Noise in Signal Sources, IEE Telecommunications Series 9*, Peter Peregrinus Ltd., p. 78, 1984.

57. X. S. Yao and L. Maleki, "Converting light into spectrally pure microwave oscillation," *Opt. Lett.*, **21**, 483–5, 1996.

58. X. Wang, W. Mao, M. Al-Mumin, S. A. Pappert, J. Hong, and G. Li, "Optical generation of microwave/millimeter-wave signals using two-section gain-coupled DFB lasers," *IEEE Photon. Technol. Lett.*, **11**, 1292–4, 1999.

59. N. G. Walker, D. Wake, and I. C. Smith, "Efficient millimetre-wave signal generation through FM–IM conversion in dispersive optical fibre links," *Electron. Lett.*, **28**, 2027–8, 1992.

60. K. E. Razavi and P. A. Davies, "Semiconductor laser sources for the generation of millimetre-wave signals," *IEE Proc.*, **145**, Pt. J, 159–63, 1998.

61. R. P. Braun, G. Grosskopf, H. Heidrich, C. Helmolt, R. Kaiser, K. Kruger, D. Rohde, F. Schmidt, R. Stenzel, and D. Trommer, "Optical microwave generation and transmission experiments in the 12- and 60-GHz region for wireless communications," *IEEE Trans. Microwave Theory Tech.*, **46**, 320–30, 1998.

62. U. Gliese, T. Norskov Nielsen, S. Norskov, and K. E. Stubkjaer, "Multifunctional fiber-optic microwave links based on remote heterodyne detection," *IEEE Trans. Microwave Theory Tech.*, **46**, 458–68, 1998.

63. U. Gliese, "Multi-functional fibre-optic microwave links," *Opt. Quantum Electron.*, **30**, 1005–19, 1998.

64. D. Roulston, "Low-noise photoparametric up-converter," *IEEE J. Solid-State Circuits*, **SC-3**, 431–40, 1968.

65. W. Kulczyk and Q. Davis, "The avalanche photodiode as an electronic mixer in an optical receiver," *IEEE Trans. Electron. Dev.*, **19**, 1181–90, 1972.

66. R. MacDonald and K. Hill, "Avalanche optoelectronic downconverter," *Opt. Lett.*, **7**, 83–5, 1982.

67. D. A. Humphreys and R. A. Lobbett, "Investigation of an optoelectronic nonlinear effect in a GaInAs photodiode, and its application in a coherent optical communication system," *IEE Proc.*, **135**, Pt. J, 45–51, 1988.

68. R. MacDonald and B. Swekla, "Frequency domain optical reflectometer using a GaAs optoelectronic mixer," *Appl. Opt.*, **29**, 4578–82, 1990.

69. Q. Liu, R. Davies, and R. MacDonald, "Experimental investigation of fiber optic microwave link with monolithic integrated optoelectronic mixing receiver," *IEEE Trans. Microwave Theory Tech.*, **43**, 2357–60, 1995.

70. T. Hoshida and M. Tsuchiya, "Broad-band millimeter-wave upconversion by nonlinear photodetection using a waveguide p-i-n photodiode," *IEEE Photon. Technol. Lett.*, **10**, 860–2, 1998.

71. H. Ogawa and Y. Kamiya, "Fiber-optic microwave transmission using harmonic laser mixing, optoelectronic mixing, and optically pumped mixing," *IEEE Trans. Microwave Theory Tech.*, **39**, 2045–51, 1991.
72. Z. Urey, D. Wake, D. J. Newson, and I. D. Henning, "Comparison of InGaAs transistors as optoelectronic mixers," *Electron. Lett.*, **29**, 1796–7, 1993.
73. Y. Betser, D. Ritter, C. P. Liu, A. J. Seeds, and A. Madjar, "A single-stage three-terminal heterojunction bipolar transistor optoelectronic mixer," *J. Lightwave Technol.*, **16**, 605–9, 1998.
74. A. Foyt, F. Leonberger, and R. Williamson, "InP optoelectronic mixers," *Proc. SPIE*, **269**, 109–14 (1981).
75. D. Lam and R. MacDonald, "GaAs optoelectronic mixer operation at 4.5 GHz," *IEEE Trans. Electron Dev.*, **ED-31**, 1766–8, 1984.
76. C. H. Cox, V. Diadiuk, R. C. Williamson, and A. C. Foyt, "Linearity measurements of high-speed InP optoelectronic switches," *Solid State Research Report, MIT Lincoln Laboratory*, vol. 3, p. 5–11, 1983.
77. J. J. Pan, "Cost-effective microwave fiber optic links using the heterodyne laser," *Proc. SPIE*, **995**, 94–8, 1988.
78. H. Ogawa and Y. Kamitsuna, "Fiber-optic microwave links using balanced laser harmonic generation, and balanced/image cancellation laser mixing," *IEEE Trans. Microwave Theory Tech.*, **40**, 2278–84, 1992.
79. E. Portnoi, V. Gorfinkel, E. Avrutin, I. Thayne, D. Barrow, J. Marsh, and S. Luryi, "Optoelectronic microwave-range frequency mixing in semiconductor lasers," *IEEE J. Sel. Topics Quantum Electron.*, **1**, 451–9, 1995.
80. J. Lasri, M. Shtaif, G. Eisenstein, E. A. Avrutin, and U. Koren, "Optoelectronic mixing using a short cavity distributed Bragg reflector laser," *J. Lightwave Technol.*, **16**, 443–7, 1998.
81. E. Eichen, "Interferometric generation of high-power, microwave frequency, optical harmonics," *Appl. Phys. Lett.*, **51**, 398–400, 1987.
82. B. H. Kolner and D. W. Dolfi, "Intermodulation distortion and compression in an integrated electrooptic modulator," *Appl. Opt.*, **26**, 3676–80, 1987.
83. S. Pajarola, G. Guekos, and J. Mørk, "Optical generation of millimeter-waves using a dual-polarization emission external cavity diode laser," *IEEE Photon. Technol. Lett.*, **8**, 157–9, 1996.
84. J. L. Hall and W. W. Morey, "Optical heterodyne measurement of Neon laser's millimeter wave difference frequency," *Appl. Phys. Lett.*, **10**, 152–5, 1967.
85. M. Sauer, K. Kojucharow, H. Kaluzni, D. Sommer, and W. Nowak, "Simultaneous electro-optical upconversion to 60 GHz of uncoded OFDM signals," *International Topical Meeting on Microwave Photonics*, Princeton, NJ, 219–21, 1998.
86. T. Kuri, K. Kitayama, and Y. Ogawa, "A novel fiber-optic millimeter-wave uplink incorporating 60 GHz-band photonic downconversion with remotely fed optical pilot tone using an electroabsorption modulator," *International Topical Meeting on Microwave Photonics*, Princeton, NJ, 17–20, 1998.
87. G. H. Smith, D. Novak and Z. Ahmed, "Technique for optical SSB generation to overcome dispersion penalties in fibre-radio systems," *Electron. Lett.*, **33**, 74–5, 1997.
88. G. P. Kurpis and J. J. Taub, "Wideband X-band microstrip image rejection balanced mixer," *IEEE MTT Symposium Digest*, 200–5, 1970.
89. L. Chao, C. Wenyue and J. F. Shiang, "Photonic mixers and image-rejection mixers for optical SCM systems," *IEEE Trans. Microwave Theory Tech.*, **45**, 1478–80, 1997.
90. R. Helkey, "Advances in frequency conversion optical links," *MICROCOLL*, Budapest, 1999.

91. G. K. Gopalakrishnan, R. P. Moeller, M. M. Howerton, W. K. Burns, K. J. Williams and R. D. Esman, "A low-loss downconverting analog fiber-optic link," *IEEE Trans. Microwave Theory Tech.*, **43**, 2318–23, 1995.

92. K. J. Williams, R. D. Esman and M. Dagenais, "Effects of high space-charge fields on the response of microwave photodetectors," *IEEE Photon. Technol. Lett.*, **6**, 639–41, 1994.

93. G. A. Davis, R. E. Weiss, R. A. LaRue, K. J. Williams, and R. D. Esman, "A 920–1650-nm high-current photodetector," *IEEE Photon. Technol. Lett.*, **8**, 1373–5, 1996.

94. L. Y. Lin, M. C. Wu, T. Itoh, T. A. Vang, R. E. Muller, D. L. Sivco, and A. Y. Cho, "Velocity-matched distributed photodetectors with high-saturation power and large bandwidth," *IEEE Photon. Technol. Lett.*, **8**, 1376–8, 1996.

95. M. M. Howerton, R. P. Moeller, G. K. Gopalakrishnan and W. K. Burns, "Low-biased fiber-optic link for microwave downconversion," *IEEE Photon. Technol. Lett.*, **8**, 1692–4, 1996.

96. K. J. Williams and R. D. Esman, "Optically amplified downconverting link with shot-noise-limited performance," *IEEE Photon. Technol. Lett.*, **8**, 148–50, 1996.

97. X. S. Yao, "Phase-to-amplitude modulation conversion using Brillouin selective sideband amplification," *IEEE Photon. Technol. Lett.*, **10**, 264–6, 1998.

98. T. Young, J. Conradi, and W. R. Tinga, "Generation and transmission of FM and $\pi/4$ DQPSK signals at microwave frequencies using harmonic generation and optoelectronic mixing in Mach-Zehnder modulators," *IEEE Trans. Microwave Theory Tech.*, **44**, 446–53, 1996.

99. K. P. Ho, S. K. Liaw, and C. Lin, "Efficient photonic mixer with frequency doubling," *IEEE Photon. Technol. Lett.*, **9**, 511–13, 1997.

100. E. Hashimoto, A. Takada, and Y. Katagiri, "Synchronisation of sub-Terahertz optical pulse train from PLL-controlled colliding pulse modelocked semiconductor laser," *Electron. Lett.*, **34**, 580–2, 1998.

101. R. Helkey, D. Derickson, A. Mar, J. Wasserbauer, J. E. Bowers, and R. Thornton, "Repetition frequency stabilisation of passively mode-locked semiconductor lasers," *Electron. Lett.*, **28**, 1920–1, 1992.

102. R. Helkey, A. Mar, W. Zou, D. Young, and J. E. Bowers, "Mode-locked repetition rate feedback stabilization of semiconductor diode lasers," *SPIE Ultrafast Pulse Generation and Spectroscopy Processings*, **1861**, 62–71, 1993.

103. J. M. Fuster, J. Marti, and J. L. Corral, "Chromatic dispersion effects in electrooptical upconverted millimetre-wave fibre optic links," *Electron. Lett.*, **33**, 1969–70, 1997.

104. J. M. Fuster, J. Marti, V. Polo, and J. L. Corral, "Fiber-optic microwave link employing optically amplified electrooptical upconverting receivers," *IEEE Photon. Technol. Lett.*, **9**, 1161–3, 1997.

105. C. H. Cox, G. E. Betts, and L. M. Johnson, "An analytic and experimental comparison of direct and external modulation in analog fiber-optic links," *IEEE Trans. Microwave Theory Tech.*, **38**, 501–9, 1990.

106. A. C. Lindsay, G. A. Knight, and S. T. Winnall, "Photonic mixers for wide bandwidth RF receiver applications," *IEEE Trans. Microwave Theory Tech.*, **43**, 2311–17, 1995.

107. J. Önnegren and J. Svedin, "Photonic mixing using a Franz-Keldysh electroabsorption modulator monolithically integrated with a DFB laser," *International Topical Meeting on Microwave Photonics*, IEEE, Kyoto, vol. WE1-4, 1996.

108. X. S. Yao, "High-quality microwave signal generation by use of Brillouin scattering in optical fibers," *Opt. Lett.*, **22**, 1329–31, 1997.

109. X. S. Yao, "Brillouin selective sideband amplification of microwave photonic signals," *IEEE Photon. Technol. Lett.*, **10**, 138–40, 1998.
110. C. K. Sun, R. J. Orazi, R. B. Welstand, J. T. Zhu, P. K. L. Yu, Y. Z. Liu and J. M. Chen, "High spurious free dynamic range fibre link using a semiconductor electroabsorption modulator," *Electron. Lett.*, **31**, 902–3, 1995.
111. G. E. Betts, F. J. O'Donnell, K. G. Ray, D. K. Lewis, D. E. Bossi, K. Kissa, and G. W. Drake, "Reflective linearized modulator," *Integrated Photonics Research, OSA Technical Digest Series*, vol. 6, 1996.
112. H. Roussell and R. Helkey, "Optical frequency conversion using a linearized LiNbO₃ modulator," *IEEE Microwave Guided Wave Lett.*, **8**, 408–10, 1998.
113. G. E. Betts, "A linearized modulator for high performance bandpass optical analog links," *IEEE MTT-S Intl. Symp. Dig.* vol. 2, 1097–1100, 1994.
114. R. Helkey, "Narrowband optical A/D converter with suppressed second-order distortion," *IEEE Photon. Technol. Lett.*, **11**, 599–601, 1999.
115. H. Ogawa, D. Polifko, and S. Banda, "Millimeter-wave fiber optics systems for personal radio communications," *IEEE Trans. Microwave Theory Tech.*, **40**, 2285–93, 1992.
116. K. Morita and H. Ohtsuka, "The new generation of wireless communications based on fiber-radio technologies," *IEICE Trans. Commun.*, **E76-B**, 1061–8, 1993.
117. J. J. O'Reilly, P. M. Lane, M. H. Capstick, H. M. Salgado, R. Heidemann, R. Hofstetter, and H. Schmuck, "Race R2005: microwave optical duplex antenna link," *IEE Proc.* **140**, pt. J, 385–91, 1993.
118. S. Komaki and E. Ogawa, "Trends of fiber-optic microcellular radio communication networks," *IEICE Trans. Electron.*, **E79-C**, 98–104, 1996.
119. T. Kuri, K. Kitayama, and Y. Ogawa, "Fiber-optic millimeter-wave uplink system incorporating remotely fed 60-GHz band optical pilot tone," *IEEE Trans. Microwave Theory Tech.*, **47**, 1332–7, 1999.
120. P. D. Biernacki, L. T. Nichols, D. G. Enders, K. J. Williams, and R. D. Esman, "A two-channel optical downconverter for phase detection," *IEEE Trans. Microwave Theory Tech.*, **46**, 1784–7, 1998.
121. J. L. Corral, J. Marti, and J. M. Fuster, "Optical up-conversion on continuously variable true-time-delay lines based on chirped fiber gratings for millimeter-wave optical beamforming networks," *IEEE Trans. Microwave Theory Tech.*, **47**, 1315–20, 1999.
122. S. A. Pappert, C. K. Sun, R. J. Orazi, and T. E. Weiner, "Photonic link technology for shipboard rf signal distribution," *Proc. SPIE*, **3463**, 123–34, 1998.
123. J. E. Roman, L. T. Nichols, K. J. Williams, R. D. Esman, G. C. Tavik, M. Livingston, and M. G. Parent, "Fiber optic remoting of an ultrahigh dynamic range radar," *IEEE Trans. Microwave Theory Tech.*, **46**, 2317–23, 1998.

11

Antenna-coupled millimeter-wave electro-optical modulators

WILLIAM B. BRIDGES

California Institute of Technology

11.1 Introduction

As the modulation frequency is increased in a traveling wave electro-optic modulator, good performance becomes more and more difficult to realize. There are generally three reasons for this: (1) velocity mismatch, (2) electrode loss, and (3) parasitic inductance and capacitance in the connections to the electrodes. Several schemes have been invented to deal with these limitations, some of which have already been discussed in Chapter 5. In this chapter we introduce a new scheme that offers improvement in all three limiting aspects, but is practical only at very high microwave frequencies. The basic scheme is simple. The transmission line electrodes of the modulator are broken into N sections, and each section is connected to an on-substrate antenna. The array of antennas thus formed is excited by a plane wave incident from the substrate side, as illustrated in Fig. 11.1. The illumination angle is chosen so that the phase velocity from antenna to antenna is the same as the phase velocity of the light in the optical waveguide. This assures velocity matching from segment to segment, even though the phase velocities of RF and optical waves are not matched within a segment. This overcomes limitation (1), so that the modulator may be made as long as desired. While the optical power is now divided by N (at best), the attenuation along each segment is now $\alpha L/N$ compared to αL for a simple modulator of length L with transmission line loss of α nepers/unit length. The length limitation (2) imposed by high α can be overcome by making N large. Finally, connections to the modulator electrodes are eliminated along with their parasitics (3).

Of course, we now have new problems to deal with. The antennas are by nature bandpass structures, so we no longer have a base-band response in such a modulator; it is inherently a bandpass modulator, with bandwidth limited by the frequency response of a single antenna-plus-transmission-line-segment response. And while we have eliminated the parasitic circuit problems of connecting the transmission

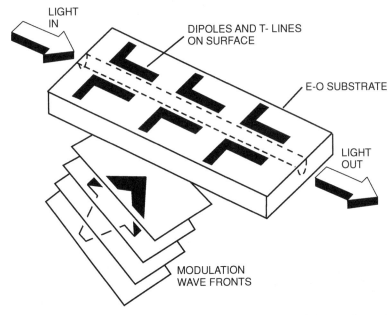

LIGHT
IN

DIPOLES AND T- LINES
ON SURFACE

E-O SUBSTRATE

LIGHT
OUT

MODULATION
WAVE FRONTS

Figure 11.1. Schematic picture of an antenna-coupled phase modulator. The optical wave is coupled into and out of an integrated optical waveguide at the surface of an electro-optic crystal. The modulating wave is incident at an angle from below the substrate, and is received by a succession of antennas on the upper surface (shown here as simple dipoles) which are, in turn, connected to short sections of transmission line. The conductor-to-conductor electric fields of the transmission lines penetrate the optical waveguide and modulate the wave via the electro-optic effect. The angle of illumination is chosen so that phase velocity from antenna to antenna is the same as the phase velocity of the light in the optical waveguide.

line to the package connectors, we have replaced them with the problem of coupling radiation from whatever transmission medium is desired at the input to the modulator (waveguide, coax, ...) to the array of antennas. Antenna-coupled modulators will be most useful at mm-wavelengths where quasi-optic feed systems can be used to supply the RF to the antennas.

To make a logical development of the quantitative behavior of such a modulator, we review briefly the problems of velocity mismatch and electrode loss and some of the schemes already discussed in Chapter 5 to deal with them.

11.2 Velocity mismatch in traveling wave electro-optic modulators

Electro-optic materials often have different relative permittivities or refractive indices in the optical range compared to their values in the range of modulation frequencies. (The refractive index is the square root of the relative permittivity, and both are real numbers for lossless or low-loss materials). For example, lithium niobate has a refractive index of about 2.2 in the optical range, but it is substantially

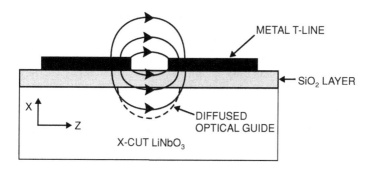

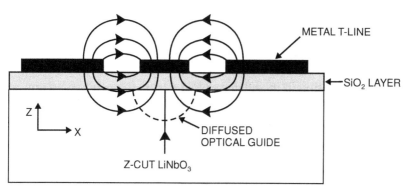

Figure 11.2. Schematic cross-sectional view of typical *x*-cut and *z*-cut modulators, showing the in-diffused optical waveguides and the transmission electrodes for the modulation. The optical waveguides can be located in regions of primarily horizontal fields (*x*-cut) or primarily vertical fields (*z*-cut).

higher in the DC to mm-wave range. The exact values depend on the polarization of the electric field with respect to the crystal axes. With the electric field parallel to the crystal's *z*-axis (extraordinary axis) the optical index is about 2.14 and the microwave index is about 5.3. With the electric field parallel to the crystal's *x*-axis (ordinary axis) the optical index is about 2.22 and the microwave index is about 6.4 [1]. In practical modulators, both the modulating field and the optical field are made parallel to the crystal's *z*-axis to take advantage of the largest electro-optic coefficient. Typical electrode arrangements for a simple phase modulator with quasi-TEM transmission lines are shown in Fig. 11.2 for both *x*-cut and *z*-cut crystal substrates. A thin (typically 0.2 μm) layer of silicon dioxide (optical refractive index of about 1.45 and microwave refractive index of 1.94 [2]) is typically used between the metal and the lithium niobate, so that the optical fields do not "see" the metal layer, which is relatively lossy at optical frequencies.

The velocity of an optical wave well-confined by an optical waveguide formed by diffusion near the surface of the lithium niobate is about $c/2.2$, while the modulating wave guided by the electrodes on the surface above the optical waveguide will travel at a slower velocity. If the electrodes were "buried" in lithium niobate, for example, by placing a lithium niobate superstrate over the electrodes and optical waveguide, then the velocity of the modulating signal traveling along the electrodes would be $c/5.3$. (Note that this is only an approximate value, since the fields of the wave traveling on the electrodes have both x and z components, so they "feel" both refractive indices. We argue that since it is the z-component of the wave on the transmission line with which the optical signal interacts, it is appropriate to use that refractive index for the velocity.) If no superstrate is used, as is usually the case, then the velocity of the modulating wave is somewhat higher, since the electric field from electrode to electrode extends through both the lithium niobate below and the air above the modulator surface, as illustrated schematically in Fig. 11.2. The approximate velocity is given by averaging the relative permittivity of lithium niobate ($5.3^2 = 28$) and air (1) to give a velocity of about $c/3.8$. However, this is still a very large mismatch in velocity, and it will limit the length of the modulator.

The relative variation in modulated output versus frequency for a simple phase modulator (or Mach–Zehnder intensity modulator) of length L, with optical and microwave refractive indices n_o and n_m is given by

$$T(f) = \operatorname{sinc}[\pi f L (n_m - n_o)/c]. \tag{11.1}$$

The first zero in this function occurs at $L_0 = c/f(n_m - n_o)$, and the -3 dB point of the modulator occurs at $L_{3dB} = 0.6034c/f(n_m - n_o)$. For a simple lithium niobate modulator, these values are $L_0 = 18.75$ mm and $L_{3dB} = 11.3$ mm at 10 GHz, and one tenth these lengths at 100 GHz. Velocity matching by some means is thus imperative if we wish to increase the sensitivity of the modulator by increasing the effective length beyond this velocity mismatch limit.

11.3 RF loss in the traveling wave electrodes

Electrode loss can also limit the effective length of a traveling wave electro-optic modulator. At microwave- and millimeter-wave frequencies, the metal electrodes (typically gold) have loss due to their finite conductivity and very small size. Electrode thickness and width as indicated schematically in Fig. 11.2 may vary a great deal. Simple evaporation may yield thicknesses of 0.2 μm to 1 μm. Still thicker electrodes are realized by electroplating on top of the evaporated layer, for thicknesses up to 5 μm or more. Evaporated gold typically exhibits a DC resistance equal to that calculated from the handbook conductivity (about 4×10^7 siemens/meter) while electroplated gold may have a resistance 50% higher [3]. For a 0.2 μm × 6 μm evaporated electrode the ideal DC resistance might be of the order of 20 Ω/mm, assuming the current is uniformly distributed over the cross-section. Increasing the evaporated thickness to 1 μm would reduce the DC resistance to 4 Ω/mm.

In reality, the RF currents on the transmission line electrodes will be not be uniformly distributed over the electrode cross-section, but will be excluded from the interior by the skin effect (Ref. 4, Section 3.16). The skin depth for gold is about 2.5 µm at 1 GHz, decreasing as $f^{-1/2}$ to 0.25 µm at 100 GHz. Thus, electroplating is important at 1 GHz, but simple evaporated electrodes should be adequate at 100 GHz. In addition to the skin effect (which affects primarily the design of the thickness of the electrodes), the RF currents will be concentrated along the edges of the transmission lines that face each other, as governed by the modal solution for propagation along the line. Increasing the width of the line much beyond the separation between adjacent lines does not decrease the resistance, since only a small fraction of the current flows in the outer regions of the increased width. Thus we are lead to conclude that increasing the thickness and width of a transmission line electrode beyond 0.5 to 1 µm thickness by 10 µm width will not improve the losses in the line at 100 GHz. This leads to effective RF resistances at 100 GHz of 2–20 Ω/mm or so.

The exponential attenuation along the transmission line can be found from the effective (RF) resistance by

$$\alpha \approx R/2Z_0, \qquad (11.2)$$

where R is the actual resistance per unit length and Z_0 is the characteristic impedance of the transmission line (see Ref. 4, p. 251, for example). For 10 Ω/mm and a 50 Ω transmission line, α is about 0.1 nepers/mm. This means the applied modulating fields will have been decreased to $1/e$ or 37% in 10 mm along the transmission line and to $1/e^2$ or 14% in 20 mm. The "effective" length of the transmission line is thus limited by the decay of the modulating fields.

11.4 "True" velocity matching

"True" velocity matching has been demonstrated by several groups (See Chapter 5 and Refs. 5–7). The approach taken is to increase the velocity on the transmission line electrodes by reducing the effective refractive index from about 3.8 (which results from the fields being half in air and half in lithium niobate with $n = 5.3$). The increase is realized (1) by increasing the thickness of the silicon dioxide buffer layer ($n = 1.94$) to a few micrometers so that a higher fraction of the electric field resides in the silicon dioxide layer, and (2) by up-plating the gold electrodes so that their height above the lithium niobate surface is comparable to or grater than the spacing between them, resulting in an increase in the fraction of the electric field that resides in air. Both these changes reduce the effective refractive index of the transmission line, which can now be made to equal the optical index, 2.2. Unfortunately, moving the electric fields out of the lithium niobate substrate (and optical waveguide) into the silicon dioxide layer or the air, also reduces the effective interaction of the modulation fields with the optical fields via the electro-optic effect. Thus, while long interaction lengths are now possible with reduced or zero velocity mismatch, the interaction per unit length is lower. Of course, the inter-action length is ultimately limited by RF losses in the electrodes. And the parasitic elements associated with the input and output transitions between the modulating electrodes and the coaxial or other transmission systems necessary to connect to the rest of the system are still a problem. "True" velocity matching by these methods is thus a trade-off.

11.5 Velocity matching "on the average" by phase shifts

Another technique, also described in Chapter 5 is "velocity matching on the average." In this method, we allow the velocity mismatch ($c/2.2$ vs. $c/3.8$) to occur over a short length of the transmission line electrodes, then we break the transmission line and re-phase the modulation so that it is again in-phase with the modulation already imposed on the optical wave.

One way of accomplishing the re-phasing is the "phase reversal" modulator of Alferness *et al.* [8], in which the modulating electrodes are transposed periodically. Such a structure is shown in Fig. 11.3. After a transposition at the input end of a transmission line segment, the modulating electrical wave leads the signal modu-lated onto the light by 90 degrees and then gradually slows down until it lags by 90 degrees at the output end of the segment. At the end of the segment, it encoun-ters another transposition which gives it a 180 degree change, so that it now leads the optical wave by 90 degrees again, and so on. The modulation coupling thus reduced by the average of the sinc function in Eq. (11.1) over ±90 degrees around the in-phase value, but the number of segments can now be increased until the mod-ulating wave is substantially attenuated by loss in the transmission structure. Note, however, that the whole structure now exhibits a "bandpass" characteristic around the modulating frequency for which the segment length is 180 degrees. Further-more, the bandwidth around this frequency is decreased by increasing the number of segments, since the overall frequency response is equal to the segment frequency response raised to the Nth power, where N is the number of segments. Figure 11.4 illustrates this bandwidth narrowing for $N = 2, 5$, and 20 sections, compared to the response for just one section. And the problem of parasitic elements in the input and output transitions is the same (although likely not as hard to "tune out" because of the narrow band nature of this modulator).

A variation on the phase reversal modulator was proposed and demonstrated by Schaffner [9]. The transmission line is again broken into short segments and re-phased at the beginning of each segment. But the re-phasing is accomplished by

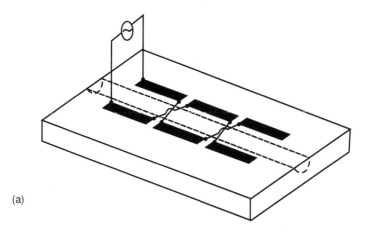

(a)

MACH-ZEHNDER WAVEGUIDES

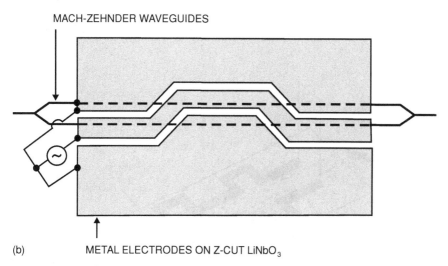

(b)   METAL ELECTRODES ON Z-CUT LiNbO$_3$

Figure 11.3. (a) Schematic representation of a phase reversal modulator, showing short electrode segments re-phased by 180 degrees by cross-over wiring at the end. (b) Actual electrode pattern that accomplishes the reversal of the electric field between segments in a z-cut Mach–Zehnder modulator.

delaying the modulating wave by 360 degrees minus the velocity mismatch phase difference suffered over the segment. Such a structure is illustrated schematically in Fig. 11.5, showing lateral transmission lines to accomplish the desired delay. By allowing shorter segments, the loss in interaction caused by averaging over the sinc function can be reduced (in principle). Like the phase reversal modulator, this modulator is now a bandpass structure, with a bandwidth that narrows as the number of segments is increased. Like the true velocity-matched modulator and the phase reversal modulator, it is ultimately limited by attenuation in the transmission

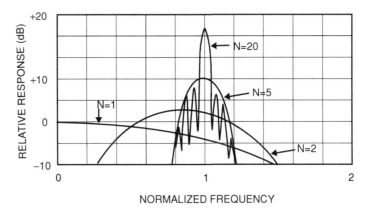

Figure 11.4. Frequency dependence for an N-section modulator where the sections are connected in series, but somehow re-phased at the beginning of each section. The bandwidth becomes quite narrow around the frequency for which the re-phasing is perfect. (Note that the phase mismatch between optical and modulating waves has been included as well, which is why the $N = 1$ curve falls off, and the other curves are not symmetrical about $f = 1$). From Ref. 11.

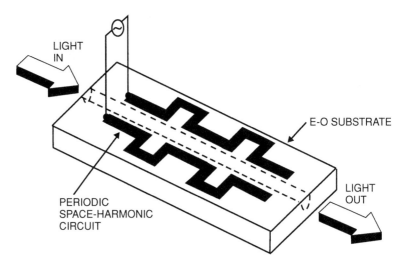

Figure 11.5. Schematic representation of a space harmonic transmission structure in which transmission line segments are phased by extra delays in transverse transmission lines.

structure. In the scheme sketched in Fig. 11.5, this loss is actually increased because the signal must propagate along the side delay lines as well as along the active modulating segments, so microwave loss is even more serious here. And of course, the transition parasitics in and out of the transmission structure must still be dealt with.

Figure 11.5 is only one possibility of what is more generally termed "space harmonic" operation of a periodic structure. The trick is finding a structure that exhibits the correct phase velocity and does not introduce too much additional transmission attenuation over a parallel strip line.

11.6 Velocity matching on the average with a corporate feed

Another variation on the "velocity matching on the average" theme was proposed by Schaffner and Bridges [10]. In this scheme, the main parallel strip transmission line is broken into short segments whose lengths permit efficient interaction even though velocity mismatched. Now, instead of feeding these segments in series as in the previous two schemes, each segment is fed from an array of strip lines slanted off to the side on the lithium niobate chip. This array of feedlines results from the successive splitting via −3 dB couplers from a single input stripline. This kind of feed structure is usually called a "corporate feed" when used in phased array radar antennas. An example is shown schematically in Fig. 11.6. The tilt angle, Θ, is chosen so that the time delay in the optical waveguide under a segment of length L, equal to $2.2L/c$, is equal to the time delay difference along successive lines in the corporate feed, $3.8L \sin \Theta / c$. Thus we have $\Theta = \sin^{-1}(2.2/3.8) = 35.4$ degrees.

There are some interesting differences from the "phase reversal" and "space harmonic" modulators previously described. Unlike the other two, the corporate feed modulator still exhibits a base-band response rather than a bandpass response, since the re-phasing is accomplished by "true time delay" means (provided the

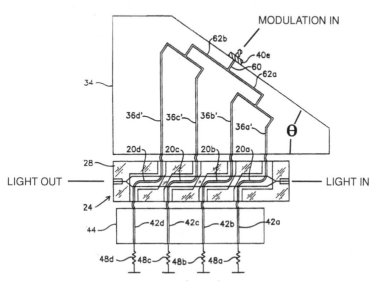

Figure 11.6. Schematic representation of a corporate feed modulator, showing the transmission line splitters and tapered line lengths necessary to feed a series of modulator segments in the proper phase. From Ref. 10.

transmission lines in the corporate feed are not dispersive) rather than phase shifting. And the bandwidth is not reduced by increasing the number of segments used. The attenuation of the transmission lines used in the modulator segments ultimately limits the length of the modulator, but in a different way than with the phase reversal and space harmonic modulators, as described below.

11.7 Effect of transmission line loss in N re-phased segments

In all modulators in which the transmission line segments are in series the voltage along the transmission structure decays as $\exp(-\alpha L)$, where α is the voltage attenuation constant and L is the overall length of the modulator. When L is made equal to α^{-1}, the voltage on the output end of the line is only 37% of the input voltage, and making the modulator longer adds little to the modulation of the optical wave. This is the consideration that limits simple traveling wave modulators and the Alferness phase reversal scheme. Schaffner's space harmonic electrode scheme suffers an additional loss, since the modulating RF traverses a longer path due to the added transverse sections of line, and the condition $\alpha L + \alpha L_{transverse} = 1$ is a more likely limit for the space harmonic scheme.

The corporate feed scheme suffers loss somewhat differently, however, as shown by Sheehy [11]. If we neglect transmission loss in the corporate feed transmission lines themselves for the moment, then we note that division of RF power into N paths to drive the N modulator segments means that the modulator suffers a loss of $N^{-1/2}$ in voltage at the outset, followed by attenuation $\exp(-\alpha L/N)$ along each segment, for a total loss in voltage of $N^{-1/2} \exp(-\alpha L/N)$ at the end of the segment. To compare performances, we calculate the total phase modulation resulting from the two methods of configuring the modulator. For a simple phase modulator of length L_0, a voltage decaying as $V(z) = V_0 \exp(-\alpha z)$ produces a total phase modulation

$$\phi = \phi_0[1 - \exp(-\alpha L_0)], \tag{11.3}$$

where ϕ_0 is the modulation we would have if we made the modulator infinitely long. If we make $L_0 = \alpha^{-1}$ then $\phi = 0.63\phi_0$. If we make $L_0 = 2\alpha^{-1}$ then $\phi = 0.86\phi_0$.

An N-segment modulator with a total length L and a lossless power-division corporate feed will have a phase modulation of

$$\phi = N^{-1/2}\phi_0[1 - \exp(-\alpha L/N)] \tag{11.4}$$

at the end of the first segment, and a total phase modulation after N segments of

$$\phi = N^{+1/2}\phi_0[1 - \exp(-\alpha L/N)], \tag{11.5}$$

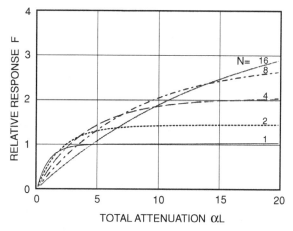

Figure 11.7. Variation of the modulation obtained in an N-section modulator in which the power is divided by N and fed to each section in the proper phase, as a function of the total attenuation αL of a modulator of overall length L. No accounting has been made for loss in the splitting and phasing means.

since the modulation from each segment adds coherently due to phase matching on the average. Taking the ratio of phase modulation for the segmented modulator to that of an infinitely long simple modulator gives the improvement factor F obtainable with the corporate feed modulator.

$$F = N^{1/2}[1 - \exp(-\alpha L/N)]. \tag{11.6}$$

This factor is plotted in Fig. 11.7 as a function of αL with N as a parameter. Values of N are taken as powers of 2, since N will likely be produced by a simple branched feed. The curve for $N = 1$ corresponds to the simple phase modulator used for comparison. All of the segmented modulators are inferior to the simple modulator at short lengths or low values of attenuation. However, all the curves eventually intersect the $N = 1$ curve, and then rise above it. For example, the $N = 4$ modulator is as good as the simple modulator at $\alpha L \approx 2.5$, then goes on to be twice as good for large values of αL. It is evident from the figure and from the equation that in the limit of large αL, the improvement factor F approaches $N^{1/2}$.

Of course, we have neglected the transmission line loss in the corporate feed structure itself. One can argue that the corporate feed transmission lines can be made much wider and thicker than the lines used in the modulator. And the corporate feed may be made on a different substrate material. Both these arguments were made in the patent by Schaffner and Bridges [10]. However, ultimately there will be loss in the corporate feed, so that the performance of this segmented modulator will be worse than that implied in Fig. 11.7.

11.8 Antenna-coupled modulators – initial experiments

The invention of antenna-coupled modulators [12] and its reduction to practice
[11, 13–16] by Sheehy, Bridges and Schaffner came about as an alternative way of
dividing the power and distributing it to the several segments of a transmission line
velocity matched on the average. As shown in Fig. 11.1, small antennas are attached
to the beginning of each transmission line segment. This linear array of antennas
is illuminated by quasi-optical means at an angle such that the time delay from
antenna to antenna is the same as that of the optical wave in the optical waveguide
underneath the electrodes. The illumination must be from the substrate side, as is
easily seen from the following simple argument. If the modulating illumination
were from the air side, then the phase velocity from antenna to antenna would vary
from c for a wave just grazing the modulator surface, to infinity for a wave incident
normal to the surface. At no angle is the velocity $c/2.2$. For illumination from the
substrate side, the phase velocity varies from $c/5.3$ at grazing incidence to infinity
at normal incidence. Thus there is an angle, $\Theta = 24.5$ degrees, where this velocity
is $c/2.2$.

Antenna coupling does give up an important feature of the corporate feed modulator, namely, the antenna-coupled modulator is no longer a low pass device. While it
is still a true time delay device, the antennas themselves have a bandwidth that does
not extend to low frequencies. While broadband antennas are possible, eventually
there will be a low-frequency roll-off when the size of the antenna is comparable to
a wavelength (corrected for the dielectric properties of the substrate). However, the
overall modulator bandwidth is that of a single antenna-plus-transmission line segment, and does not narrow because there are N segments, unlike the phase reversal
and space harmonic modulators.

It is important to ask how the dielectric interface modifies the radiation pattern of
an antenna located on that interface. Since we must illuminate the dipole from the
substrate side, it is important to know whether the antenna pattern is strong or weak
in that direction. (Without the interface, the dipole would be omni-directional in
the plane perpendicular to its conductor.) Fortunately, nature is kind to this scheme.
The antenna response is actually greatly enhanced in the substrate direction. Several
workers have treated this problem [17–20] but give calculations for only relatively
low dielectric constant materials. In his thesis (Ref. 11, Appendix A), Sheehy
devised a clever and intuitively satisfying technique for calculation, involving the
superposition of direct radiation and radiation reflected from the interface for an
antenna immersed in the higher dielectric constant material, but very near the
interface. By taking the limit as the antenna approached the interface, he obtained
results that were identical to those in Ref. 18 in the case of an infinitesimal dipole. He
was then able to calculate the antenna patterns for high dielectric constant material.

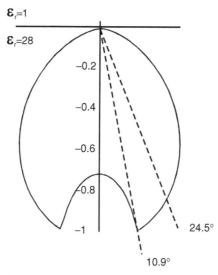

Figure 11.8. Antenna radiation pattern from a dipole antenna on a dielectric substrate with $\varepsilon_r = 28$. The radiation into the air region is zero on the scale of this drawing. The cusps in the pattern occur at the critical angle for LiNbO$_3$, 10.9 degrees, and the radiation in the desired direction for segment phasing, 24.5 degrees, is 90% of the maximum value. From Ref. 11.

Sample results are shown in Fig. 11.8 for two half-waves in phase on an air–lithium niobate ($\varepsilon_r = 28$) surface. the "cusps", while not seen in ordinary antenna patterns, are real, and occur at the critical angle (10.9 degrees here). The response at 24.5 degrees is approximately 90% (in power) of the value at the peak. The response to the air side is essentially zero on the scale of this figure.

Ideally, the modulating-wave power would be divided equally among the N antennas, with no loss. This is not possible in practice, since the transverse shape of the illumination beam of modulation radiation will not likely have the exact size and shape of the array of antennas, nor will it be uniform over whatever shape it has. The situation is improved however, if a waveguide transition is used to shape this illumination function to fit the array. The first experiment done with antenna coupled modulators took advantage of this fact. Sheehy *et al.* [11,13] used the radiation from the standard WR-90 waveguide (0.900 inches by 0.400 inches) to illuminate a modulator substrate at X-band (8.2–12.4 GHz), where the five-antenna array had approximately these same dimensions when tilted at the proper angle. The arrangement is shown in Fig. 11.9. The layout of antennas on the modulator chip is shown in Fig. 11.10. The antennas extend 2.8 mm on either side of the transmission line, and the transmission line is 3.5 mm long. The wavelength at 12 GHz in a line with effective index 3.8 is 6.6 mm, so that the sum of the length of one antenna arm plus that of the transmission line, equaling 6.3 mm, would be

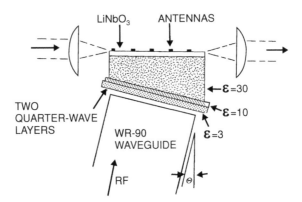

Figure 11.9. Side view of an X-band wave coupled modulator, showing the entering wedge and the matching layers. The broad (0.900 inch) dimension of the WR-90 waveguide is shown. From Ref. 13.

nearly resonant. The resonant length for "fat" antennas is always somewhat shorter than exact wavelength values, but the exact correction factor for flat antennas at a dielectric interface is not known explicitly, so we "guessed" at 5%.

The spacing between the transmission line conductors was 8 μm and the conductors were 50 μm evaporated aluminum stripes. Using data from Rutledge *et al.* [21], the calculated characteristic impedance of this line was 35 Ω. A 1500 Å thick silicon dioxide buffer layer was used between the aluminum and the substrate. An x-cut substrate was used since the fringing fields from the transmission line strips are largely horizontal through the optical waveguide (Fig. 11.2, A).

Two quarter-wave dielectric plates of $\varepsilon_r = 3$ and 10 were used to match the waveguide impedance of 450 Ω (for WR-90 at 12 GHz) to that of the wave impedance in a space of $\varepsilon_r = 30$, about 70 Ω. A wedge-shaped block of Stycast® $\varepsilon_r = 30$ material, with a wedge angle of 23 degrees was inserted to couple the wave in at (almost) the correct angle. The critical angle for lithium niobate is 10.9 degrees, so it is impossible to couple in to a surface parallel to the antenna array and produce an interior angle of the correct value. We used the wedge to couple into the high dielectric constant material at normal incidence, then change the direction of the wave incident on the modulator within the high dielectric constant region. The Stycast® $\varepsilon_r = 30$ material was the closest to $\varepsilon_r = 28$ that we could obtain without special order.

Sheehy used several wedges at different angles to confirm that there was actually an optimum angle as predicted by the simple theory. Figure 11.11 shows the strength of the modulation as a function of angle, with a maximum at approximately the correct angle. The solid curve is the simple theory for five antennas with the sine-squared power distribution that would be expected from the TE_{10} mode in the WR-90 waveguide.

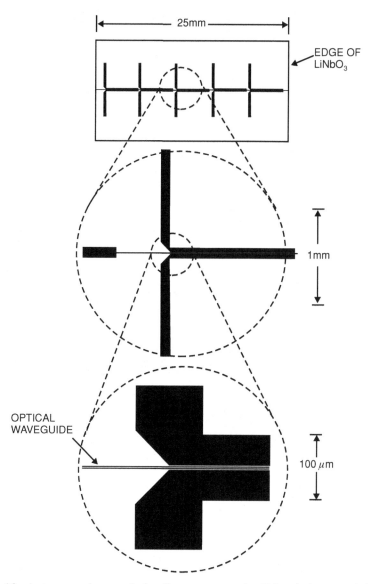

Figure 11.10. Antenna and transmission line pattern on the X-band phase modulator chip. The single optical waveguide was positioned in the 8 μm gap between the transmission line conductors. From Ref. 13.

A novel method of measuring modulation performance was used in these initial experiments. A simple single-waveguide phase modulator was fabricated instead of the more usual Mach–Zehnder intensity modulator. Since phase modulation is not detected by a simple photodetector, a scanning Fabry–Perot (SFP) interferometer was used as an optical spectrum analyzer to measure the phase modulation sidebands on the modulated output. While photodetectors with X-band response

Figure 11.11. Experimentally determined response of the X-band phase modulator as a function of the modulation wave illumination angle (changed by using different wedge angles in Fig. 11.9). Also shown is a theoretical response (solid line) expected for illumination of a five antenna array with a sine-squared illumination over the same range of angles. From Ref. 13.

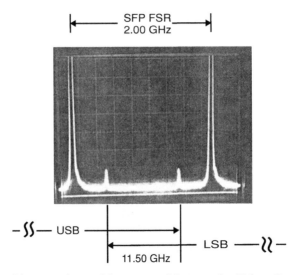

Figure 11.12. Oscilloscope photo of the output of the scanning Fabry–Perot modulator with a free spectral range (FSR) of 2.00 GHz. The large signals are the optical carrier and its alias 2.00 GHz away. The small signals are the phase modulation sidebands 11.5 GHz away from the optical carrier aliases 10 GHz "outside" this picture. From Ref. 13.

are readily obtainable, photodetectors useful at 100 GHz were not available at the time of Sheehy's work. The SFP sideband detection method is independent of modulation frequency, working just as well at 100 GHz as at 10 GHz, and this was the direction in which Sheehy's modulator research was headed. The only drawback of the SFP method is that modulation at frequencies that are integral multiples of the free spectral range (FSR) of the Fabry–Perot are masked by aliases of the carrier. Sheehy's SFP had a FSR of 2.00 GHz, so that modulation at 8, 10, and 12 GHz was not detectable, for example. Figure 11.12 shows the spectral display of the SFP, showing the carrier (and its alias, 2.00 GHz away). Also shown are the phase modulation sidebands from 11.5 GHz modulation. While they appear to be 1.5 GHz from the carrier, they are actually $1.5 + 5 \times 2.00$ GHz $= 11.5$ GHz due to the aliasing from the SFP. The actual phase modulation may be obtained from the equation

$$P_{\text{sideband}} = \phi^2 P_{\text{carrier}}/4 \tag{11.7}$$

or

$$\phi = 2(P_{\text{sideband}}/P_{\text{carrier}})^{1/2}. \tag{11.8}$$

Figure 11.13a shows the modulator performance obtained in radians/W$^{1/2}$ as a function of frequency. The "gaps" in the data are around integral multiples of

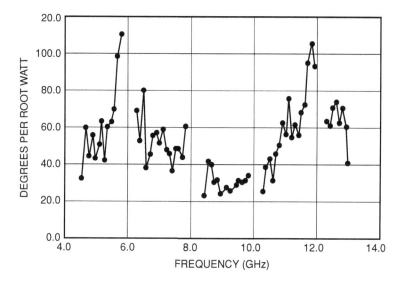

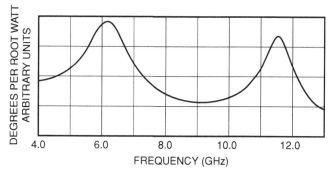

Figure 11.13. Phase modulation as a function of modulation frequency for the five antenna X-band modulator. The design frequency was 12 GHz, where the antennas are two-half-waves in phase and the transmission line sections are a half wavelength long. The response at 6 GHz is the resonance where the antennas are simple half-wave dipoles and the transmission line segments are one-quarter wave long. The upper plot shows experimental data and the lower curve is from a simple transmission line theory. From Refs. 11, 13.

2.00 GHz, where the "skirts" of the carrier signals caused by the finite resolution of the SFP mask the sidebands. As described above, the original modulator design had been for the antennas to be two half-waves in phase connected to a half-wave transmission line at 12 GHz. The antenna-plus-transmission-line segment forms a resonant structure, as the antenna should exhibit a high drive-point impedance at resonance, and the open-circuited transmission line a half wavelength long transforms this high impedance to a high impedance where it connects to the antenna. The data in Fig. 11.13a show just such a resonant behavior. A second resonant peak was also measured at about one-half the frequency, or 6 GHz, where the antenna now looks like a center-fed half-wave dipole (low drive-point impedance), and the transmission line becomes a quarter wavelength long (transforming the open circuit into a short circuit where it connects to the antenna). To measure these lower frequency points, the WR-90 feed structure was replaced by a WR-137 waveguide feed, which no longer matched the modulator size and aspect ratio.

Figure 11.13b shows a theoretical prediction from a simple transmission line model of the antenna's drive-point impedance coupled to the transmission line segment [11]. Since there is 100% reflection from the open circuit on the end of the transmission line segment, there are traveling waves in both directions on the segment. The wave traveling in the same direction as the optical signal interacts more strongly than the wave traveling in the opposite direction, since the velocity mismatch for that wave is much greater. However, the interaction with the backward-traveling wave is not zero. Sheehy included the total modulation from both traveling waves [11]. The qualitative agreement with experiment seems quite good. The modulation decreases to a minimum of about -7 dB at 9 GHz. Based on the agreement, we could project at least octave bandwidth performance with more sophisticated broadband antenna designs replacing the simple dipole.

These experiments used a 633 nm helium–neon laser as the optical source. The optical waveguides were single mode. The laser power was kept in the order of 100 to 200 microwatts to preclude photorefractive effects (which were seen at the milliwatt level).

11.9 Millimeter-wave modulator experiments at Caltech

11.9.1 A 60 GHz phase modulator

Armed with confirmation of the basic idea of antenna-coupled modulators, Sheehy and his co-workers scaled the X-band experiment described above to 60 GHz [15]. Figure 11.14 shows the antenna and transmission line segment layout and dimensions used on the lithium niobate modulator chip. Again, a simple

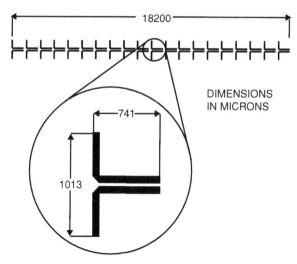

Figure 11.14. Antenna array used on the 60 GHz phase modulator. Twenty antennas were used with a total overall length of 18.2 mm. From Ref. 15.

single-waveguide phase modulator was fabricated. The transmission line spacing and conductor width remained the same, 8 μm and 50 μm, for a Z_0 of 35 Ω. The dipole antenna was 1.013 mm end-to-end and the transmission line segments were 0.741 mm long. The length of half the dipole plus the transmission line was thus 1.25 mm, while the wavelength at 60 GHz and $n_m = 3.8$ was 1.31 mm. Twenty antenna-plus-segment elements were used, covering a length of 18 mm along the waveguide. Evaporated aluminum was used for these electrodes.

A different feed structure was designed for this modulator. Whereas the WR-90 waveguide dimensions had nicely matched the X-band modulator chip dimensions, now the WR-15 waveguide dimensions (3.76 mm by 1.88 mm) did not match the modulator antenna array dimensions (18 mm by 1 mm). Accordingly, a new feed was developed and is shown schematically in Fig. 11.15. A tapered slab dielectric waveguide 1 mm thick and made of Teflon ($\varepsilon_r = 2.1$) was used to couple the modulation wave. A 1 mm thick wedge of lithium niobate was cut to allow the wave to enter the lithium niobate region at normal incidence. A thin quarter-wave matching section of Stycast® $\varepsilon_r = 9$ was interposed between the tapered slab and lithium niobate wedge. With this arrangement, the modulation signal had transverse dimensions of approximately 18 mm by 1 mm at the bottom surface of the modulator substrate (also 1 mm thick).

The results for this phase modulator are shown in Fig. 11.16. The highest value obtained was about 80 degrees per root watt, which was about twice what was expected from scaling the dimensions and losses from the X-band modulator. This was attributed to better coupling from the feed. The large variations in performance

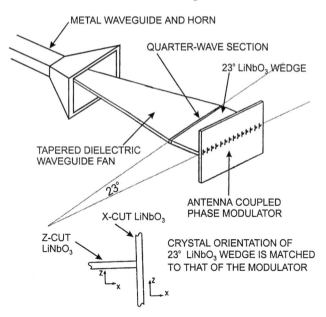

Figure 11.15. Feed arrangement for the 60 GHz modulators, showing the tapered dielectric fan (Teflon) from the WR-15 waveguide and metal horn antenna. A quarter-wave slice of $\varepsilon_r = 9$ material was used to reduce reflections at the interface with the lithium niobate wedge. From Ref. 15.

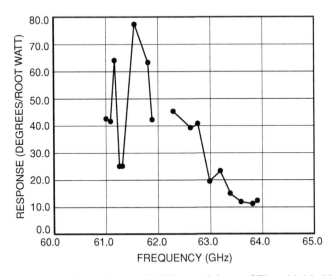

Figure 11.16. Phase modulation for the 60 GHz modulator of Figs. 11.14, 11.15. The data gap at 62 GHz results from masking the sideband by the carrier alias every 2.00 GHz. The extreme variation with frequency is thought to be an artifact introduced by the E-H tuner in the waveguide feed. From Ref. 15.

with frequency were unexpected. After the experiment was torn down for the next measurement, it was determined that the rapid variations were very likely due to an E-H tuner that had been used to "tune out" the reflections from the pointed end of the tapered dielectric waveguide at one frequency, but not re-optimized for each measurement. Unfortunately, the 60 GHz klystron used to make these measurements burned out during the next experiment, so it was impossible to go back and repeat the measurements.

A second 60 GHz modulator was also reported by Sheehy *et al.* [15]. This modulator used Mach–Zehnder optical waveguides, designed for 1.3 μm laser operation, and was similar to the 94 GHz Mach–Zehnder modulator described below. While awaiting delivery of a 1.3 μm laser, the modulator was initially tested at 633 nm, where the optical waveguides were multimode, and good intensity modulation contrast was impossible to obtain. It was during these measurements that the 60 GHz klystron burned out. Sheehy was able to recover some phase modulation results from the limited data (since only one arm of the MZ was modulated), and the results were consistent with the previous data from the 60 GHz phase modulator.

11.9.2 A 94 GHz Mach–Zehnder modulator

A modulator was designed for 94 GHz employing the electrode mask shown in Fig. 11.17. Bow-tie dipole antennas are coupled to short segments of the transmission line. The dipole antennas have a tapered length of 0.69 mm side to side, and are coupled to transmission line segments 0.4 mm in length. The same transmission

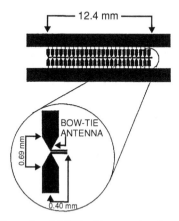

Figure 11.17. Electrode mask used at 96 GHz. Twenty-five bow-tie antennas are distributed over a total length of 12.4 mm. The effective radiating length of the antennas is thought to be only the "bow" portion, with the wide extensions "cold". Connection to these cold extensions was made through thin traces to wide busses to which bias voltage could be applied. From Ref. 11.

line widths of 50 μm separated by 8 μm were used. Twenty-five of these antenna-plus-transmission line segments were spread out over a 12 mm active length along one arm only of the MZ optical waveguide, since the two-conductor transmission line used had a spacing much less than the spacing between optical waveguides. (A three-conductor line like those commonly used with MZ modulators would have required some sort of insulated conductor crossover at the connection to the antenna, and we did not wish to tackle that particular problem in this first MZ experiment.) An *x*-cut substrate was used. This modulator was designed for 1.3 μm laser operation. A reduction in performance was anticipated because the bow-tie antennas are not supposed to be resonant and are matched (approximately) to the transmission line over a wide range of frequencies. Thus the antenna re-radiates the wave reflected from the open circuit at the far end of the transmission line segment independently of the modulation frequency.

According to the accepted theory of bow-tie antennas on dielectric substrates [22] the ends of the bow tie are electrically "cold" at all but the lowest operating frequency, so that they may be extended without changing the radiation characteristics at higher frequencies. This was done here, and narrow traces added to connect all the dipoles on each side of the transmission line segments together in a DC fashion so that the proper DC bias could be applied to reach the linear intensity modulation point of 50% transmission, which results in simultaneous phase and intensity modulation of the output.

The feed structure for this antenna was similar to that shown in Fig. 11.15, except that the feeding waveguide was WR-10 (2.54 mm by 1.27 mm). A photograph of this modulator (actually, two modulators on the same lithium niobate wafer) is shown in Fig. 11.18. An optical fiber is butt-coupled to provide the input, and a microscope objective is used to focus the output on the SFP. The small coils of wire are the DC bias connections. Figure 11.19 shows the feed structure more clearly.

The results for this modulator are shown in Fig. 11.20, given in terms of m^2/W, where m is the usual intensity modulation index. Again, there is missing data at 96 GHz because the SFP used with the 1.3 μm laser had a free spectral range of 8.00 GHz, and $96 = 12 \times 8.00$. Otherwise, the response seems reasonably flat over the tuning range of the klystron, 91 to 98 GHz. (No E–H tuner was used in the feed waveguide). The performance of this modulator agrees well with scaled performances of the 10 and 60 GHz modulators, accounting for increased skin-effect loss in the electrodes and the smaller length.

11.9.3 A 94 GHz directional coupler modulator

We wished to see if the antenna coupling scheme would work with directional coupler modulators as well as simple phase modulators and Mach–Zehnder intensity modulators. An optical directional coupler consists of two optical waveguides side

Figure 11.18. Photograph of the 94 GHz Mach–Zehnder modulator set up. Two modulators are on this lithium niobate chip. The optical input is from the left through a butt-coupled fiber. The microscope objective couples the output to the scanning Fabry–Perot. The wire coils are the bias connections.

by side and spaced very closely, so that the optical fields that die exponentially outside of the guide actually penetrate the other guide slightly. (See, for example, Ref. 23.) The optical wave in one guide can then "transfer" power to the other guide by gradually exciting the mode in that guide. If the two guides have exactly the same propagation constant, β, then the transfer can be 100% after a distance L_{transfer} down the pair of guides. If the length of the coupler is made still longer, then the optical power can transfer back to the original guide, also up to 100%. Thus by controlling the length and the optical field overlap (expressed as a coupling coefficient), power at the input to one guide may be split in any proportion between the outputs of the two guides. To make a modulator (or a switch) out of such a coupler, we place electrodes over the two waveguides so that we can induce a small change in the refractive index within the guides and hence a small change $\Delta\beta$ in the β's of the two guides. We arrange the electrodes so that they induce opposite signs of this change in the two guides, so there is a net $\Delta\beta_e$ resulting from the applied voltage. This change will change the transfer of power between the two guides. If the directional coupler is designed (or biased) to produce a 50%–50% split at the output, then an applied modulation voltage will produce complementary increases and decreases in the transmitted light intensity. This device is known as a directional coupler modulator (DCM) or a "$\Delta\beta$" modulator (or switch, if the applied

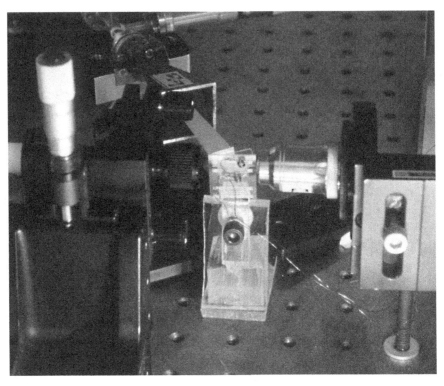

Figure 11.19. Photograph of the same setup as Fig. 11.18, but showing the feed horn and fan dielectric waveguide behind the modulator.

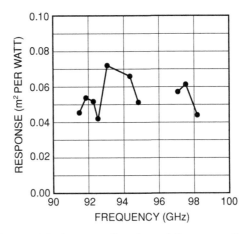

Figure 11.20. Intensity modulation as a function of frequency for the 94 GHz Mach–Zehnder modulator with bow-tie antennas of Fig. 11.18. The data gap at 96 GHz results from the 6.00 GHz free spectral range of the SFP used in this measurement.

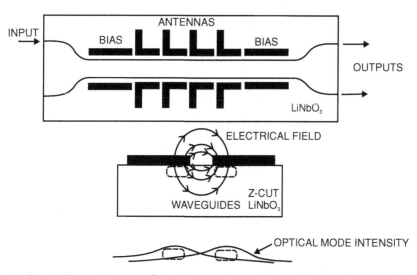

Figure 11.21. Schematic layout of the directional coupler modulators, showing the directional coupler waveguides, two DC bias sections, and four antennas and transmission line segments. The figure also shows the electric field orientation through the waveguides and the optical field intensity across the two coupled waveguides.

voltage is large enough to change the output completely from one guide to the other).

Sheehy designed and fabricated such a modulator [11] using dimensional and coupling coefficient data from Gaeta at the Hughes Research Laboratories [24]. The antenna and transmission line layout is shown schematically in Fig. 11.21. Simple resonant dipoles plus transmission line segments were used for this modulator experiment rather than the bow-tie antennas. DC electrodes upstream and downstream from the antenna array were used to fix the proper operating point of the directional coupler at a 50%–50% split of optical power between the two output waveguides. Since the two optical waveguides of the directional coupler were so close together (6 μm spacing) the transmission line electrodes could be placed directly on top of them. A z-cut substrate had to be used in this case, since the fields directly under the transmission line are roughly vertical (Fig. 11.2b).

Again, two-half-waves-in-phase dipole antennas were used, 0.656 mm tip to tip. The transmission lines were 0.44 mm in length, as shown in Fig. 11.22. The same conductor width of 50 μm and transmission line spacing of 6 μm were used. Figure 11.23 shows the complete mask for 12 modulators, with the antenna array and the DC electrodes. The modulators were fabricated at HRL by Schaffner.

Sheehy made preliminary measurements on this modulator, but was unable to make sense out of the transfer function obtained by applying DC to the bias electrodes. As he states in his thesis, "Since we did not understand its DC bias behavior, it was hopeless to attempt to measure its amplitude modulation performance at

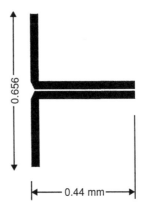

Figure 11.22. Detail of the dipole and transmission line segments used in the 94 GHz directional coupler modulator. From Ref. 11.

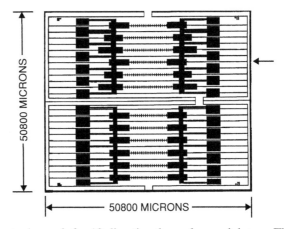

Figure 11.23. Electrode mask for 12 directional coupler modulators. The arrow indicates the particular modulator used in the experiment. The large blocks at either side are the bias pads for the DC electrodes. Adjacent modulators share bias connections. The arrow indicates the particular modulator used to obtain the results reported here.

94 GHz." Sheehy received his Ph.D. in June 1993, and his modulator chips were given to a Caltech undergraduate, Uri V. Cummings, to unravel the mystery as a summer research project. After considerable measurement, he did. The "mystery" is illustrated in Fig. 11.24, the transfer function showing the output of the crossover arm of the directional coupler as a function of the DC bias voltage on the input DC electrode section. The solid curve shows the transfer function expected from theory, which is symmetrical about zero bias. If the directional coupler length, or coupling coefficient is changed, the maxima and minima and shape of this function will change, but it will always be symmetrical about zero bias voltage. The experimental points show a function that is quite asymmetrical, however. Cummings was able to fit the experimental points with a theoretical transfer function as shown in

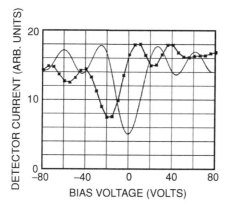

Figure 11.24. Data points measured by applying DC to one set of bias electrodes on the 94 GHz directional coupler modulator. The solid curve is a theoretical calculation for one set of assumed directional coupler design parameters. The important feature here is that the data points are asymmetrical about zero volts, while any assumed design parameters would result in a symmetrical curve about zero volts.

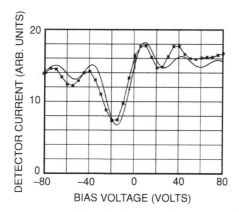

Figure 11.25. The same data as in Fig. 11.24, but with a curve fit by assuming that the two optical waveguides are not identical, so that there is a "built in" $\Delta\beta$.

Fig. 11.25 by assuming that the directional coupler was itself asymmetrical. That is, there was a built-in $\Delta\beta$, not just an electro-optically induced $\Delta\beta$. We did not know the origin of the asymmetry at the time, but it came to light later, when we were inspecting a copy of the optical waveguide mask used in fabrication. The original mask had developed defects, in which semi-circular regions of metal at the sides of the waveguide had chipped or peeled off. These are called "mouse-bite" defects, since the mask looks like a tiny mouse had taken a bite out of the metal coating. We then noted that these defects had been faithfully replicated in the modulator waveguides. Thus, there were several "bulges" in both optical guides along their length. This was the source of the asymmetrical β in the two guides, and also the source of the rather poor optical transmission of the modulator (about -20 dB optical loss).

Armed with an understanding of what the transfer function of the DC bias electrodes was in actuality, Cummings was able to determine the correct biases to apply
to place the transfer function for the antenna array at the 50%–50% split point, and
make a mm-wave measurement. The SFP method was used to detect the modulation
sideband to carrier ratio of 1:600 (power) with 37 mW of power out of the klystron
at 91.2 GHz. This gives a value of m^2/W of 0.18 [16,25]. We may compare this
result with ordinary modulators by noting that m^2/W for a Mach–Zehnder biased
at its linear point is

$$m^2/W = \left(\pi^2 Z_0 / 2 \, V_\pi^2\right). \qquad (11.9)$$

This gives an equivalent V_π of about 31 V for an assumed Z_0 of 35 Ω. Of course,
this modulator still had -20 dB optical transmission. The "mouse bite" origin
of the directional coupler asymmetry was not known at the time the mm-wave
performance was obtained. Cummings remained at Caltech as a graduate student
and set out to improve the performance of mm-wave DCMs as a portion of his
doctoral research. However, the subsequent history of this modulator is a tale of
frustration in several aspects, too sad to recount here.

11.9.4 The slot Vee Mach–Zehnder modulator

In order to improve the performance of the antenna-coupled modulator, Sheehy
considered the possibility of using antennas with more directional gain than simple
dipoles could offer. In free space, the gain of a simple half-wave dipole is about
2.1 dBi (dB with respect to a fictitious isotropic radiator), and the gain of a two-
half-waves-in-phase dipole is about 3.9 dBi, or about 1.8 dB better than a simple
dipole. While the corresponding gains are not known for antennas on a dielectric
half-space, we imagine they will have about the same relative ratio. Increasing the
length of the dipole elements beyond 1.25 wavelengths overall actually reduces
the gain in the plane normal to the wire (Ref. 4, Section 12.6). However, the gain
may be increased for longer dipoles by bending them in to a "V" shape (Ref. 4,
Section 12.10). Using the superposition method he had developed, Sheehy analyzed several cases of "V" (or Vee) antennas of different lengths and angles on a
dielectric interface, and calculated the directional performance [11] at the preferred
direction into the substrate (24.5 degrees). While the results were encouraging, no
modulators were fabricated with Vee antennas because Sheehy had also hit upon yet
another idea for an antenna with gain, one that changed the whole geometry of the
modulator.

It is easy to understand why antennas at a dielectric interface have a null in their
radiation pattern in the directions parallel to the interface. Imagine a plane wave

radiated in that direction. Part of the wave would have to travel with a phase velocity c, appropriate to the air side, while the other part would travel at a phase velocity c/n, where n is the refractive index of the dielectric. Such a situation is impossible. For the phases to match at the interface, the waves must travel in different directions on the two sides of the interface, as given by Snell's law. However, if the interface between the two dielectric regions is coated with a metal, then the waves on the two sides are isolated from one another, and waves can propagate that have finite amplitude in the direction parallel to the metal surface. In fact, many antennas will have a maximum response in this direction. The only question now is how to make a planar antenna in the metal layer between the two dielectrics. The answer is to make a "slot" or cut in the metal. One can think of making a long, thin cut in the metal to form a "slot dipole." Such an antenna has characteristics that are the complement or dual of the more usual dipole (Ref. 4, Sections 12.16, 12.24).

Rather than design a modulator with slot dipole antennas, Sheehy envisioned an array of "slot Vee" antennas, that would be directive parallel to the modulator sufrace and would also have significant gain. (See Section 11.10 below for modulators employing slot dipoles.) He envisioned using the lithium niobate modulator substrate as a dielectric slab waveguide to illuminate an array of such slot Vee antennas. To realize the proper time delay from antenna to antenna, he proposed tilting the illuminating modulation wavefront in the plane of the modulator substrate at the proper angle to the optical waveguide axis, this time using the effective phase velocity of the slab dielectric waveguide to find the proper angle. If we assume that the modulation wave is "well confined" to the slab waveguide, then it is travelling at $c/5.3$, so the proper tilt angle is $\sin^{-1}(2.2/5.3) = 24.5$ degrees, as before. If the wave is not well confined to the slab waveguide, then the phase velocity will be greater than $c/5.3$, and the tilt angle will be greater than 24.5 degrees.

Sheehy realized he had to orient the slot Vee antennas in the direction of the incoming signal. The resulting configuration for antennas and a transmission line segment is shown in Fig. 11.26. (The second antenna on the other end of the transmission line re-radiates the signal and becomes, in effect, a matched load). Of course, the antennas are now actually open spaces in a continuous metallic coating on the lithium niobate substrate. Figure 11.27 compares the electric field distribution and radiation direction for the Vee and slot Vee antennas. The Vee is connected to a two-strip transmission line, while the slot Vee requires a "two-gap" line, or a single strip line with ground planes on either side. Note that a single optical waveguide could be excited with either z-cut or x-cut substrates by placing it where the vertical or horizontal electric fields are strongest.

Sheehy's original modulator concept is shown in Fig. 11.28, with a half-horn antenna exciting the dielectric slab waveguide through a tapered fan. (Dielectric slab waveguide on a metallic surface is usually called "image line," since only the

20 GHz SLOT VEE ANTENNA ELEMENT

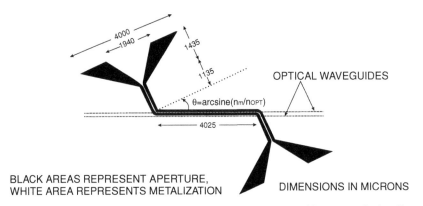

Figure 11.26. Design and dimensions of a slot Vee antenna and its transmission line segment. A second slot Vee antenna is used to re-radiate the power, thus acting as a matched termination for the transmission line segment. The dark areas are openings in a continuous metal surface. From Ref. 28.

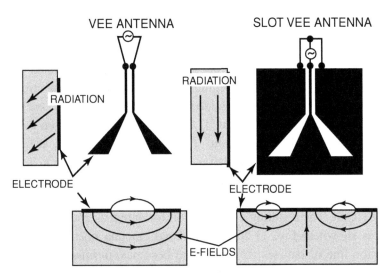

Figure 11.27. Comparison of the electric field distribution around the antenna elements in a conventional Vee antenna and a slot Vee antenna on a dielectric substrate. The conventional Vee is horizontally polarized and radiates down into the substrate at an angle. The slot Vee is vertically polarized and radiates primarily parallel to the surface. From Ref. 28.

slab waveguide modes with mirror symmetry are allowed.) The metal plate must be machined away under the slot antennas so that it does not short out the fields that penetrate into the air side of the evaporated metal coating. By using tapered slots in the slot Vee elements, it was presumed that the antenna bandwidth would be large (since this is, in a sense, the bent dual of the "bow-tie" antennas discussed

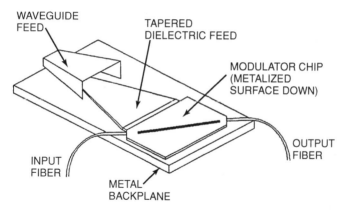

Figure 11.28. Overall concept of the slot Vee modulator by Sheehy. The modulator and its tapered dielectric feed are mounted on a metallic ground plane, thus forming image guide. The metal coated surface of the lithium niobate is on the bottom side, making contact with the metal backplane. There is a hole cut in the metal backplane under the antenna array, so the fields on the air side of the antennas are not shorted out. A broadband waveguide horn excites the tapered feed, and an absorbing material along the lower right edge of the modulator chip absorbs the radiation. From Ref. 11.

above). This broadband performance had already been confirmed experimentally by Moussessian [26]. There remained the problem of the reflection from the open end of the transmission line segment. This Sheehy proposed to solve by terminating the line in another slot Vee antenna, which would re-radiate the modulating signal back into the dielectric image line, with an eventual absorption at the far edge of the modulator substrate chip.

However, Sheehy went only as far as describing this modulator in his thesis [11]. An incoming graduate student, Lee Burrows, took up the challenge of actually building it and making it work, a task that turned out to be much harder than anticipated. Burrows' design for the actual antenna and transmission line segment is shown in Fig. 11.26 for a slot Vee antenna at approximately 20 GHz. The dimensions were determined to provide an antenna radiation resistance that matched the transmission line impedance. Note that the conductors of antenna and transmission line are all connected and thus at the same DC potential. No DC bias can be applied to the transmission line connected to the slot Vee antenna. A separate, isolated transmission line segment must be added somewhere along the transmission line for the DC bias. Figure 11.29 shows that 50 mm square electrode mask used for the fabrication of five modulators, two each at 94 and 44 GHz, and one at 20 GHz. The bias connection pads to short lengths of isolated center conductor can be seen near the left end of each modulator. Mach–Zehnder waveguides were fabricated on z-cut lithium niobate by Schaffner at HRL Laboratories, and the electrode mask in Fig. 11.29 was used by Burrows to define the antennas and

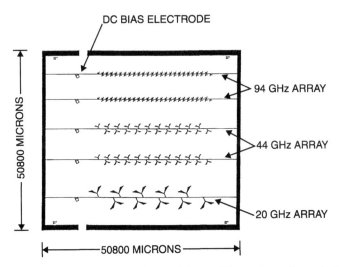

Figure 11.29. Mask used to make slot Vee modulators. The regions shown in black become the spaces in the otherwise gold coated surface of the lithium niobate chip. Bias electrodes approximately 6 mm long are connected to the square pads shown. From Ref. 28.

transmission line segments in evaporated gold. The transmission line is oriented so that it overlays both arms of the Mach–Zehnder to provide the desired anti-symmetrical excitation.

The 20 GHz modulator was chosen for the initial experiments. Several attempts were made at realizing the image-guide fan transition envisioned by Sheehy and shown schematically in Fig. 11.28. This proved impractical for an unanticipated reason: it seemed practically impossible to insert the thin quarter-wave matching section on the image guide. What had been relatively easy to do in a full dielectric slab guide (where the edges of the quarter-wave section were allowed to extend beyond the transverse dimensions of the slab) proved difficult here, since the quarter-wave section had to fit tightly against the metal image plane as well as against the Teflon and lithium niobate slabs. After trying several other schemes (including making a fan out of lithium niobate, another unsuccessful idea), we finally decided to make the transition from standard waveguide (in this case, WR-42) to lithium niobate inside the metal waveguide, then use a lithium niobate image-guide fan to expand the wave to the length of the array on the modulator. A long taper in empty metal guide was used to reduce the height of 4.32 mm for WR-42 to 1.00 mm to match the thickness of the lithium niobate guide. Then a tight-fitting plug of $\varepsilon_r = 4$ Stycast[®], cut to five quarter-wavelengths in the reduced height metal waveguide was inserted in the guide. Finally, a close-fitting piece of lithium niobate was inserted to extend to the end of the guide, where it was butt coupled to the end of the lithium niobate fan. Two such sections were fabricated so that back-to-back transmission

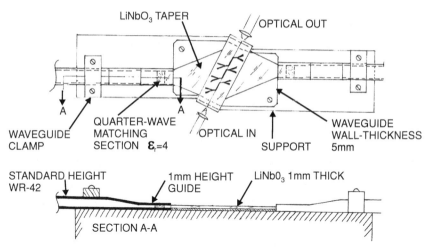

Figure 11.30. Overall assembly drawing of the slot Vee modulator and its feed system. Standard WR-42 waveguide is reduced in height to 1 mm through a taper. A plug of $\varepsilon_r = 4$ material five quarter-waves long (measured in the reduced height, dielectric loaded guide) is inserted, followed by a plug of lithium niobate. At the end of the guide, a lithium niobate fan serves as a dielectric image guide to expand the radiation in width to match the modulator chip, which has the slot Vee antennas on the side in contact with the metal ground plane. (There is a hole in the ground plane under the antennas.) A mirror image transition couples back into WR-42 waveguide so that RF loss measurements in this entire structure could be made. During operation, this guide is terminated in a matched load. From Ref. 28.

measurements could be made to determine the loss due to the various components. The overall assembly is shown in Fig. 11.30.

RF transmission measurements were made flange-to-flange over the range 18 to 26 GHz with a microwave network analyzer. With the two metal waveguides butted together, omitting the fans and the modulator, we measured 2–8 dB flange-to-flange loss through the entire system over the range 20–26 GHz. With the two fans added but the modulator omitted, the loss increased to 12–23 dB. Finally, with the modulator inserted the loss increased to 17–30 dB. We speculate that the major cause of the increase with the modulator was a poor contact between the evaporated gold coating on the modulator and the aluminum base plate on which it was mounted, since the loss seemed quite sensitive to vertical pressure.

The loss decreased to 16–24 dB when a metal plate was placed on top of the image guide. The presence of this metal plate effectively converted the image guide into a parallel-plate waveguide loaded with lithium niobate. In theory, this should increase the loss, since the fields now see losses in two metal surfaces instead of one, compared to the "lossless" upper dielectric surface. The rapid variations of transmission with frequency indicated that the main problem was reflections from the various dielectric boundaries interfering, rather than the waveguide tapers or five-quarter-wave matching plugs. We concluded that our losses also included radiation from the discontinuities (e.g., the butt joints) and that addition of the second metal plate reduced these radiation losses more than they increased the resistive losses in the metal. Clearly, one should be able to reduce these losses with more care in piecing the various sections of lithium niobate together. In any case, we estimate that the transmission loss from the WR-42 input flange to the antennas on the modulator was 11.3 dB at 23.47 GHz, the frequency where the modulator was eventually evaluated.

Modulation measurements were made over a much narrower range of frequencies, since we needed more power to detect optical sidebands than we could get from a laboratory signal generator. A Varian VA-98M reflex klystron was used at 23.47 GHz to drive the modulator with 290 mW. A resulting modulation sideband is shown in Fig. 11.31, on the skirt of a carrier alias. The resulting value of m^2/W is 0.17, which is equivalent to a V_π of 38 V in a conventional modulator with an impedance level of 50 Ω. (Refs. 27, 28. The value of 76 V cited in Ref. 27 is in error.) Unfortunately two of the five slot Vee antennas shown on the 20 GHz mask of Fig. 11.29 actually had shorted transmission lines. If we correct the measured results by assuming the active length was only 3/5 of 20mm, then the comparable V_π would be 23 V. If we further correct for the estimated flange-to-modulator RF transmission loss of 11.2 dB, then m^2/W becomes 2.2, and the equivalent V_π becomes 10.6 V. This number is quite reasonable compared to the DC V_π of about

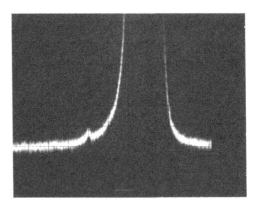

y = 5 mV/div   x = 50 MHz/div

Figure 11.31. Oscilloscope photo of the scanning Fabry–Perot output, showing one sideband and one carrier alias. Since the exiting Nd-YAG laser was not single mode, the free spectral range of the SFP was quite cluttered. For that region, only a small fraction of the FSR is shown so that the sideband is seen clearly. The sideband to carrier amplitude ratio was 1:325 in this picture. From Refs. 27, 28.

20 V measured for the DC bias section, which is only 6 mm long. The length of transmission line connected to three antennas is about 12 mm, so that a V_π of 10 V is about what we would expect. The indication is that the antennas and modulator are working properly, and are properly phased, and the RF input loss in the coupling is the source of poorer than expected performance. Of course, any real coupling scheme will have some RF loss.

In addition to all the other problems in getting this slot Vee modulator to work, it turned out that the Mach–Zehnder waveguides were multimode at 1310 nm, so that only about a 6 dB contrast ratio could be obtained. We anticipate that the modulator would have had a better performance if the waveguides had been single mode.

No measurements were made with the 44 and 94 GHz modulator chips (which, unfortunately, also have multimode waveguides).

11.10 Other antenna-coupled modulators

An antenna-coupled modulator of a somewhat different type was developed and demonstrated by workers at ATR Optical and Radio Communication Laboratories in Japan. Penard, Matsui, and Ogawa described their work at an SPIE meeting in 1994 [29]. Their modulator is similar to the Caltech approach in that a linear array of dipoles and transmission line segments is deployed along the integrated optical waveguide to obtain phase-matched operation. They use slot dipoles rather than ordinary dipoles, so that the upper surface of the lithium niobate is largely covered by a conducting layer, the same as for the slot Vee modulator described in Section 9.4.10. The difference is that the array is illuminated by a plane wave *perpendicular* to the array rather than at an angle. This results in infinite phase velocity from antenna to antenna, so that the antenna array is phase-matched to the modulation wave only for frequencies where the modulation phase shift from antenna to antenna at the optical waveguide velocity is an integral multiple of 360 degrees. Thus the spacing from antenna to antenna is fixed at d, given by

$$d = c/f_m n_o, \tag{11.10}$$

where f_m is the modulation frequency. The bandwidth of the response around this frequency is given by the familiar antenna array factor:

$$\text{Response}(f) = \sin(N\pi f/f_m)/\sin(\pi f/f_m). \tag{11.11}$$

The bandwidth around the modulation center frequency thus decreases as the number of segments N increases. A sample design is given in Ref. 29 for a modulator with a center frequency of 25 GHz with $N = 6$, $d = 5.6$ mm and $L = 2.6$ mm.

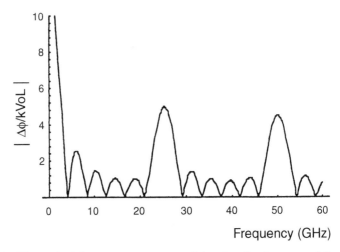

Figure 11.32. Theoretical frequency response of the slot dipole array of Penard, Matsui, and Ogawa, due to the array illumination factor alone. The frequency response of the slot dipoles alone is not included. From Ref. 29.

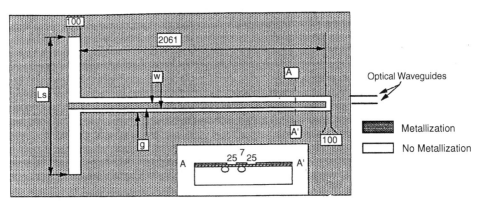

Figure 11.33. A simple linear slot dipole and its transmission line segment. Dimensions are in micrometers. The inset shows the orientation of the transmission line over the optical waveguides. From Ref. 29.

(Note that their segment length is considerably shorter than the distance between antennas; less than 50% of the array length is used for modulation.) The theoretical frequency dependence of this array factor is shown in Fig. 11.32. The frequency dependence of the individual antenna-plus-transmission-line-segments is not included.

Several different slot antennas are described in Ref. 29, and their electrical characteristics are discussed. Figures 11.33–11.35 show three of these antennas: a simple slot dipole, a square patch antenna, and three coupled slots of slightly different length. Experimental results were given for modulators using the simple slot dipole and the coupled slots dipole, as well as a "phase reversal" array of simple slot

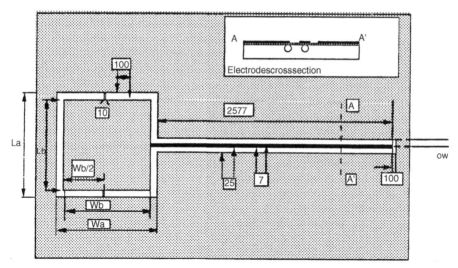

Figure 11.34. A square slot dipole and its transmission line segment. Dimensions are in micrometers. The inset shows the orientation of the transmission line over the optical waveguides. From Ref. 29.

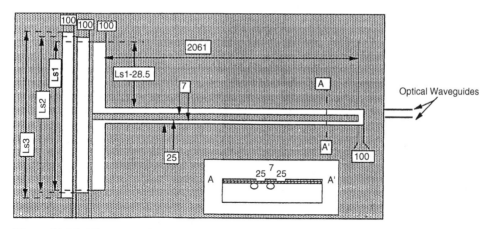

Figure 11.35. Three coupled slot dipoles with a transmission line segment attached to one slot. Dimensions are in micrometers. The inset shows the orientation of the transmission line over the optical waveguides. From Ref. 29.

dipoles that are staggered over the optical waveguides as shown in Fig. 11.36. In this case, the dipoles should be spaced at half the distance given in Eq. (11.10).

The measured frequency response is shown in Fig. 11.37. The coupled slots antenna array seems to give about 10 dB better performance than either the array of simple slot dipoles ("Linear slot 1") or the phase reversal array using simple slot dipoles ("Linear slot 2"). In all cases, the slot array was illuminated by a horn antenna with 23 dB gain located 35 cm from the modulator, with a transmitted power level of +25 dBm. The authors estimate that the power actually delivered to

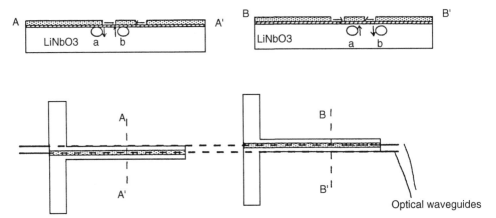

Figure 11.36. Three antenna-plus-transmission line segments arrayed along the optical waveguides in a Mach–Zehnder modulator. The segments are staggered with respect to the optical waveguides to effect a phase reversal modulator configuration. From Ref. 29.

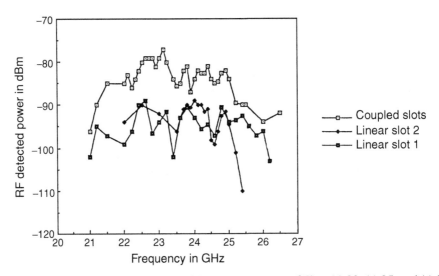

Figure 11.37. Measured performance of the antenna arrays of Figs. 11.33, 11.35, and 11.36. The curve marked "Linear slot 2" was the phase reversal modulator. From Ref. 29.

the modulator was about −6 dBm. The modulator was driven by 4 mW of optical power and exhibited 8 dB optical loss. A high speed photodetector (40 GHz) was used as the intensity detector. The simple slot dipole array used 6 antennas spaced 5.6 mm, the phase reversal array used 11 antennas spaced 2.8 mm, and the coupled slots antenna array used 6 antennas spaced 5.6 mm.

In light of Fig. 11.8, one might wonder how you could illuminate this modulator from the air side and have any sensitivity at all, since the antennas are strongly directional into the dielectric rather than into the air. In this experiment, the modulator substrate was placed on a second conducting plane, which acts as a mirror. Thus the dipoles have no choice but to be sensitive to radiation from the air side. Figure 11.8 does not apply in this case. Of course, there is now an additional phase shift from the double transit through the thickness of the modulator substrate, but this is the same phase shift for all antennas. This phase shift will have a strong influence on the impedance of the slot dipole, however.

11.11 Summary and suggestions for future projects in antenna-coupled modulators

Antenna coupling has been successfully demonstrated in a variety of modulator configurations with a variety of antennas. Whether or not this technique will prove valuable in applications in the future will depend on the specific requirements of the system that uses the modulator and the state of the art for competing alternatives. Antenna coupling is inherently a band-pass technique, but should result in bandwidths greater than the serial periodic techniques such as phase reversal and space harmonic electrode techniques. And it is clear that as low pass modulators using true velocity matching are further developed for high microwave and mm-wave use that the bar will be raised for applications of antenna coupling. To be fair, additional engineering development of antenna-coupled modulators, particular the wave coupling feeds should be undertaken to see how good such modulators can become. One fruitful direction might be to extend Sheehy's work on conventional Vee antennas, or other higher gain antennas for use in the array.

The primary advantage in the antenna coupling schemes described here is in overcoming high transmission line loss at millimeter frequencies. And the higher the modulation frequency, the bigger the advantage of antenna coupling. It would be nice to demonstrate this by making a modulator with a modulation frequency in the sub-millimeter wave or far infrared (FIR) range. Lithium niobate remains a low-loss material well below 1 mm. And the photolithography of the electrodes presents no problems. One could envision an experiment with a sub-millimeter or FIR gas laser source of 1 mm to 100 μm wavelength radiation as the modulation source, with simple cylindrical optics to spread the beam in one dimension to the 10–20 mm length of an antenna array. Detection of the optical sidebands with an SFP would work as well as it does at lower frequencies (except that the aliases are still determined by the free spectral range of the SFP, while the sidebands would be several hundred GHz from the carrier, so the frequency stability of the gas laser might become an issue). Thus all elements are available for the experiment, except the motivation. What use is a FIR-modulated optical signal? And who would pay to see it demonstrated or developed? These are questions to be answered by the next generation of researchers.

Acknowledgements

The author wishes to acknowledge the many contributions over the past decade by his students, Finbar T. Sheehy, Lee J. Burrows, and Uri V. Cummings, his technical assistant, Reynold E. Johnson, his associates James H. Schaffner and Adrian E. Popa at the Hughes Research Laboratories (now HRL Laboratories LLC), and Charles H. Cox, III at the MIT Lincoln Laboratories. The late Brian H. Hendrickson fought the budget battles for this work at Rome Laboratories and DARPA, and Norman H. Bernstein oversaw the technical ups and downs of this work through the years.

References

1. C. J. G. Kirkby, "Refractive index of lithium niobate, wavelength dependence", Sections 5.1 and 5.2 in *Properties of Lithium Niobate*, EMIS Datareviews Series No. 5, INSPEC, The Institution of Electrical Engineers, London, 1989, pp. 131–42.
2. A. R. Von Hippel, *Dielectric Materials and Applications*, The MIT Press, Cambridge MA, 1954, pp. 311, 402.
3. J. H. Schaffner, HRL Laboratories, LLC, private communication.
4. S. Ramo, J. R. Whinnery, and T. Van Duzer, *Fields and Waves in Communication Electronics*, 3rd Edn., J. Wiley, New York, 1994.
5. G. K. Gopalakrishnan, W. K. Burns, R. W. McElhanon, C. H. Bulmer, and A. S. Greenblatt, "Performance and modeling of broadband $LiNbO_3$ traveling wave optical intensity modulators," *J. Lightwave Technol.*, **12**, 1807–18, 1994.
6. O. Mitomi, K. Noguchi, and H. Miyazawa, "Design of ultra-broad-band $LiNbO_3$ optical modulators with ridge structure," *IEEE Trans. Microwave Theory Tech.*, **QE-43**, 2203–7, 1995.
7. D. W. Dolfi and T. R. Ranganth, "50 GHz Velocity-matched broad wavelength $LiNbO_3$ modulator with multimode active section," *Electron. Lett.*, **28**, 1197–8, 1992.
8. R. C. Alferness, S. K. Korotky, and E. A. J. Marcatili, "Velocity-matching techniques for integrated optic traveling wave switch modulators," *IEEE J. Quantum Electron.*, **QE-20**, 301–9, 1984.
9. J. H. Schaffner, "Analysis of a millimeter wave integrated electro-optic modulator with a periodic electrode," *Proc. SPIE OE-LASE Conference*, Los Angeles, CA, January 16–17, 1990, **1217**, 101–10.
10. J. H. Schaffner and W. B. Bridges, "Broad band, low power electro-optic modulator apparatus and method with segmented electrodes," U.S. Patent 5,291,565, March 1, 1994.
11. F. T. Sheehy, "Antenna-coupled mm-wave electro-optic modulators and linearized electro-optic modulators," Ph.D. Thesis, California Institute of Technology, June 1993.
12. W. B. Bridges, "Antenna-fed electro-optic modulator," U.S. Patent 5,076,655, December 31, 1991.
13. F. T. Sheehy, W. B. Bridges, and J. H. Schaffner, "Wave-coupled $LiNbO_3$ electrooptic modulator for microwave and millimeter-wave modulation," *IEEE Photon. Technol. Lett.*, **3**, 133–5, 1991.
14. W. B. Bridges and F. T. Sheehy, "Velocity matched millimeterwave electro-optic modulator," Final Technical Report RL-TR-93-57 on contract F30602-88-D-0026 with U.S. Air Force Rome Laboratories, May 1993.

15. F. T. Sheehy, W. B. Bridges, and J. H. Schaffner, "60 GHz and 94 GHz antenna-coupled LiNbO₃ electrooptic modulators," *IEEE Photon. Technol. Lett.*, **5**, 307–10, 1993.
16. W. B. Bridges, L. J. Burrows, U. V. Cummings, R. E. Johnson, and F. T. Sheehy, "60 and 94 GHz wave-coupled electro-optic modulators," Final Technical Report RL-TR-96-188 on contract F30602-92-C-0005 with U.S. Air Force Rome Laboratories, September 1996.
17. C. R. Brewitt-Taylor, K. J. Gunton, and H. D. Rees, "Planar Antennas on a Dielectric Surface," *Electron. Lett.*, **17**, 1981.
18. N. Engheta, C. H. Papas, and C. Elachi, "Radiation patterns of interfacial dipole antennas," *Radio Sci.*, **17**, 1557–66, 1982.
19. G. S. Smith, "Directive properties of antennas for transmission into a material half-space," *IEEE Trans. Antennas Propag.*, **AP-32**, 232–46, 1984.
20. M. Kominami, D. M. Pozar, and D. H. Schaubert, "Dipole and slot elements and arrays on semi-infinite substrates," *IEEE Trans. Antennas Propag.*, **AP-33**, 600–7, 1985.
21. D. B. Rutledge, D. P. Neikirk, and D. P. Kasilingam, "Integrated circuit antennas," in *Infrared and Millimeter Waves*, Volume 10, Academic Press, 1983, Chapter 1.
22. R. C. Compton, R. C. McPhedran, Z. Popovic, G. M. Rebeiz, P. P. Tong, and D. B. Rutledge, "Bow-tie antennas on a dielectric half-space: Theory and experiment," *IEEE Trans. Antennas Propag.*, **35**, 622–31, 1987.
23. R. V. Schmidt, "Integrated optics switches and modulators," in *Integrated Optics: Physics and Applications*, eds. S. Martellucci and A. N. Chester, Plenum Press, New York, 1981, pp. 181–210.
24. C. Gaeta, Hughes Research Laboratories, private communication.
25. U. V. Cummings, W. B. Bridges, F. T. Sheehy, and J. H. Schaffner, "Wave-coupled LiNbO₃ directional coupler modulator at 94 GHz," *Photonic Systems for Antenna Applications Conference* (PSAA-4), Monterey, CA, 18–21 Jan. 1994, Paper 4.3.
26. A. Moussessian and D. B. Rutledge, "A millimeter-wave slot-V antenna," IEEE AP-S International Symposium, July 18–25, 1992, Chicago IL, *Conference Digest*, Vol. 4, pp. 1894–1897.
27. L. J. Burrows and W. B. Bridges, "Slot-vee antenna-coupled electro-optic modulator," *Proc. SPIE Conference on Photonics and Radio Frequency II*, 21–22 July 1998, Vol. 3463, pp. 56–65.
28. W. B. Bridges, L. J. Burrows, and U. V. Cummings, "Wave-coupled millimeter-wave electro-optic techniques," Final Technical Report AFRL-SN-RS-TR-2001-37 on contract F30602-96-C-0020 with U.S. Air Force Rome Laboratories, March 2001.
29. E. Penard, K. Matsui, and H. Ogawa, "Intensity modulation of LiNbO₃ electro-optic modulator by free space radiation coupling," *Proc. SPIE Conference on Optoelectronic Signal Processing for Phased-Array Antennas IV*, 26–27 Jan. 1994, Vol. 2155, pp. 55–66.

12

System design and performance of wideband photonic phased array antennas

GREGORY L. TANGONAN, WILLIE NG, DANIEL YAP,
AND RON STEPHENS

HRL Laboratories, LLC

12.1 Introduction

The application of RF photonics to RF antenna systems is a thread of development that runs parallel to the development of high performance analog links and components. This chapter deals with the promise of high performance analog photonic links applied to the field of phased arrays or RF manifolds.

In the early 1980s RF design engineers began to view photonics as a promising system option because of the possibility of modulating RF signals onto an optical carrier and the advantage of fiber optic transmission for low loss over large distances. As a cable replacement, fiber optics offers extremely wide bandwidth with no dispersion, significant weight reduction, loss reduction over a distance greater than 100 m, and immunity from electromagnetic interference or cross coupling. The challenges for achieving cable replacement have been lowering the overall conversion loss, and achieving high spur-free dynamic range, and high dynamic range.

In the late 1980s the focus shifted to applying photonics to antenna systems, as the development of optical networking techniques suggested the possibilities of routing optical modulated carriers to perform different RF functions. In particular the systems designers were interested in the possibility of performing several RF functions with photonics, namely beam steering, null steering, channelization, and local oscillator distribution over an optical manifold. To implement a photonic architecture, we assume the existence of high fidelity links and networks and device schemes for signal gathering, beam control, phase steering and true time delay steering, multichannel remoting, and pre-processing in the optical domain.

The overall phased array architecture being considered is shown in Fig. 12.1, where an array of elements is depicted with several processing functions that rely on photonic techniques for signal remoting and multiplexing onto a single fiber, for beam formation and steering, for optical filtering or channelization, and for local

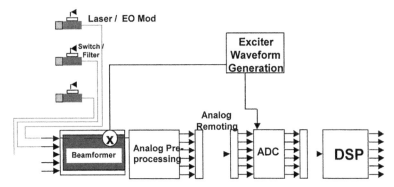

Figure 12.1. Photonic architecture for a digital-on-receive array with the different functions of optical processing – beam steering, optical channelization, LO distribution, and antenna remoting.

oscillator (LO) distribution. These functional blocks form the analog front-end processing that precedes the digital signal processing blocks that perform the standard functions of analog to digital conversion, signal detection, and digital signal processing for radar and communications. This architectural picture is a functional diagram that shows how photonics can be used in analog signal processing of RF inputs from an array. The reality facing system designers is that, depending on the application, only a few of these functions may require a photonic approach.

This chapter describes various demonstrations of the functionality of photonics in several of the processing blocks. Most notably these demonstrations deal with the development of fiber optic beam steering techniques using optical true time delay and phase steering approaches.

12.2 Modern wideband arrays

Phased array antennas are composed of three major parts – antenna elements, phase and amplitude controllers, and the manifold that includes the receivers. As shown in Fig. 12.2, the beams can be steered electronically through the piece-wise control of the phase of each element, and the array elements are typically separated a half-wavelength apart.

In radar applications, phased arrays with electronic beam steering replace the mechanical beam steering. The beam-switching agility of an electronically scanned antenna is crucial for search and track radars that must handle multiple targets. The operating bandwidth of the array may be very high for this class of radar, since the search mode is commonly done at low frequencies (say L-band or lower) and high resolution tracking is best done at higher frequencies (say S- or X-band). The need for wide operation bandwidth and the use of wideband waveforms press the system designers to specify wideband TR modules, phase shifters and manifolds.

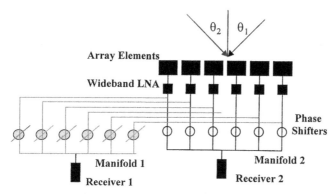

Figure 12.2. Phased receiver array for two independent beam operations using two sets of phase shifters.For broadband operation the LNA, phase shifters, and manifolds must support the overall frequency of operation.

Depending on the instantaneous bandwidth required and the angle of scan, true time delay steering may be dictated.

Historically this was the point where photonics first began to make inroads into beam steering of phased arrays. Initial work on optically controlled phased arrays focused on the need for high instantaneous bandwidth over a wide operating band.

As part of the coherent combining of the elements, one can use different summing algorithms to adapt the pointing – to form beams with nulls, to detect the direction of arrival, to null out jamming signals, or to reduce sidelobe strengths. These signal processing functions are considered to be part of the smart manifold functions in modern radar design. Here again there is a growing literature on the use of photonics for active beam steering and sensing of the direction of arrival for wideband signals.

In mobile communications an array of receive antennas can help system providers overcome the problem of supplying service over a limited bandwidth to as large a number of mobiles as possible. In fact modern mobile communication systems rely heavily on the directionality of the antennas to provide enhanced performance. Figure 12.3 shows a case encountered in mobile communications, where several potential listeners (L1 and L2) and the desired receiver are within the radiation pattern of the transmitter X. Further complicating the situation is the presence of two jammers (J1 and J2) that are radiating into R. In interference-limited systems like modern wireless systems, this is a common problem. Here the antenna pattern is modified to place a null in the direction of the transmission of X and the receive pattern of R. Clearly in a situation where maximal frequency reuse leads to optimum system bandwidth efficiency, smart antennas like this can radically improve system performance. A phased array may also help in improving wireless system performance by reducing fading, controlling multiple beams from satellite to the ground, increasing channel bandwidth, reducing co-channel interference,

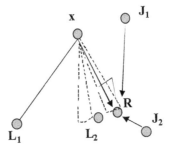

Figure 12.3. Smart antenna system for a wireless communication system with L1 and L2 as listeners and J1 and J2 as jammers of the communications between X and R. The antenna patterns of the transmit antenna of X and the receive antenna of R are modified to minimize the co-channel interference.

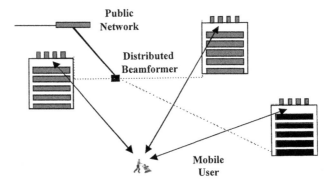

Figure 12.4. Large distributed antenna system with coherent summing over a large geographic area. Fiber optic remoting will play an enabling role in making this approach a reality.

increasing spectrum efficiency, and extending range. At the base station X, that array may consist of several subarrays separated by several wavelengths, as is typical of modern cellular deployments.

As wideband communications channels and the connection to mobile users over large acquisition angles increases, the simple technique of phase steering may no longer suffice. Also, forming a wideband null in a specific direction will become even harder to realize. Thus the very natural tendency of communication channels to grow in bandwidth and the number of users to increase within a given cell will cause designers to reach for new wideband beam steering technologies.

Of particular interest to the research community is the connection of several antennas over a large geographical area (several kilometers radius) to form a distributed antenna system. The large aperture of this distributed system and the wideband processing will allow spatial focusing to pinpoint the transmitter and increase the receiver signal-to-interference ratio. Of course what is required is a means of collecting the signals over large areas into a common receiver that is broadband and can compensate for the propagation delay differences. Clearly this is an area where RF photonics can play a crucial role in the very hot commercial field of wireless networking.

For both radar and communications the trend is clearly towards phased array designs that are considerably more complex than the first generation of active arrays. Wideband radars and receive-only arrays require multiple beams operating over multiple octaves. A high performance listening array may require 50 pencil beams over UHF to 18 GHz with instantaneous bandwidth in excess of 2 GHz. For LEO, MEO and GEO satellite communications may involve several hundred spot beams operating over Ka-, Ku-, or V-band with instantaneous bandwidths of over 200 MHz.

Thus, multibeam, multiband beam steering techniques are being explored in several research and development laboratories. Many teams agree that digital signal processing will be the ultimate winner – *all* the processing (wideband reception and conversion to digital signals, beam forming, null steering, and array calibration) will be done digitally. Short of attaining this final vision are the teams of researchers searching for unique ways to address the fundamental problem at hand – ultra-wideband beam of multiple beams with a major reduction in system complexity.

Research on photonically steered antennas with multiband, multibeam perfor-mance is just appearing in the literature. It has only recently been understood that this level of functionality is of crucial importance to several military and commercial applications.

12.3 LO distribution as an example of RF photonic signal remoting

An excellent example of the use of photonics in a phased array system is the distribution of LO signals. A photonic network can be used to switch LO signals from high-performance synthesizers to frequency converters in remotely located receiver and exciter modules. As shown in Fig. 12.5, this is a powerful alternative to having a dedicated LO synthesizer at each of possibly many frequency converters, which would lead to undesirable high cost as well as increased size, weight and power consumption. Distribution of LO signals from centralized synthesizers by means of coaxial microwave cables or waveguides is limited by the bulk, dispersion, and transmission losses of these devices, especially for large airborne or space based arrays. Photonics offers an attractive alternative because optical fiber cables are compact, lightweight, and provide low transmission loss and large bandwidth with minimal RF dispersion.

In an integrated sensor system, the required LO signals are generated by centrally located frequency synthesizers that are shared by the various sensors via the distri-bution network. The LO signals are used to modulate the optical transmitters and the

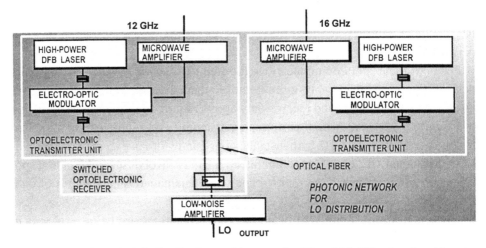

Figure 12.5. Photonic LO distribution architecture for 12 and 16 GHz signals with opto-electronic receiver selection for high isolation (>80 dB).

modulated lightwaves are distributed by the optical fiber network to the destination sensors where they are detected and selected by means of opto-electronic switches. The photonic architecture has the beneficial characteristic of versatile connectivity, supporting both broadcast and point-to-point delivery of the LO signals. The capability of selecting a given LO signal from an array of LO signals optically delivered to a given frequency converter is especially useful for EW functions and for integrated sensor systems (ISSs) that combine multiple sensor functions.

The task of distributing LO signals for multiple RF sensor functions imposes severe demands on the performance of the network. One requirement is high channel isolation. In the ISS, the specified channel isolation is that all non-selected LO signals must be at least 80 dB below the selected LO signal. Also, the network must add no more than 1 dB of amplitude or phase noise to the original LO signal. Because of this requirement the network must have a residual phase noise below −120 dBc/Hz at 10 kHz offsets from the carrier for radar functions, and below −147 dBc/Hz at offsets above 25 MHz for some EW and communications functions.

The reverse isolation of the photonic network is essentially perfect, since the reverse optical signal path from a frequency converter does not contain an RF-to-optical transducer to generate the interfering signal. This is in contrast to a coaxial or conducting waveguide medium that is susceptible to impedance mismatch and other discontinuities. The measured channel isolation and phase noise performance of a switched photonic link are described later in this chapter. As expected, the phase noise performance depends an the choice of laser type (either a distributed feedback (DFB) semiconductor laser or a diode-pumped solid-state laser (DPSSL)), the modulation efficiency of the photonic transmitter, the optical power level of the laser, the responsivity of the photodetector, and the losses in the network.

A two-channel photonic unit was built to demonstrate the performance of a switched photonic link for distributing LO signals. This unit consists of two separate photonic transmitter channels that are connected by optical fibers to a single 2-to-1 opto-electronic switched receiver. The unit was constructed from commercially available components to provide an indication of the capability of the deployable photonic technology. The unit is self-contained and operates from externally supplied AC power. The desired LO is chosen through mechanically operated switches located on the front panel of the unit. The unit is accessed through three RF connectors – two for the LO inputs and one for the selected LO output.

Each of the external-modulation transmitters uses a DFB laser driving an electro-optic modulator whose RF input is an amplified LO signal. The amplified input compensates transduction losses and minimizes noise contribution of the link. The link is provided with control electronics to stabilize laser operating current and temperature and modulator bias voltage. The opto-electronic switch consists of two photodetectors whose microwave outputs are combined by means of a 2-to-1 Wilkinson combiner. The bias voltage of each photodetector is switched by CMOS circuits to enable the photodetector to output the LO signal ("on" state) or to suppress that signal ("off" state). A low-noise microwave amplifier is placed after the optoelectronic switch to achieve the desired net link gain of +2 dB.

The switched photonic link is set to have an overall gain of +2 dB at 12 GHz by adjusting the output powers of the lasers and the gains of the microwave amplifiers in the two transmitter channels. Both channels are adjusted to have the same gain. The intrinsic link gain consisting of the photonic components alone is −27 dB. The modulator input amplifier provides 15 dB of amplification. The low-noise, post-detector amplifier provides the remaining amplification needed to achieve the desired link gain. The measured gain rolls off as the frequency is increased. This reduced response with frequency is due to the commercial photodetectors. Photodetectors that have both a flat frequency response (to beyond 20 GHz) and the ability to handle high levels of optical power (generating photocurrents > 10 mA) are being produced at HRL, although they are not yet available commercially.

The input and output ports of the photonic unit are matched quite well to a 50 Ω impedance. The magnitude of S_{11} and S_{22}, the input and output reflection coefficients, is less than −10 dB over the 8–16 GHz frequency range. To achieve low phase noise, it is important to minimize the reflections that may occur at any of the microwave or optical junctions in the unit.

The isolation of the optoelectronic switch has been measured by applying a 12 GHz LO signal to one input port of the photonic unit and a 16 GHz LO signal to the other input port. The output of the photonic unit was then observed with a microwave spectrum analyzer. The level of the 16 GHz output is somewhat lower than that of the 12 GHz output, because of the reduced photodetector response (as seen from the lower left plot). The optoelectronic switch isolation was tested by alternately switching off the bias voltage to the photodetector of each channel by means of a CMOS circuit. When a channel is switched off, its associated LO signal is not observable above the noise floor of the measurement setup, which is more than 80 dB below the signal level (as seen from the two plots to the right). The measured isolation of the optoelectronic switch compares well with that of a PIN-diode microwave switch, whose best performance likewise is approximately 80 dB.

The phase noise performance of the photonic unit was measured using custom, low-noise synthesizers built by Raytheon. The absolute phase noise was measured with a Hewlett-Packard phase noise test set. One synthesizer was connected to the photonic unit. A second, essentially identical synthesizer provided the reference signal. The resultant phase, shown in Fig. 12.6, was measured for offsets as large as 40 MHz from a 12 GHz LO signal. The absolute phase noise at an offset of 10 kHz is only −125 dBc/Hz. For offsets greater than 3 MHz, the phase noise is approximately −147 dBc/Hz. This performance is sufficient for the combined radar, EW and CNI functions.

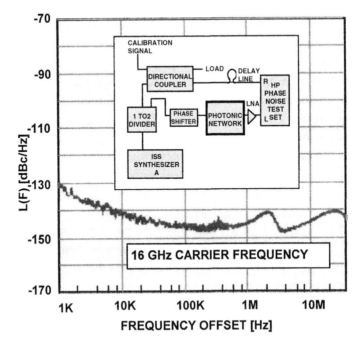

Figure 12.6. The measured residual phase noise of the photonic LO distribution system is <145 dBc/Hz, limited by the post detector amplifier.

12.4 Demonstrations of wideband photonically controlled phased arrays

12.4.1 Phase steering and true time delay (TTD) steering for wideband arrays

The majority of phased array antennas in use today and in design use phase shifting to accomplish the steering. As shown in Fig. 12.7, individual phase shifters are placed at the radiating aperture to control the phase of the individual radiator (or subarray). The steering angle θ for an angle ϕ is given by the grating equation:

$$\phi_i = 2\pi f/c \sin(\theta).$$

This basic design works very well for narrow band apertures where the instantaneous bandwidth is a small percentage of the operating frequency.

The performance of conventional phased arrays, albeit satisfactory for many applications, is perceived to have ultimate limitations set by the bandwidth constraints and mechanical rigidity of the microwave components employed to form its beam. In particular, conventional arrays steered by phase shifters suffer from the well-known phenomenon of beam squint – the drifting of the antenna's beam-scan angle (θ_o) with frequency. When normalized by the beam half-width $\theta_{1/2}$, the beam squint $\Delta\theta_o$, is given by [1]:

$$\frac{\Delta\theta_o}{\theta_{1/2}} = 1.13 \sin\theta_o \frac{L}{\lambda}\frac{\Delta f}{f} \tag{12.1}$$

From Eq. (12.1), we see that the penalty due to beam-squint increases with: (i) $\Delta f/f$, the array's fractional bandwidth of operation, (ii) $2L/\lambda$, the number of radiating elements in the array of length, and (iii) the array's angular scan range (θ_{max}).

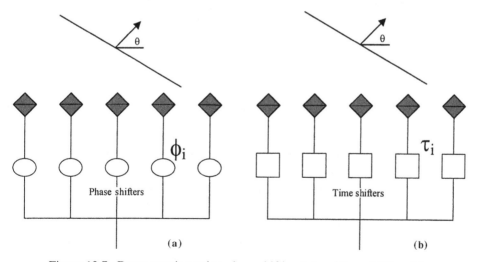

Figure 12.7. Beam steering using phase shifting (a) and time shifting (b).

For example, the normalized beam squint doubles as one increases the beam scan angle (θ_o) from 28° to 70°. At 60°, a small variation in frequency of 5% will cause a 3 dB drop in array gain for an array 7λ long.

This can be a severe problem in many broadband applications. A ground-based L-band radar may require 20–30% instantaneous bandwidth so the radar can use the lower frequencies for detecting low observable targets and the upper band to track the targets once found. For communications, consider an airborne SHF satellite receive antenna that must receive wideband (100 MHz) signals over a large scan angle (say ±70°).

Two different approaches have been applied towards the development of wideband arrays that exploit photonic architectures. The first approach has been true time delay (TTD) beam steerers that exploit the basic insight that the wideband delay line can compensate for the propagation delay from different points on the array. The basic challenge for a photonic implementation of TTD is that a photonic delay several wavelengths long perform better than a strip line delay line or coax of the same length. Several approaches for photonic delay lines have been developed, but only a few illustrative examples will be given here.

The second approach relies on phase steering but limits the scan to small angular excursions about a multiplicity of beam positions. The effect of squinting is markedly reduced, allowing wideband performance. Crucial to this approach is the development of a beam steerer with multiple fixed beam positions and a means of phase scanning about these positions. By limiting the scan angle about these positions, one can indeed achieve wideband performance.

For antenna system designers, the main issue in exploiting photonics remains the trade-off between complexity, cost and risk, and performance, cost and risk. While it is fair to say that the development of photonically steered antennas has achieved several successful demonstrations of wideband performance, it is also fair to say (and humbling to be reminded) that the systems designers remain skeptical.

12.4.2 True time delay demonstration systems

The HRL team achieved its first major demonstration under an EHF Study Program with Rome Laboratories: [2], the wide instantaneous bandwidth of photonic TTD steering in a transmit array [3]. We demonstrated that a single, common photonic beamformer could enable squint-free operation from L-band (1–2.6 GHz) to X-band (8–12 GHz). Meanwhile, great improvements were made in the NF and SFDR of analog fiber optic links with the development of: (i) directly modulated links using low noise (RIN $\sim$ −160 dB/Hz) DFB lasers with high (fiber-coupled) slope efficiencies ($\sim$0.15 mW/mA), and (ii) externally modulated links [4] using high power ($P_o > 100$ mW) $\lambda = 1.3$ μm Nd:YAG lasers. Furthermore, a spreadsheet analysis [5] of ours showed that the array-NF (with some trade-off in SFDR) could be significantly improved via coherent signal summation from multiple antenna elements. Therefore, we proceeded to demonstrate, in a second program [6], photonic beamforming for a 96-element *radar* that would be fully equipped to perform transmit and receive operations at L-band. We describe an ongoing program [7] where we will demonstrate part of an SHF-band satellite communication (SATCOM) antenna specified to scan (with a resolution of ∼4°) over an ultra-wide angular range of 140° (±70°).

12.4.3 Dual band transmit array

The EHF-array (Fig. 12.8) was a dual-band (L- and X-band) transmit antenna designed to scan from −28° to +28°. Consisting of 8 elements for L-band and 32 elements for X-band, it was divided into four subarrays. The length of the array was ∼53 cm (for both L- and X-band). The identical aperture sizes for both bands enabled one BFN to feed two different sets of radiating elements. Each of the four subarrays was steered by a 3-bit photonic time shifter whose delay times were varied by selectively "biasing-up" a bank of directly modulated diode lasers. By implementing TTD at the subarray level, we reduced the number of photonic time shifters needed – hence the complexity and cost of the optical BFN. Inside each X-band subarray we incorporated, in addition, electronic phase shifters with 4-bits of resolution to fine steer the beam. It should be noted that the mating of

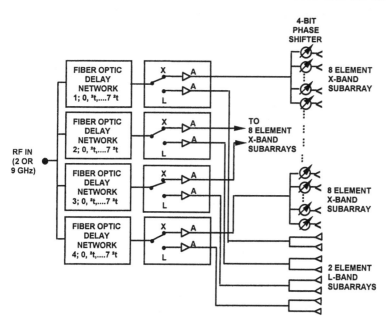

Figure 12.8. Architecture of EHF dual band array.

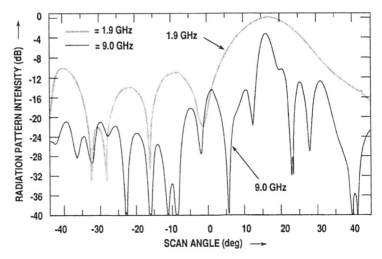

Figure 12.9. Transmit pattern of EHF array at $\theta_0 = +16°$ for L- and X-band.

photonic TTD steering at the subarray level with electronic phase steering *inside* the subarray is an effective architecture that allows one to enhance *substantially* the array performance (instantaneous bandwidth or scan angle for a given array size) with only a moderate increase in cost. In addition, it enables us to take advantage of the sophisticated T/R modules developed (for conventional arrays) to fine steer the beam from the much *smaller* subarrays. Figure 12.9 shows the measured beam pattern of the EHF array at L- and X-band (for the scan angle of $\theta_0 = +16°$). As shown, we demonstrated [3] the achievement of squint-free transmission patterns in a frequency range that measured almost one decade wide.

In association with the EHF-array, we also developed a monolithic photonic time shifter (Fig. 12.10) that utilized detector switching to vary the delay times. Specifically, the time delays in this optical circuit were generated by an optical carrier propagating through rib-waveguides fabricated from GaAlAs/GaAs epitaxial layers. Aside from compactness, we expected that the waveguide-delay-lines fabricated by photolithography would offer better differential phase accuracy than their fiber-based counterparts, whose lengths were typically trimmed individually to meet specific time delay requirements. By measuring the differential RF insertion phase between the waveguide-delay-lines, we obtained [8] time delays that fell within 4 ps of their designed values. Finally, we demonstrated [9] an on/off ratio that was larger than 40 dB (at 10 GHz) for the waveguide-coupled MSM detectors integrated to switch the delay lines.

12.4.4 L-band conformal radar with 96 elements

As mentioned earlier, rapid improvements in the NF and SFDR of analog photonic links occurred in the early 1990s, and the time was ripe to demonstrate an array that

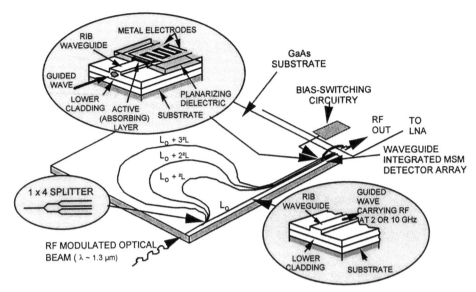

Figure 12.10. Monolithic time-delay network integrating rib-waveguide delay lines and waveguide-coupled MSM detectors on GaAs.

would incorporate photonic TTD on receive. We chose to demonstrate this photonic radar at L-band where – stimulated in part by the commercial CATV industry – we saw the most significant improvements in analog transmission. Designed to scan ±60°, the 96-element array (Fig. 12.11) was divided into eight subarrays, each consisting of three columns of radiating elements. As shown, it was arranged geometrically to form an arc with a radius of ~3.05 m. Each of the subarrays was steered by a 4-bit/5-bit photonic time shifter module. The large aperture size and wide angular scan requirement of this radar imposed a step size (Δt) of 0.25 ns for its quantized delay times. We used a combination of laser and detector switching to switch its optical delay lines. Inside each of the eight subarrays additionally we incorporated an electronic time shifter that consisted of microstrip lines to obtain six more bits of resolution. By combining photonic time steering for the major bits, and electronic time steering for the less significant bits, we achieved TTD beamforming with an overall precision of 10/11 bits. Finally, by judiciously setting the first LNA at the receive path to have a gain of ~48 dB, we accomplished [5] an overall system noise figure of 3.44 dB with an accompanying SFDR of 96 dB/Hz$^{2/3}$.

As reported previously [5,10], we demonstrated squint-free beam patterns for both transmit and receive from 0.85–1.4 GHz, which corresponded to an instantaneous bandwidth of ~50%. Figure 12.12 shows the receive beam patterns at $\theta_0 = 30°$ and 60°.

We also developed, within this program, a new approach that took advantage of pulse synthesis techniques [10,11] to characterize the photonic time shifters. Taking advantage of the wide operational bandwidth intrinsic to these time-delay

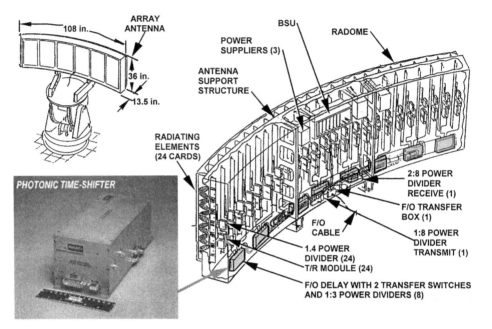

Figure 12.11. The 96-element L-band conformal radar with TTD beamforming. The inset shows one of the photonic time-shifters used to steer its subarray.

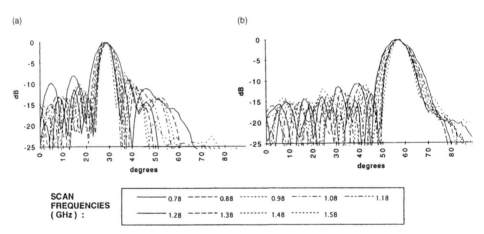

Figure 12.12. Receive patterns obtained for (a) +30° and (b) +60° scan.

modules, we can synthesize their *pulse* response from experimental data measured over a bandwidth of ∼1 GHz in the frequency domain. When an inverse Fourier transform is exercised, we obtain the module's time-domain response due to a pulse excitation that is ∼2 ns wide. Compared with characterization techniques based on measurements of the module's RF-insertion phase, the time-domain approach holds the following advantages. First, from the separation between the module's pulse response and its input excitation, one can read off directly the time delay incurred

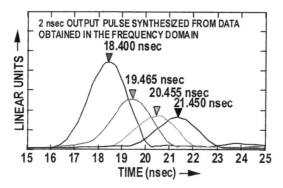

Figure 12.13. Synthesized pulse response of four delay lines from silica-waveguide module. The arrows indicate the measured time delays.

by a specific delay path. Second, any undesirable RF/optical crosstalk between the RF/optical components in the module would be discerned as a distortion in the synthesized output pulse, making it easier to diagnose.

Finally, we accomplished system insertion [10] of the waveguide-delay-lines in this L-band array. Using silica waveguides with a propagation loss of only 0.1 dB/cm, we successfully integrated eight delay lines with lengths as long as 67.2 cm on two waveguide-chips connected by a 4 × 4 star coupler. With a fabrication process that is based on photolithography, we demonstrated a viable technology that allows us to reproduce identical sets of delay lines efficiently and with high precision. Figure 12.13 shows the pulse responses of four delay paths designed to generate time delays in step increments of 1 ns. As shown, the output pulses from the silica-waveguide module suffered almost no distortion.

12.4.5 SHF SATCOM array for transmit and receive

The SHF SATCOM band consists of an uplink frequency band (7.9 to 8.4 GHz, for transmitting to the satellite) and a downlink frequency band (7.25 to 7.75 GHz, for receive). An important performance goal of this airborne array is the formation of a beam with a half-power beamwidth ($\theta_{1/2}$) of ~4° or less. Since the radiating elements of the array are typically spaced half a wavelength apart (to avoid the formation of grating lobes), the above beamwidth requirement implies that the array aperture needs to be populated by at least 32 × 32 radiating elements. Finally, because of the aircraft's motion and the variability of its position with respect to the satellite, the antenna needs to cover an angular scan range of 140° (i.e., the maximum scan angles (θ_{max}) of the array are as large as ±70°). Although the fractional bandwidth ($\Delta f / f$) for the transmit/receive band ($\Delta f = 500$ MHz) is only ~7%, the large scan angle requirements of ±70° for the array could lead to unacceptable beam squint if *true time delay* (TTD) were *not* pursued to steer its beam. A detailed discussion of the array's design is given in Ref. 12.

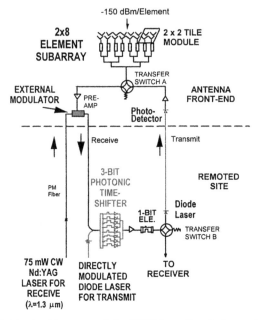

Figure 12.14. Architecture of the BFN for a 2 × 8 element subarray.

Figure 12.14 shows the architecture of one of the subarrays we are building for a system demonstration. Specifically, we use a directly modulated RF photonic link for transmit, and an externally modulated link for receive. By choosing an externally modulated link in the array's receive path, we were able to achieve a point-to-point SFDR of 112.6 dB/Hz$^{2/3}$ (for a receiver photocurrent of 9.5 mA). The fundamental response and third order intermodulation product of this link is plotted in Fig. 12.15.

The time-delay network for the array consists of a 3-bit photonic time-shifter module that is cascaded with a 1-bit electronic (microstrip) time-delay module. Together, they offer 4-bits of resolution for the quantized time delays that we will use to steer the array. By cascading our 3-bit photonic time-shifter module with a 1-bit microstrip module, we accomplish a better link insertion gain for the BFN, and yet maintain the merits of compactness and high EM-isolation offered by photonics. Specifically, the time-delay increment in our 3-bit photonic time shifter is generated from one silica-waveguide chip in steps of $2\Delta t_d \sim 0.243$ ns. A prototype waveguide chip integrating eight delay lines has been demonstrated. In particular, the percentage time-delay error of the longest delay line on the chip was measured to be only 0.95%. Finally, our 4-bit time-shifter module is designed to provide sufficient resolution so that one can match (to within a wavelength) the delay length needed to steer a 32 × 32 element array to its θ_{max} of ±70°.

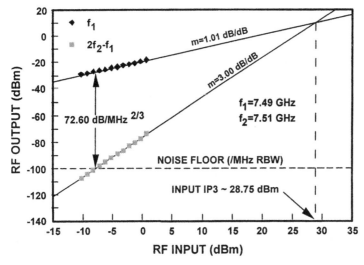

Figure 12.15. Third order intermodulation product and SFDR of externally modulated link at 7.5 GHz.

12.4.6 Multibeam Rotman lens array controlled by an RF-heterodyne photonic BFN

In this section, we briefly describe a photonic beamforming architecture that combines an RF-heterodyne concept with the classical Rotman lens feed. A detailed description of this BFN is given in Ref. 13. As we shall see, a combination of these two concepts enables the formation of multiple beams with the following characteristics. At specific Rotman angles (θ_R), TTD is implemented and the beams thus formed possess, in principle, unlimited instantaneous bandwidth. With an RF-heterodyne photonic feed to be described below, we now fine scan the beam over narrow angular ranges about each of these Rotman angles. Although the fine scanning (about θ_R) is accomplished via phase steering, a minimal level of beam squint is incurred because θ_{max} is typically only $\pm10°$. In this manner, we can accomplish *continuous* scanning over the whole "visible space" of $\pm90°$. It is worth mentioning that this is a significant improvement over the original [14] Rotman lens feed that was designed to generate only discrete beams. In Fig. 12.16 we show a schematic of the RF-heterodyne photonic feed.

As shown, two RF frequencies f_1 and f_2 are modulated onto two optical carriers generated, respectively, by laser 1 and laser 2. These two optical carriers then propagate through two optical manifolds that are designed to have a differential length ΔL between adjacent feeds to the radiating elements. As in the TTD BFN described in Section 12.4.1, each of the RF signals suffers a phase shift proportional to the product f_i and its time delay (Δt_d) generated by the optical manifold. As

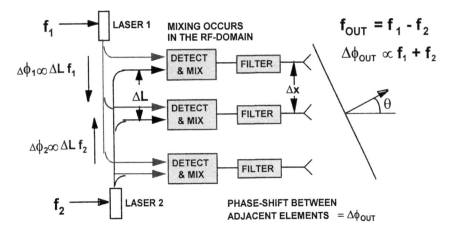

$$\Delta t_d = (n\Delta L)/c \; ; \; \Delta\Phi_{OUT} = \Delta t_d \cdot 2\pi(f_1 + f_2)$$

Figure 12.16. Schematic of RF-heterodyne photonic feed (*n*-effective refractive index of optical waveguide in manifold).

the optical carriers are detected and mixed in the feeds to the radiating element, we obtain a relative phase $\Delta\phi_{\text{out}}$ given by:

$$\Delta\phi_{\text{out}} = \Delta t_d \, 2\pi(f_1 + f_2) \tag{12.2}$$

for the mixed product f_{out} ($f_{\text{out}} = f_1 - f_2$) of the two RF frequencies. Notice that the relative phase $\Delta\phi_{\text{out}}$ is proportional to the sum of f_1 and f_2. If we maintain the difference f_{out} to equal the desired transmit frequency of the array (e.g., with the help of a phase-locked loop), we can scan the array's output beam by varying the sum $(f_1 + f_2)$. If λ_o and ΔX are, respectively, the RF-wavelength of the radiated beam and the spacing between the radiating elements, the scan angle θ_o of the transmitted beam is given by:

$$\sin\theta_o = \frac{\lambda_o}{\Delta X}[\Delta t_d(f_1 + f_2) - m], \tag{12.3}$$

where m is an integer. Typically, the ratio of ΔX to λ_o is 1/2. Experimentally, we have verified the above relationship (Eq. 12.3) for an X-band array ($f_{\text{out}} = 9$ GHz), with f_1 and f_2 varying, respectively, over the frequency ranges of 10–12 GHz and 1–3 GHz. Finally, we also demonstrated the formation of two beams (at $f_{\text{out}} = 9.38$ GHz and $f_{\text{out}} = 8.57$ GHz) and their scanning via RF-heterodyning. The details of these experiments are described in Ref. 12. In Fig. 12.17 we show the schematic of a photonic Rotman lens feed that incorporates the RF-heterodyning concept to achieve continuous scanning.

The RF-heterodyne Rotman lens BFN depicted here consists of two photonic manifolds (1 and 2) with differential lengths ΔL_1 and ΔL_2. In particular,

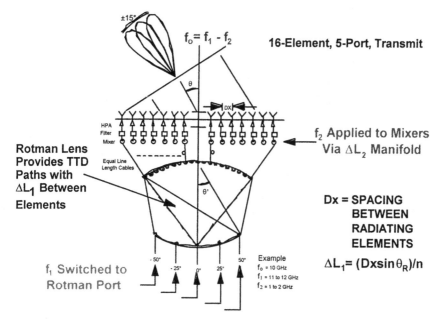

Figure 12.17. Schematic of RF-heterodyne Rotman lens BFN. Arrows at the bottom arc of the figure designate the five Rotman beam-ports.

$\Delta L_1 (= (Dx)(\sin \theta_R)/n)$ is designed to satisfy the TTD beamforming condition at the Rotman angle θ_R. In a transmit array, each of the Rotman beam ports is equipped with an optical source designed to generate the optical carrier for the RF input f_1. If we only utilize manifold 1, the BFN functions as a photonic implementation of the classical Rotman lens feed. Thus, excitation of the beamport (with a subtended angle θ') generates a beam with TTD characteristics in the direction designated θ. When we introduce the optical carrier (modulated by the RF f_2) from the manifold 2 at the mixer, a differential phase $\Delta\phi$ (between adjacent radiating elements) given by:

$$\Delta\phi = 2\pi(\Delta L_1 f_1 + \Delta L_2 f_2)\frac{n}{c} \qquad (12.4)$$

is generated for the difference frequency $f_0 = f_1 - f_2$. From Eq. (12.4), we see that the beam scan angle can now be fine tuned about θ by varying f_1 and f_2, but keeping f_0 constant. In particular, ΔL_2 can be chosen to provide good scan sensitivity and be made compatible with a given phase-locked loop design. A calculation performed for this design indicates that we can keep the normalized beam squint of a 16-element array to less than 0.1 if the fine-scanning range can be kept below $\pm 10°$ (for an instantaneous bandwidth Δf of $\sim$1.5 GHz within the operation frequency band of 6–18 GHz). Further details for this BFN are given in Ref. 13.

12.5 New architectures for photonic beam steering

While the results described above show definite progress towards systems implementation, the simple broadcast and select architecture suffers from the inherent
1 to N fanout loss. In a systems implementation this loss can cause severe penalties
for receive manifolds, such as higher noise figure, poorer spur free dynamic range,
and extremely high-gain LNA requirements.

Several new schemes have been developed that address this major difficulty
with the opto-electronic architecture. In general these new architectures exploit
the new capabilities of dense wavelength division multiplexing applied to analog
systems. Goutzoulis first proposed a WDM approach to manifold to attain a major
reduction in complexity [15]. The fundamental problem is the larger the array the
greater the range of delays needed for steering, and each part of the array requires
different delays. Thus the complexity grows very rapidly. By partitioning the delay
into binary stages with short delay segments and long delay segments cascaded,
Goutzoulis demonstrated that major savings in the total parts count can be achieved
using wavelength multiplexing.

More recent architectures rely on wavelength switchable laser sources and wavelength routing components to eliminate the 1 to N fanout losses. Several schematic
diagrams are shown in Fig. 12.18 through Fig. 12.22 that illustrate the new architectural designs that have been demonstrated. In Fig. 12.18 we show a fiber-grating-
based true time delay element developed by Ming Wu *et al.* of UCLA [16]. In
this architecture a wavelength tunable source coupled with a modulator provides a
wavelength switched RF modulated light. This light is fed into a wavelength dependent cross-connect system that incorporates 2×2 spatial switches, fiber circulators,
and fiber gratings. The laser is set to a particular color and then the modulated light
is switched into a particular path (by the 2×2 switch). A specific wavelength is
reflected off a particular grating after a specific delay. The net effect is to have a
true time delay line that can be switched very rapidly with low loss.

Obviously what makes this a reliable and practicable way to implement true time
delay is the wavelength tunable source. There has been tremendous progress in this

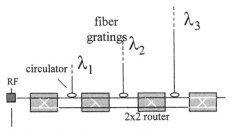

Figure 12.18. A wavelength routing architecture for true time delay beam steering (after
Tong and Wu [16]).

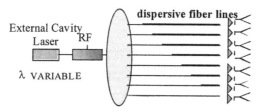

Figure 12.19. True-time-delay beam steering concept using dispersive prism for highly dispersive waveguides (after Esman *et al.*[19]).

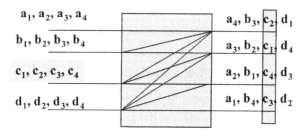

Figure 12.20. Wavelength routing chart for a Dragone wavelength crossconnect router. a_i, signal at port a with frequency i.

area as part of the drive to make practical dense WDM for the telecom industry. A good example is the development of a 40-wavelength laser that can be digitally tuned with wavelength spacing of 100 GHz.(17) The advent of these new sources makes this approach viable for RF photonics applications.

Researchers at the Naval Research Laboratories led by Ron Esman were the first to implement tunable lasers for beam steering applications. They combined a tunable laser with the dispersive delay lines to actually build working radar [18–20]. The dispersive delays provide a convenient way to vary the delay continuously by simply varying the wavelength, as shown in Fig. 12.19. Highly dispersive fibers can be used to reduce the actual total delay through the system to minimize the problem of slow beam switching speeds.

The second architecture that has been developed for delay applications builds on the wavelength routing properties of the Dragone coupler [21], shown in Fig. 12.20. The connection diagram indicates how the different channels are routed through the coupler. This is essentially a wavelength routing map that in the hands of a delay network designer can become a delay module. Jalali *et al.* [23] designed this type of delay line element for true time delay applications; the design is shown in Fig. 12.21. Notice that Jalali's design makes full use of routing capabilities of the array waveguide router. The optical energy is routed to the particular path of delays. This approach exemplifies the power of the newer routing schemes for true time delay beam steering: several bits of resolution can be attained without an optical switch, since the routing is done passively through the wavelength cross-connect.

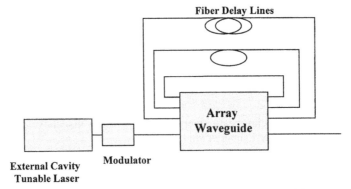

Figure 12.21. True-time-delay module based on the array waveguide router (after Jalali *et al.* [23]).

Acknowledgements

DARPA and the Air Force Research Laboratories in Rome, New York support this work. The authors wish to acknowledge the encouragement of N. Bernstein, J. Hunter, and the late B. Hendrickson. In general the RF photonics community benefited tremendously from the visionary contributions of Brian Hendrickson.

References

1. R. Tang and R. W. Burns, "Phased arrays," in *Antenna Engineering Handbook*, 2nd Edn., New York, McGraw-Hill, 1984, Ch. 20.
2. USAF Rome Labs. Program (F30602-87-C-0014), "EHF optical fiber based subarrays."
3. W. Ng, A. Walston, G. Tangonan, J. J. Lee, I. Newberg, and N. Bernstein, "The first demonstration of an optically steered microwave phased array antenna using true-time-delay," *J. Lightwave. Technol.*, **LT-9**. 1124–31, 1991.
4. C. Cox, III, G. Betts, and L. Johnson, "An analytic and experimental comparison of direct and external modulation in analog fiber-optic links," *IEEE Trans. Microwave Theory* Tech., **38**, 501–9, 1990.
5. J. J Lee, R. Loo, S. Livingston, V. Jones, J. Lewis, H. W. Yen, G. Tangonan, and M. Wechsberg, "Photonic wideband array antennas," *IEEE Trans. Antennas Propag.*, **43**, 966–82, 1995.
6. DARPA/USAF Rome Labs. Program (F30602-91-0006), "Optical control of phased arrays."
7. AFRL (Rome) Program (F30602-96-C-0025), "SHF SATCOM array hardware."
8. W. Ng, D. Yap, A. Narayanan, and A. Walston, "High-precision detector-switched monolithic GaAs time-delay network for the optical control of phased arrays," *IEEE Photon. Technol. Lett.*, **6**, 231–4, 1994.
9. W. Ng, A. Narayanan, R. R. Hayes, D. Persechini, and D. Yap, "High efficiency waveguide-coupled $\lambda = 1.3$ μm $In_xGa_{1-x}As/GaAs$ MSM detector exhibiting large extinction ratios at L and X Band," *IEEE Photon. Technol. Lett.*, **5**, 514–17, 1993.
10. W. Ng, R. Loo, V. Jones, J. Lewis, S. Livingston, and J. J. Lee, "Silica-waveguide optical time-shift network for steering a 96-element L-band conformal array," in *Proc. SPIE Conf. for Optical Technology for Microwave Applications VII*, Vol. 2560, pp. 140–7, 1995.

11. J. J. Lee, S. Livingston, and R. Loo, "Calibration of wideband arrays using photonic delay lines," *Electron. Lett.*, **31**, 1533–4, 1995.
12. W. Ng, R. Loo, G. Tangonan, J. J. Lee, R. Chu, S. Livingston and F. Rupp, "A photonically controlled airborne SATCOM array designed for the SHF band," in *Proc. SPIE Conf. for Optical Technology for Microwave Applications VIII*, Vol. 3160, pp.11–16, 1997.
13. J. J. Lee, R. Stephens, W. Ng, H. Wang, and G. Tangonan, "Multibeam photonic Rotman lens antenna using an RF-heterodyne approach," in *Proc. SPIE Conf. for Photonics and Radio Frequency II*, San Diego, CA, July 21–22, 1998.
14. See for example, J. Ajioka and J. McFarland, "Beam-forming feeds," in Chapter 19, *Antenna Handbook, Vol. III, Applications*, Chapman and Hall, 1993.
15. A. Goutzoulis and D. K. Davies, "Hardware compressive 2-D fiber optic line architecture for time steering of phased-array antenna," *Appl. Opt.*, **29**, 5353–9, 1990.
16. D. T. K. Tong and M. C. Wu, "Common transmit/receive module for multiwavelength optically controlled phased array antennas," Optical Fiber Communication Conference and Exhibit, 1998. *OFC '98, Technical Digest*, 354–5.
17. C. R. Doerr, C. H. Joyner, and L. M. Stulz, "40 wavelength rapidly digitally tunable laser," *IEEE Photon. Technol. Lett.*, **11**, 1139, 1999.
18. J. L. Dexter, R. D. Esman, M. J. Monsma, and D. G. Cooper, "Continuously variable true time delay modulator," *OFC/IOOC'93 Technical Digest*, Paper ThC6, p. 172.
19. R. D. Esman, M. Y. Frankel, J. L. Dexter, L. Goldberg, M. G. Parent, D. Stilwell, and D. G. Cooper, "Fiber optic prism true time delay antenna feed," *IEEE Photon. Technol. Lett.*, **5**, 1347–9, 1993.
20. M. Y. Frankel, R. D. Esman, and M. G. Parent, "Array transmitter/receiver controlled by a true time delay fiber optic beamformer," *IEEE Photon. Technol. Lett.*, **7**, 1216, 1995.
21. C. Dragone, "A $N \times N$ optical multiplexer using a planar arrangement of two star couplers," *IEEE Photon. Technol. Lett.*, **3**, 812–15, 1991.
22. D. T. K. Tong and M. C. Wu, "A novel multiwavelength optical controlled phased array antenna with programmable dispersive matrix," *IEEE Photon. Technol. Lett.*, **8**, 812, 1996.
23. B. Jalali and S. Yegnanarayanan, "Advances in recirculating photonic true time delay for optically controlled phased array antennas," *Photonic Systems for Antenna Applications* (PSSA-9), 17–19 Feb. 1999, Technical Digest.

Index

401